T0215559

ARM®
Microprocessor Systems

Cortex®-M Architecture, Programming, and Interfacing

ARM®
Microprocessor Systems

Cortex®-M Architecture, Programming, and Interfacing

Muhammad Tahir and Kashif Javed

CRC Press
Taylor & Francis Group
Boca Raton London New York

CRC Press is an imprint of the
Taylor & Francis Group, an **informa** business

CRC Press
Taylor & Francis Group
6000 Broken Sound Parkway NW, Suite 300
Boca Raton, FL 33487-2742

First issued in paperback 2020

© 2017 by Taylor & Francis Group, LLC
CRC Press is an imprint of Taylor & Francis Group, an Informa business

No claim to original U.S. Government works

ISBN 13: 978-0-367-57391-1 (pbk)
ISBN 13: 978-1-4822-5938-4 (hbk)

Library of Congress Cataloging-in-Publication Data

Library of Congress Cataloging-in-Publication Data
Names: Tahir, Muhammad (Electrical engineer), author. | Javed, Kashif, author.
Title: ARM microprocessor systems : cortex-M architecture, programming, and interfacing / Muhammad Tahir and Kashif Javed.
Description: Boca Raton : Taylor & Francis, CRC Press, 2017.
Identifiers: LCCN 2016038555| ISBN 9781482259384 (hb : alk. paper) | ISBN 9781482259391 (electronic)
Subjects: LCSH: ARM microprocessors--Programming.
Classification: LCC QA76.6 .T337 2017 | DDC 005.1/8--dc23
LC record available at https://lccn.loc.gov/2016038555

Visit the Taylor & Francis Web site at
http://www.taylorandfrancis.com

and the CRC Press Web site at
http://www.crcpress.com

Contents

Preface

This book is the outgrowth of our experience in teaching the Microprocessor Systems course in the Electrical Engineering Department, University of Engineering and Technology, Lahore. The book contents are divided into three parts: architecture, programming, and interfacing. In architecture, we discuss the ARM processor and interrupt architecture. The programming part introduces the assembly language in a fair bit of detail. Once the reader is familiarized with the ARM processor architecture and its programming-using assembly language, then we describe interfacing various input/output devices with the microprocessor using different peripheral interfaces. Over the years, we have observed that teaching these three contents prepares the reader for attractive jobs in the embedded market. When an embedded engineer encounters an unknown architecture, his first job is to understand its architecture. The processor architecture knowledge equips the programmer to develop efficient programs. This is what we learned during our tenure as embedded engineers at Mentor Graphics (www.mentor.com).

Prior to 2013, we were teaching the course using the 80x86 microprocessor family. Because of the maturity of the 80x86 family, a number of good books was available. Architecture, programming, and interfacing can be covered from a single book. However, the interfacing aspect covered in most of the 80x86 texts is outdated. This can partly be attributed to the fact that the 80x86-based architecture has become too complex over the years and does not support many of the interfaces. For instance, the 80x86 used to have a parallel peripheral interface for printer interfacing that has become obsolete. It was not possible to conduct a good laboratory session based on the 80x86 architecture, which is highly important for such a practical subject. In the fall semester of 2013, we decided to switch to a different microprocessor architecture to improve the laboratory experience. After a thorough survey, we found ARM Cortex-M-based microprocessor to be the most suitable candidate. Afterward, the search for a suitable textbook began. We extensively searched but couldn't find a good textbook that incorporates architecture, programming, and interfacing aspects. So, we decided to write our own book. Our goal was to write a textbook that is easy to read and covers these three topics in fair detail. We have chosen ARM Cortex-M processor based microcontroller, TM4C123, from Texas Instruments for the hardware platform. The material is presented in a logical order along with interesting and practical examples to maintain the reader's interest.

Chapter Summary

The book expects understanding of some basic concepts such as representation of numbers in binary and hexadecimal forms. We also assume that the reader is familiar with the basic digital systems. Knowledge of C programming language as a prerequisite is also a must to understand different example programs.

There are 12 chapters in the book. Chapter 1 of the book is an introduction to embedded systems and its building blocks. Specifically, we discuss different integral components of embedded systems, such as microprocessors and their architectures, memories, software system components, and debugging interfaces. The software tools such as compilers and debuggers are also introduced briefly. The remaining 11 chapters are organized in three parts, namely, architecture, programming, and interfacing. Part I comprise two chapters related to the architecture. In Chapter 2, we discuss the ARM Cortex-M architecture in detail. Its register set, operating modes, reset sequence, pipelined architecture, memory map model, and the bus system architecture are presented. Chapter 3 is dedicated to discussing the architectural aspects of exceptions and interrupts. We introduce the usage of Cortex-M nested vector interrupt controller (NVIC), different steps required to perform interrupt configuration, how to set up an interrupt vector table, and interrupt masking and handling. Toward the end of this chapter, advanced concepts such as interrupt tail chaining and interrupt nesting are presented.

Part II of the book is about Cortex-M programming, and is split into four chapters. Chapter 4 teaches the basics of Cortex-M assembly programming. The fundamentals such as the syntax format and the Thumb2 Instruction set are discussed. The reader will learn how to write the first assembly program and its compilation in this chapter. The KEIL μVision4 Tools are used throughout the book for writing programs, building their executables, and programming and debugging of the hardware platform. The instruction encoding used to map the assembly language mnemonics to machine language codes is discussed at the end of this chapter. Chapter 5 presents instructions such as shift, rotate, basic math, bit operations, etc. In Chapter 6, assembly instructions used to perform memory-related operations are discussed. These instructions allow the processor to read and write memory locations. This chapter also introduces assembly instructions used to interact with the stack memory. Chapter 7 describes program flow control instructions, including conditional and unconditional branch instructions, combined compare and branch, and IF-THEN (IT) block. The use of IT block allows conditional execution of many assembly instructions other than branch instructions.

The material covered in the first two parts doesn't require a specific hardware platform. Rather, one can use a simulator to learn the workings of different assembly programming instructions. However, Part III is all about interfacing different real-life hardware devices to the ARM Cortex-M controller. For this part we have selected the Texas Instruments TM4C123-based microcontroller hardware platform, which is also called Tiva^{TM} C LaunchPad. Using this hardware platform, we introduce simple input/output interfacing in Chapter 8. Chapter 8 discusses the workings of general purpose input-output (GPIO) pins, their features, possible alternate functionalities, and interfacing of output (LED, LCD displays) as well as input (switches and keypads) devices. The first interfacing example that toggles an on-board green LED is developed using both assembly and C programming languages to elaborate the relation between C and assembly programming. Chapter 9 discusses different techniques to achieve synchronization between hardware and software components of an interface. We also provide a comparison of different methods employed for synchronization with the main emphasis on interrupts. In Chapter 10, timing interface-related concepts are introduced. We discuss how timers work in input and output modes. The working of the Systick timer is also illustrated. In addition, various clock sources available on TM4C123 are also listed, which show the available flexibility to clock the microcontroller. Chapter 11 discusses different serial communication interfaces involving UART, SPI, I^2C, and CAN bus. Both the communication protocol-related details along with the microcontroller configuration steps required to use any serial interface are outlined. Chapter 12 of the book is related to analog interfacing. In this chapter, we present the basics related to analog-to-digital conversion (ADC) and then detail the various capabilities and configuration steps required to use the ADC module available on the TM4C123 microcontroller.

Audience

This textbook is aimed for the first course on *microprocessor systems* or *embedded systems*. The book is equally useful for electrical/computer engineering as well as computer science departments. The book assumes that the students should have experience with C programming language and digital logic design. The book is appropriate for undergraduate students. It can also be used to teach a first-year graduate course related to embedded systems. In addition to the theory related to ARM Cortex-M, the book has been written in such a manner so that one should be able to design one's own embedded systems after going through it. So, it should be useful to practitioners in the embedded industry working on ARM Cortex-M-based platforms.

One Final Note

We would like to encourage students and instructors: while using this book, if you have any suggestions or comments, please e-mail us. We will certainly incorporate your useful ideas to make the book better in any subsequent edition. You can contact us using our e-mail ids mtahir@uet.edu.pk and kashif.javed@uet.edu.pk.

Acknowledgments

Since we began writing this book in 2013, many people, though mostly our students, kept on providing their help and support. We are grateful to all of them. Special thanks go out to the following who have reviewed and provided us with their invaluable feedback:

Muhammad Abdullah, lab instructor, Embedded Systems Lab, Al-Khawarizmi Institute of Computer Science (KICS), University of Engineering and Technology (UET),

Lahore; and Zeeshan Altaf, Senior Engineering Operations Manager at Mentor Graphics Pakistan.

We would also like to express our gratitude to the entire team at CRC Press. In particular, we are thankful to Richard O'Hanley for his support. In the end, we are grateful, too, for the patience, ongoing support, and encouragement of our family members. Visit the eResource at: https://www.crcpress.com/9781482259384

Muhammad Tahir and Kashif Javed

Department of Electrical Engineering
University of Engineering and Technology, Lahore

Chapter 1

Introduction

Overview

This chapter provides a general introduction to embedded systems and its building blocks. In particular, basic concepts related to microcontrollers, microprocessors, different types of memories, and the software tools required to work with microcontrollers are introduced. Different attributes that are the basis for microprocessor architecture classification are also discussed. We adopt a top-down approach in explaining these concepts.

1.1 What's the Book About

If we look around at our surroundings for a short while, we will encounter systems being controlled by tiny computers everywhere. Air-conditioners, digital clocks, washing machines, cell phones, point-of-sale systems, and blood sugar measuring equipment are a few to name. These are examples of embedded systems. A system that performs a dedicated task, with the help of a computer embedded inside it, is named an embedded system. The tiny computer, which is comprised of components such as memory, *central processing unit* (CPU)[1], and programmable input/output (I/O) interfaces is located on a single integrated chip and is termed a microcontroller. Since the advent of the first microcontroller in the early 1970s, more sophisticated architectures and organizations of the internal components have been developed in order to meet the needs of the advanced and complicated application requirements. A microcontroller or a microprocessor, at the heart of an embedded system, is integrated with other physical devices or modules to develop a standalone system.

The overall objective of this book is to familiarize the readers with the effectiveness of microcontrollers so that they can design and develop their own embedded systems for many different appealing applications. We have chosen ARM® Cortex®-M

---

[1] We will use the words CPU and microprocessor interchangeably throughout this book.

1

family based microcontroller, which is based on a 32-bit ARM processor core from ARM Holdings. ARM Holdings designs a family of reduced instruction set computing (RISC) based processors as well as software development tools, but does not manufacture processor chips. Rather, it licenses the chip designs as well as ARM instruction set architectures to third parties to develop their own products. The Cortex-M processors are enriched with advanced features from the latest ARMv7-M architecture and have become the basis for many industry-leading 32-bit microcontrollers.

The dilemma in learning as well as teaching such a practical subject is that it is very difficult to find a resource that provides a unified treatment of the subject, covering details related to processor architecture, assembly language programming as well as interfacing of external devices using different peripherals. There are few good books available for the Intel x86 architecture, which cover the above-mentioned three aspects. However, we could not find a similar treatment in the context of ARM architecture and this book is just the first step in that direction. The general concepts related to microcontrollers are described in this introductory chapter. The next two chapters provide details about the architecture of Cortex-M while Chapters 4 to 7 will deal with the ARM assembly language programming. In Chapters 8 to 12, the reader will learn how to interface external devices using different peripheral interfaces. We will use Cortex-M4 based microcontroller platform from Texas Instruments (TI). There are many other Cortex-M based microcontrollers available in the market from different vendors, e.g., ST Microelectronics and NXP, to name a few.

This chapter provides a general introduction to embedded systems and its building blocks. In particular, basic concepts related to microcontrollers, microprocessors, different types of memories, and the software tools required to work with microcontrollers are introduced. We adopt a top-down approach in this chapter in explaining these concepts.

Most of us use digital computers everyday. Broadly, a computer is a combination of two components namely hardware and software. Hard disk, CPU, memory (volatile also called RAM), CD-ROM and printers are a few examples of the hardware to name, which are considered part of a computer. We use software such as applications (one of the well-known applications these days is SKYPE, which allows different users to communicate over the Internet with each other) for various purposes and operating systems (Windows and Linux are two highly used operating systems) to manage the applications and computer hardware efficiently. Desktops and laptops are examples of such digital computer systems. A number of tasks can be accomplished with these general-purpose digital computers. Figure 1.1 shows the block diagram of a typical computer system.

Input devices, such as keyboard and mouse, allow a user to provide input to the computer system. The users can see useful information on output devices such as monitors and printers. These input/output (I/O) devices communicate with the CPU through different types of interfaces. The CPU carries out all kinds of processing and comprises three basic building blocks, namely, arithmetic logic unit (ALU), control unit and set of registers. The main memories are random access memory (RAM) and read-only memory (ROM). These memories are accessible directly from the processor. Different types of these memories and their related features will be discussed later in

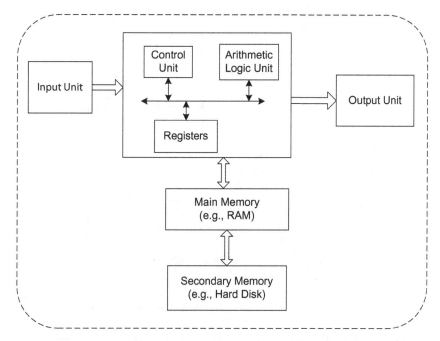

Figure 1.1: Block diagram representation of a general purpose computer.

this chapter. Low access-time and random access are their major attributes. Hard disk is an example of the secondary memory and does not interact with the CPU directly. Its information access-time is relatively higher than that of the primary memories. The memory space requirements to store a large number of different applications makes it almost mandatory to use secondary memories, which usually have a relatively large amount of space compared to primary memories.

1.2 Embedded Systems

An embedded system is based on almost the same concepts as those employed by digital computers. A generic block diagram of an embedded system is shown in Figure 1.2. However, an embedded system is normally developed for a specific application to perform dedicated task(s). Their small size, low cost, and low power requirements have resulted in their widespread use. Examples include cell-phones and other handheld devices, blood pressure monitoring equipment used by doctors, digital multi-meters in the hands of electrical engineers, temperature and humidity measuring devices used by weather stations, to name a few. The requirements of these systems make processor, memory, and other interfaces to play a key role and are the building blocks for the tiny computer (also known as microcontroller) embedded inside a specific embedded system. The software is designed to handle a range of different tasks and is normally programmed to ROM memory, which is not accessible to the user of the embedded system. Similar to the hard disk, the ROM in embedded systems is employed to store programs permanently even when the power is turned off.

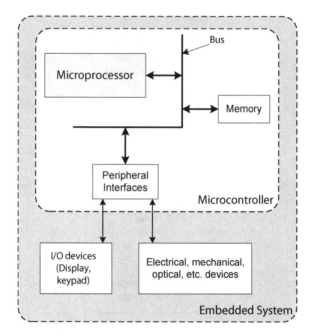

Figure 1.2: A generic block diagram of an embedded system.

To further elucidate the relationship among embedded systems, microcontrollers, and microprocessors, let's refer to Figure 1.3. From the system hardware perspective, an embedded system embeds a microcontroller inside it, which in turn contains a microprocessor core. This is a generic hierarchical view and will be elaborated further throughout this book.

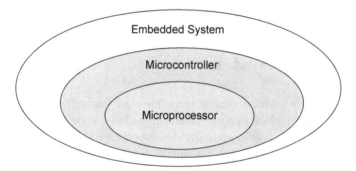

Figure 1.3: Embedded system hierarchy from a hardware design perspective.

1.2.1   Examples of Embedded Systems

No one can deny the importance of embedded systems. They are ubiquitous in our everyday lives. Our houses, cars, toys as well as offices are all equipped with these embedded systems making our lives more comfortable. We list some of the common application areas of embedded systems next.

1. Applications in communication: Radios, telephones, cellular phones, answering machines, fax machines, wireless routers.

2. Consumer electronics: Washing machine, clocks and watches, games and toys, remote controls, audio/video electronics.

3. Automotive systems: Braking system, electronic ignition, locks, power windows and seats, collision avoidance.

4. Commercial usage: ATM machines, bar code readers, elevator controllers.

5. Medical treatments: Cancer treatments, dialysis machines, blood pressure measuring equipment, electrocardiography (ECG), etc.

6. Industrial: Process automation, oil refineries, food processing plants, paper and board mills, etc.

7. Military use: Missile guidance systems, global positioning systems, surveillance systems.

1.2.2 Design Parameters of Embedded Systems

In the design of an embedded system, there are many key factors that need to be considered. Broadly speaking, a good embedded system design will be the one that can achieve better trade-off between price and performance. While designing an embedded system, the factors that are of utmost importance are listed below.

1. Power consumption

2. Speed of execution

3. System size and weight

4. Performance accuracy

An embedded system's processing speed, the power it consumes, its size, and of course how accurately it performs the task assigned to it are those parameters which help a user in the selection process. Furthermore, we can also chalk down the parameters that are commonly used in the selection of a microcontroller and are listed below:

1. Processing rate and processor size

2. Different types of I/O devices that can be interfaced

3. Memory size (RAM and ROM)

Now we introduce the microcontroller, which is next in the system hierarchy.

1.3 Microcontrollers

A *microcontroller* combines a microprocessor, read only memory (ROM), random access memory (RAM), and input/output (I/O) peripheral devices on a single chip. The microprocessor is similar to a human brain, which sends signals to control as well

as information exchange with various subsystems. For this communication among
subsystems, different buses are used. A bus is a collection of wires over which digital
signals propagate in the form of 0s and 1s. Buses internal to the chip can be broadly
categorized in three groups namely data, address, and control. Figure 1.4 shows
the basic organization of a microcontroller, depicting how buses are used to connect
memory and peripherals to the microprocessor. For the system shown in Figure 1.4,
any memory location inside RAM or ROM memories as well as registers associated
with different peripheral devices are identified by unique numbers called addresses.
Address signals flow over the address bus. The size of the address bus decides the
unique addresses generated by a microprocessor. The data bus carries the data.
Increasing the size of the data bus allows more number of bits to be communicated
between two subsystems. Control signals such as read/write inform whether the CPU
is interested in reading some information or wants to write some information to the
location whose address was generated over the address bus.

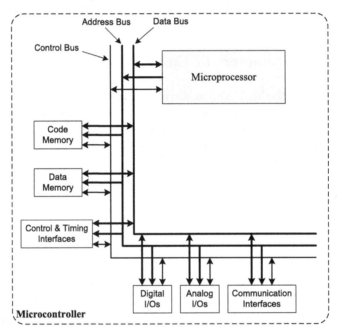

Figure 1.4: Block diagram illustrating the basic building blocks of a microcontroller.

The capability of a microprocessor to process simultaneously a certain number of bits
defines the type of microcontroller in terms of processing bits. For example, the first
generation microcontrollers were based on 8-bit CPUs, which means that the processors inside those microcontrollers can process 8-bits simultaneously. The need for
more processing power and advancements in technology have led to microcontrollers
that can process 32 bits simultaneously. In addition, the reduction in the price gap
between 8-bit and 32-bit microcontrollers has been a key factor in the popularity of
32-bit microcontrollers.

Another major component of a microcontroller is the general purpose peripheral device also called *port*, which provides a physical connection between the microcontroller
and the outside world devices. Based on the type of the device being connected with

the microcontroller, ports can be configured as input or output. Light emitting diodes (LEDS) and seven segment displays are examples of output devices and are connected with output ports of the microcontroller. On the other hand, a switch is an example of an input device and is interfaced with an input port of the microcontroller. Ports typically have registers associated with them, which are used for sending and receiving data to and from a communicating device. In addition, some of the port registers are used to monitor the device status, while other registers are used to configure the direction of the port. Obviously, the microcontroller needs a software program for proper working of ports and to read/write to the ports to exchange information with the outside world.

Information exchange between a microcontroller and a device can take place in parallel or serially. Broadly, I/O ports can be classified in two categories as discussed below.

- *Parallel Port*: When two communicating entities are connected with a group of lines, a number of data bits can be exchanged simultaneously.

- *Serial Port*: In this case, there is only one line for the data to travel between two communicating parties.

Analog-to-digital converters (ADC) and digital-to-analog converters (DAC) are also extensively used, which provide interfacing means between microcontrollers and devices generating analog signals such as a temperature sensor. We discuss them in detail in Chapter 12.

One of the key differences between microcontrollers, employed by embedded systems and microcomputers (or personal computers) is that RAM, ROM, CPU, and I/O interfaces in a personal computer are typically integrated in the form of multiple different chips on a motherboard, while all of these components in the case of a microcontroller are embedded inside a single chip. In addition, the memory and peripherals embedded in a microcontroller are of reduced speed and size compared to their counterparts used in a personal computer. Next, we start our discussion on memory and its different types, which is an important building block of a microcontroller.

1.4 Memory: Information Storage Device

Digital computers including general-purpose as well as embedded systems process and store information in binary digits called *bits*. These bits are used for representing operators, operands, and addresses. There are two possible states for a binary bit. A '1' represents the presence of a voltage and a '0' represents the absence of voltage. This representation is called *positive logic*.

To store binary information, digital electronic circuits such as complementary metal oxide semiconductor (CMOS) can be built and a collection of a large number of such circuits is named memory. Memory stores data sequentially and is normally depicted as shown in Figure 1.5. Memories are made up of data storage locations, which are uniquely addressable, and as a result accessible, by the processor. In most of the memories, each storage location can hold 8 bits of information. To uniquely address different memory locations, addresses are assigned to them. Figure 1.5 shows a few memory locations with their corresponding addresses. Since each byte is uniquely addressable, such memories are called *byte addressable* memories. Other possibilities are bit addressable as well as word (32-bit) and halfword (16-bit) addressable memories. During a write cycle, data is transferred from processor to memory. Data flows in the opposite direction while a read operation is carried out. Based on read and write capabilities, memories are grouped in two categories namely read-only memory (ROM) and random access memory (RAM). The former allows read operations only while in the case of RAM both read as well as write operations can be performed by the microprocessor. In the case of ROM, write operation requires special hardware capability. How many times the write operation can be performed depends on the type of ROM memory. Information can be accessed randomly from both memories.

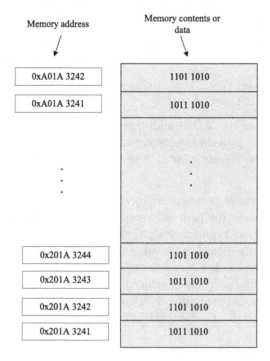

Figure 1.5: Memory locations hold data and are identified by their unique addresses.

1.4.1   Read Only Memory

The memory that allows the processor to only read its contents is read only memory (ROM). Another attribute of ROM is that they can store information permanently even when no power is applied. The information in the ROM is therefore nonvolatile. Instructions or a program code are the information that is normally stored in a ROM. For example, the boot sequence with which an operating system is loaded is a piece of code, which does not vary over time and is stored in a ROM. This is a common usage of ROMs in desktop and laptop computers. For embedded systems that have attributes such as low power, low cost, and small size, ROMs are handy in storing application program codes. Based on how a ROM is programmed, ROMs can be of different types and are listed as follows.

- *Programmable ROMs (PROMs)* can be programmed only once. Any change to the contents will require the replacement of the chip.

- *Erasable Programmable ROMs (EPROMs)* can be reused by erasing and re-programming. Conventionally, information was erased and written to with the help of ultraviolet (UV) light. The changes cannot, however, be made while the chip is installed in the system. We need to remove it with the help of dedicated equipment and reinstall it whenever we need to change its current contents. Another drawback of EPROM is that the whole chip needs to be completely erased. The EPROMs have become almost obsolete and are replaced by EEPROMs and flash memories discussed next.

- *Electrically Erasable PROMs (EEPROMs)* are EPROMs but their contents can be erased and written to by applying electric signals to the storage cells. Furthermore, erasure and writing can take place while the chip is installed in the circuit. Conventionally, EEPROMs were limited in their capability and only allowed single byte read and write operations making them slow speed devices. However, modern EEPROMs are capable of multi-byte based page operations.

- *Flash memory* is a type of EEPROM that allows read and write operations to be carried out in large multi-byte blocks. In general, the erase cycles for non-volatile memories are slow. Flash memory allows erasing large block sizes, which provides these memories a significant speed advantage compared to the EEPROM when dealing with large amounts of data. In addition, the cost of flash memory is also quite low compared to byte-programmable EEPROM. As a result flash memories have become the dominant memory type for those applications which require large volumes of non-volatile and solid-state storage. One of the limitations of flash memories as well as EEPROMs is their limited number of read and write cycles.

1.4.2 Random Access Memory

The memory which allows the processor to read from and write to its locations is named random access memory (RAM). However, the capability of randomly accessing a memory location is not limited to RAM, rather ROM-type memories can also be accessed randomly. One limitation of RAM is that information stored in it is lost as soon as the power applied to it is removed. On the other hand, RAM is not limited by the number of read and write cycles and is more suitable for storing data that is updated frequently. The read and write operations of RAMs are faster than those of ROMs. There are different types of RAMs, which are briefly described here.

- *Dynamic RAM (DRAM)* is the most commonly used type of RAM these days. Each memory cell of a DRAM, which can store one bit of information, is made up of two transistors and a capacitor. The transistor acts as a switch while the capacitor holds the charge. As the capacitor charge leaks, the voltage representing '1' needs to be refreshed. This refresh operation is performed a number of times in one second and results in reducing the memory operating speed.

- *Static RAM (SRAM)* employs a flip-flop for storing a bit in a memory cell. A flip-flop requires 4 to 6 transistors and does not require refreshing circuitry.

As a comparison of DRAM and SRAM, the latter is faster than the former because of the absence of refresh cycles. However, on a given chip SRAM yields less memory space as compared to DRAM. This is attributed to more numbers of transistors required by an SRAM cell. This also makes them more expensive. Because of their features, SRAM is normally used for building cache memories while DRAM is employed for RAM chips. In comparison to ROM, the access time of RAMs is less.

There are other types of RAMs such as synchronous dynamic RAM (SDRAM), double data rate SDRAM (DDR SDRAM), and rambus dynamic random access memory (RDRAM). We will not discuss them in this book. For further details, we refer the interested readers to [Messmer(2001)]. For byte addressable memories, with 16/32 bit data bus interface, the memory accesses can be either aligned or unaligned as discussed next.

1.4.3   Aligned and Unaligned Memory Accesses

For a byte addressable memory, each memory access of size byte is inherently aligned. A 32-bit processor will most probably have a 32-bit data bus to access the memory. For such a scenario, it is possible to perform memory read/write operations of halfword (16-bit) and word (32-bit) size. If a memory access, of size halfword, is performed from an even address, then this memory access is an aligned access. On the other hand, if an odd address is used, then the resulting memory access is an unaligned access.

Similarly, for a 32-bit word memory access, word aligned means the data is stored on a word boundary, i.e., the memory address accessed is divisible by 4. A word memory access from an address not divisible by 4 is termed unaligned word access. Next, we turn to the general concepts about the working of microprocessors, the heart of microcontrollers.

1.5   The Microprocessor

A microprocessor, also known as a central processing unit, has a capability of executing instructions at an extremely high speed. Broadly, a microprocessor has five basic components, which are listed below:

- arithmetic logic unit (ALU),

- control unit,

- bank of registers,

- interconnection buses, and

- timing unit.

The block diagram of a microprocessor is shown in Figure 1.6. The job of the arithmetic logic unit (ALU) is to perform arithmetic and logical operations such as addition, subtraction, and logic operations such as AND, OR operations. Because the microprocessor can work with only binary digits, each operation is encoded by a unique binary combination, which is termed the operational code or briefly the opcode. Various steps required for the execution of an operation are performed using binary signals, which are known as control signals. These control signals are generated by the control unit. The operations and operands are initially stored in the memory. They are brought inside the microprocessor using buses and are stored temporarily in microprocessor registers. Data transfers among different processor registers as well as ALU are performed using the internal interconnection buses inside the microprocessor and are different from the data, address, and control buses used by the processor to communicate with the external memories as well as peripheral devices.

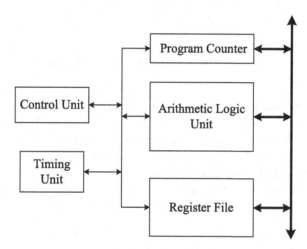

Figure 1.6: Block diagram of a microprocessor.

The job of a microprocessor is to execute a user program. The executable program is stored in memory initially and is executed instruction by instruction. During program execution, the microprocessor performs different operations including fetching instructions and data from memory, communicating with input/output ports, etc. The instructions from memory and data from memory or I/O ports is stored temporarily inside the microprocessor using processor registers. For instance, the address of the next instruction to be fetched from the memory is contained in the program counter (PC) register. Once the instruction is brought from memory into the microprocessor, it is stored in yet another register. The control unit extracts the opcode and decodes it or generates control signals so that the operation can be carried out. It may bring some variables, which are the operands of the instruction, from the memory into the registers of the microprocessor. The CPU therefore needs a large set of registers. For example, when the control unit is required to add two numbers stored in two different registers, values from the register will travel via internal buses to the input of the ALU unit. The processor carries out all the activities according to a timing sequence. This is where a processor needs a clock or timing unit to synchronize its tasks.

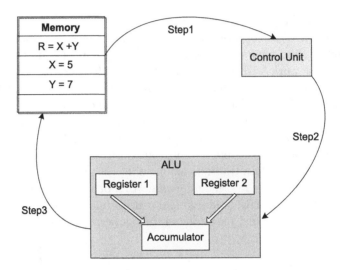

Figure 1.7: The working of a microprocessor.

Figure 1.7 shows the working of the microprocessor when it executes the $R = X + Y$ addition instruction. Let's assume that this instruction is compiled and is present in the memory. In a basic form, this instruction will consist of three fields. One of them will contain the operational code representing addition, while the other two fields will contain the memory addresses of the locations where variables X and Y are stored. We further assume that the PC register contains its memory address. The major steps that are carried out by the microprocessor are listed below.

- The microprocessor fetches the instruction by applying PC contents on the address bus, a read signal on the control bus
- The control unit decodes its opcode
- Processor executes $R = X + Y$ by
  - fetching the current value of X from the memory
  - fetching the current value of Y from the memory
  - instructing the ALU to add these two numbers
  - writing the sum back to the memory address of R

These are the steps carried out by the processor to execute one instruction. These days, processors can execute billions of such instructions in one second, which determines its true processing capability.

1.6   Microprocessor Architecture Classification

The microprocessor architecture can be classified using different aspects. Figure 1.8 shows two different possible classifications of the processor architecture.

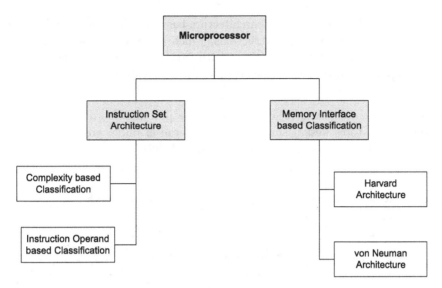

Figure 1.8: Classification of microprocessor architecture.

1.6.1 Instruction Set Architecture

Now let us discuss the processor architecture classification based on instruction set architecture (ISA). There are two important aspects that can be used to define ISA classification. One classification can be done based on the complexity of instructions, while the other possibility is based on the instruction operands. Next we discuss each of these ISA classifications.

Complexity-Based ISA Classification

Using the ISA complexity as the classification measure, we can categorize the microprocessors into two groups: complex instruction set computer (CISC) and reduced instruction set computer (RISC). In reality, there is a spectrum of architectures that we can classify as CISC or RISC. Figure 1.9 depicts the major differences between the two architectures. We make the following general observations when deciding whether to call a computer CISC or RISC.

Complex instruction set computers (**CISC**): The key features of the CISC architecture are discussed below highlighting the advantages and disadvantages of this architecture.

- One possible use of complex instruction set architecture was in early computers where the processors were much faster than available memories. Fetching an instruction from memory used to become the performance bottleneck.

- One of the advantages of CISC architecture is that a single complex instruction can perform many operations. For instance, find the zeros of a polynomial. However, complex instructions require many processor clock cycles to complete and most of the instructions can access memory.

- A program running on a CISC architecture based machine involves a relatively small number of complex instructions, which can provide high code density. In addition, many instruction types with varying instruction length are available supporting different addressing modes while requiring fewer and specialized registers.

- In CISC the complexity is embedded in the processor hardware, making the compilation tools design simpler. Some of the example processors based on CISC architecture are Intel (x86) and Freescale 9S12.

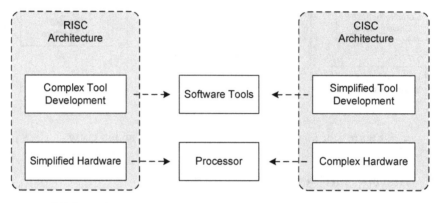

Figure 1.9: CISC emphasizes hardware complexity. RISC emphasizes compiler (software tools) complexity.

Reduced Instruction Set Computers (**RISC**): We outline the main features of the RISC architecture below and discuss both the advantages as well as disadvantages of this architecture.

- This architecture is suited for those scenarios where the processor speeds match that of memories. This speed matching reduces the penalty for instruction or parameter fetching from the memory. Another associated advantage with RISC is the simplified instructions used by this architecture.

- Each complex operation is broken into multiple simplified operations and dedicated instructions are provided for these operations. For instance, load and store instructions are provided for memory read and write operations. The other instructions cannot access memory directly.

- The simplified instructions make it possible to execute an instruction in a single processor clock cycle. However, this is not true in general and some of the instructions in RISC may require more than one clock cycle for its execution completion.

- A task when run on a RISC computer requires a relatively larger number of simplified instructions and results in low code density. An associated advantage is fewer memory addressing modes resulting in reduced complexity.

- The simplicity in the hardware architecture in RISC is complemented by the increased complexity in the generation of assembly code by the tools (compiler) or by the programmer. Some of the processor architectures based on RISC are MIPS, ARM, SPARC, and PowerPC.

The answer to the question of which architecture is best suited for a specific application is very much situation dependent and is beyond the scope of this book. However, it is important to recognize the terminology. In addition, it is very difficult to compare the execution speed of two computers, especially between a CISC and a RISC. One way to compare is to run a benchmark program on both and measure the time it takes to execute. For example, the 50 MHz ARM Cortex-M processor has one bus cycle every 20 ns, while on average it may require 1.5 cycles per instruction. If the benchmark program executes 10,000,000 assembly instructions, then the time to execute the benchmark will be 0.33 seconds. This issue will be discussed further later in this section.

Instruction Operand-Based ISA Classification

Instructions in an assembly language program in general have multiple operands. At this time it is not important how many operands an instruction has. Rather we are interested in classifying different possibilities of specifying these operands. The operands for an instruction can be specified either using memory or registers or combination of both. The ISA classification based on how the instruction operands are specified can be categorized in the following groups.

- *Memory-memory*: This type of ISA allows more than one operand of most instructions to be specified in memory. VAX and PDP series are examples of this type of architecture.

- *Register-memory*: These architectures allow one operand of an instruction to be specified in memory, while the other operand is in CPU register. The well-known x86 and Motorola 68k are examples of this architecture. In this ISA the individual instructions execute faster, compared to memory-memory based ISA, due to fewer memory accesses. However, this case may require more number of instructions to complete the same task.

- *Register-register*: This ISA classification is also called load-store architecture. Direct access to the memory is not allowed to most of the instructions in this ISA. Rather specific instructions, named as load and store instructions, are responsible for any data movement between registers and memory. All instructions other than load and store instructions get their operands from and store their results to registers. The execution of most of the instructions in load-store ISA is very fast, in many cases single clock cycle. This is due to the fact that most of these instructions operate within the processor using registers. However, it is worth mentioning that load-store based ISA requires the most number of instructions to complete a given task. But since in many cases the immediate results are not stored to the memory, rather they are temporarily placed in the registers, it might be used by the subsequent instructions, saving extra memory load and store operations. ARM and MIPS belong to this class of ISA.

1.6.2   Memory Interface-Based Architecture Classification

There are two widely used memory interface architectures, namely, von Neumann architecture and Harvard architecture. Both of them are shown in Figure 1.10. The von Neumann architecture uses a common bus for both data as well as code memory. As a result either an instruction can be fetched from memory or data can be read/written to/from memory during each memory access cycle. Instructions and data are stored in the same memory subsystem and share a common bus to the processor.

The Harvard architecture, on the other hand, utilizes separate buses for accessing code and data memories. Using separate buses for code and data memories allows instructions and data to be accessed simultaneously. In addition, the next instruction may be fetched from memory at the time when the previous instruction is about to finish its execution, allowing for a primitive form of pipelining. Pipelining decreases the per instruction execution time; however, main memory access time is a major bottleneck in the overall performance of the system. Mostly, microprocessors are implemented using von Neumann architecture, while microcontrollers use Harvard architecture. ARM-based microcontrollers use Harvard architecture.

1.6.3   Performance Comparison of Different Architectures

A microprocessor's architecture can be compared against other architectures by running benchmark programs. The following performance evaluation equation is used to quantify the execution speed of a microprocessor.

$$\text{Execution Time} = T_c \sum_{i=1}^{N} C_i \qquad (1.1)$$

In (1.1), T_c is the cycle time, C_i represents the number of cycles required for i^{th} instruction execution, and N is the number of instructions in the test program. If each instruction requires the same number of cycles for execution, then the expression in (1.1) is simplified to

$$\text{Execution Time} = T_c N C. \qquad (1.2)$$

In (1.2) we have used the fact that $C_i = C \; \forall i$. From the expression in (1.2) we observe that there are three different possible ways to speed up the execution of a microprocessor and are listed below:

1. Use fewer instructions for a given program. In other words, it is possible to improve the execution speed by efficient programming. This aspect will be highlighted in later chapters while implementing different tasks.

2. Reduce the number of cycles for the instructions. This is mainly dependent on the instruction set architecture of the microprocessor.

3. Speed up the clock frequency of the microprocessor or equivalently reduce the cycle time. This refers to the maximum clock frequency of the processor. The maximum clock frequency depends on many different aspects and a few are discussed below.

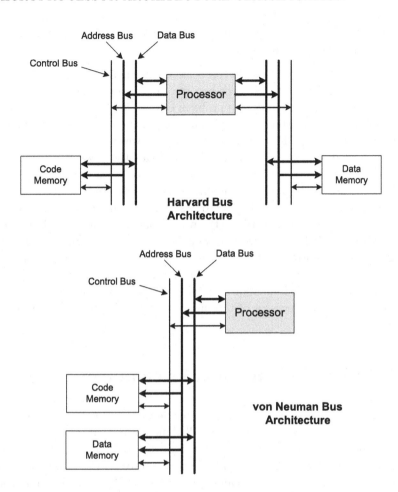

Figure 1.10: Difference between Von Neumann and Harvard Architectures.

There can be multiple different reasons for clock frequency limitation of a microprocessor. Some of those are physical and some others are purely technical. We discuss two key reasons here, first one is physical and the second is technical.

The physical limitation of processor clock frequency is based on the fact that the electrical signals in a circuit approximately travel at the speed of light. Let us consider a scenario, assuming a microprocessor is running at 3 GHz clock frequency. In other words, one clock cycle takes $\frac{1}{3}$ ns (nanosecond). Now the electrical signals travel a physical distance of approximately 10 cm during $\frac{1}{3}$ ns. To assume that a circuit is working synchronously, we need to ensure that a clock signal is available at different parts of the circuit (almost) simultaneously. This can be approximately ensured by requiring the propagation delay much less than clock cycle time. Let us use a factor of 10, which will require the size of the circuit be no more than 1 cm. The CPU core die size of Intel Core2 Duo Processor [Intel(2014)] is approximately 1 cm. Now if we want to double the speed of this processor, the size of the processor core should become one-half. If this size limitation is violated, then what happens is that some parts of the processor circuit are operating in the current clock cycle, while some other parts of the circuit are still not done with the previous clock cycle. This type of distributed system is very hard to deal with in practical systems.

Now let us discuss a technical limitation associated with increasing the clock frequency of a microprocessor. As the CPU cores become smaller and smaller due to reduced fabrication process dimensionality, the problem of heat dissipation starts arising. As you may already be aware, processors are made up of transistors and these transistors can be switched ON and OFF at the clock frequency. Now we know that each transistor dissipates power when switched from one state to another, switching at fast speeds leads to more power dissipation. When the processor core size is reduced to increase the clock speed, the corresponding amount of heat generated by these millions of switching transistors couldn't be dissipated easily due to the reduced chip size. First the heat sinks and fans are connected to the processors and sometimes liquid nitrogen is also used in some high end servers as well for the cooling purpose. The amount of speed gain and the corresponding increase in the cost of associated cooling mechanism are some of the practical barriers in increasing the processor speeds further. One natural solution to this limitation is to use multiple processors of moderate speed rather than to have one processor of too high speed, leading to multi-core processor architecture.

1.7   Software System and Development Tools

The intelligence of a microcontroller-based embedded system lies in its software, which is a collection of instruction sequences, also called functions, that are stored in memory (either ROM or RAM). These functions are implemented to perform certain tasks and are fundamental building blocks of a software system. Based on the collective functionality of a set of functions, we can further differentiate among different software sub-systems. Certain rules are followed using well-defined interfaces to integrate these sub-systems to construct a complete software system. An arbitrary software system does not necessarily require all the software sub-systems.

1.7.1   Software Sub-Systems

Some of the important software sub-systems include operating system, device drivers, libraries, and user applications. However, it is worth mentioning that this list of sub-systems is not by any means exhaustive. The word *operating system* is a standard nomenclature but we are calling it a sub-system in the context of our software system hierarchy. In the following, we discuss a brief introduction of these sub-systems and also highlight their interfaces with other sub-systems.

- *Operating system*: It is hard to define an operating system encompassing all of its different aspects. The definition of an operating system may change depending on one's needs as well as one's view of the system. Broadly speaking, operating system is the in-charge of a collection of resources including the microprocessor, memory, Input/Output devices. On one hand, the operating system manages these resources using the hardware architecture dependent OS components, while on the other hand, it allows user applications to use these resources in a systematic way through OS call interface for applications. These interfaces provided by the OS can be observed in Figure 1.11.

- *Device drives or drivers*: A device driver is a software sub-system, which allows the operating system to communicate with the hardware devices. Drivers for some of the standard devices are sometimes integrated as an essential component of the operating system. However, there are many hardware devices, which require a driver and are provided by the vendor of the device.

- *Libraries*: The software library like other sub-systems is a collection of function calls developed for specific job and is made available to the user. By using a library, the software application developer (to be discussed next) reuses the specific functionality already implemented by the library and does not need to implement it again. Libraries allow sharing the code in a modular fashion, as well as easing the distribution of the code.

- *Applications*: Application software is at the top of the hierarchy in the software system architecture. It is a collection of one or more programs, which are designed to perform operations for a specific application requirement. Application software cannot run (or execute) on its own, rather it is dependent on the OS and possibly libraries.

The integration of different sub-systems discussed above is shown pictorially in Figure 1.11.

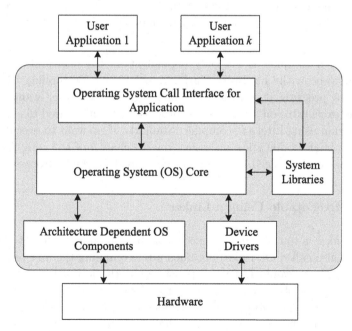

Figure 1.11: Software system architecture.

1.7.2   Software Development Tools

Next we start discussion about the software tools used to convert the user programs to executables, which can be interpreted and run by the microprocessor/microcontroller. The process of converting user program written in a high level programming language can have multiple intermediate steps with corresponding outputs. There are two main steps involved in the conversion of a user program to an executable: (1) compilation and (2) linking.

Compilation Process

The tool used in the compilation process is called either a compiler or an assembler depending on the type of the user program source file. If the user program is written in assembly language, then we use an assembler to convert it to an *object code*, which is also called *machine code*. In this process, an assembly instruction is converted to its equivalent *opcode* or machine code. Assembly language programs are also called low level programs. If the user program is written in a high level language, we will take the example of a C program, then we use a compiler to convert it to the machine code. Another possibility is that the compiler converts the code to an equivalent assembly program, which is converted to machine code by the assembler. However, in general, the compilers give the machine code as an output. Furthermore, most of the compilers can also provide the assembly equivalent code, in addition to machine code, as a side output. This can be done by configuring the appropriate compiler options (also called compiler switches) at the compilation time. The process of converting a user program to machine code using either a compiler or an assembler is depicted in Figure 1.12.

The selection of a compiler is not only dependent on the choice of the high level language, but also on the hardware platform to be used for running the program. For example, a program is written in C language and the user wants to run its corresponding executable on an Intel microprocessor. Then we need to compile the C language program using Intel's C compiler. Similarly, if we want to execute the same code (with minimal modification necessary due to change of the compilation tools) on an ARM machine, we need a C compiler developed for ARM processor.

Building an Executable Using a Linker

The job of a linker is to construct an executable by combining different object codes (or object files also called OBJ files) obtained after compiling the user program codes or files. However, we would like to emphasize that the job of linker is not merely combining of different object codes. In the process of linking, the linker has to decide what will be the locations (addresses) of different object codes and data segments. In addition, some of the functions used in the program are not implemented by the programmer, rather they are provided by a library. For instance, taking the log of a variable will require the use of a math library. It is also the job of a linker to get the desired modules from different libraries and integrate them in the process of

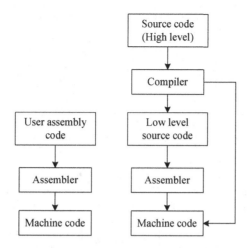

Figure 1.12: Software system architecture.

constructing the executable. A possible depiction of the functioning of the linker is shown in Figure 1.13. It should not be difficult to realize that the contents of a library are a collection of different object codes and the library itself is also constructed using a linker. The linker, in addition to building the executable, can generate, optionally, many other files containing useful information for the programmer. For instance, the linker can generate a map file that provides an overall summary of memory usage for code space as well as global data variables. Similarly, the list file can also be generated by the tools, which provides the disassembly of the code that might be helpful to the programmer in debugging. The blocks in Figure 1.13 showing the linker generated MAP and list files are colored to mark that they are optional outputs from the linker.

The above-mentioned use of libraries in the process of linking at the time of building an executable is called *static linking*. In this context, another important related concept is the *dynamic linking*, where the libraries are not used at the time of building an executable, rather they are used during the execution at runtime and are called dynamic link libraries or in short DLLs.

1.8 Debugging Tools and Techniques

The development of a microcontroller-based embedded system requires both hardware design as well as software programming. The successful design requires the hardware and software components work together seamlessly in an effective manner. The use of debugging tools helps to rectify the errors quickly and the effort required to make the system functional is reduced. Specifically, a debugging tool:

- Helps to reduce the system prototyping time,

- Can help identify both hardware and software problems in the prototype as well as the software application, and

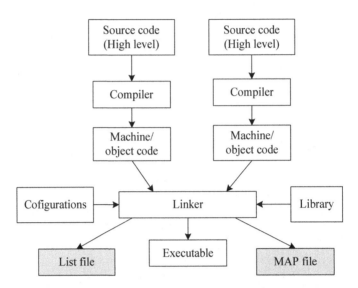

Figure 1.13: Software system architecture.

- Can assist in the process of fine tuning certain parameters of the system.

Elaborating further on the above-mentioned points, one of the key reasons to use debugging tools is based on the fact that development of an embedded system using a high performance microcontroller is a complex undertaking, which involves interfacing the hardware and software with the physical environment. It is very unlikely that the system works the first time as intended. The software may have many performance related issues and it is quite possible that the timing performance, of the properly executing code, is inadequate. It is also possible that the real time execution leads to an unpredictable behavior when exposed to external physical interactions in the form of I/Os and interrupts. Furthermore, verification of each functionality may be a mandatory requirement and the debug functionality allows verifying proper functioning of various modules for a class of test conditions. From the available large collection of debugging tools and techniques, the choice is dictated by the following three key factors:

- Cost,

- Ease of use, and

- Set of features offered by each tool to help analyze the system design.

In general, higher level of integration with the target hardware can take greater benefit of a tool which in turn can result in shorter development time. This, however, requires higher cost. Broadly, the debugging techniques and tools used for testing microcontroller-based systems can be divided into the following groups.

1. Manual methods: These techniques are not sophisticated but can be readily used for locating and correcting the program errors.

2. Software-only: The tools in this category come in the form of simulators and monitors and do not have any hardware dependency.

3. Software-hardware: This type of debugging tool is hardware dependent and spans a large range from highly expensive in-circuit emulators to relatively less expansive in-circuit debuggers.

Non-intrusiveness is a highly desirable attribute of a debugger. It allows the software-hardware system to operate in such a manner as if the debugger is not present. The level of intrusiveness is a measure of degree of perturbation that is caused to the program execution and as a result to its true performance by the debugging instrument. For instance, a print statement added to the program is highly intrusive since it significantly affects the real-time interaction of the hardware and software. In addition, for a real microcomputer system, the use of breakpoints and single-stepping is also intrusive because the hardware real-time state interacting with the physical phenomenon might change, while the software is not running. Thus, the use of intrusive debugging tools or techniques can cause significant alteration in the performance as the program interacts with real-time events. On the other hand, there are some debugging methods which are relatively less intrusive. For example, using LED-based indicators, LCD, or other external hardware displaying the status will be less intrusive. A logic analyzer, oscilloscope, etc., which passively monitor the activity in response to software execution, by monitoring the physical I/O signals is completely non-intrusive. Now we will discuss each of the above-mentioned methods and classes of tools.

1.8.1 Manual Methods

Most of the manual methods are quite simple to implement and this is the main reason for their popularity. Some of the widely used manual methods are listed below.

- Print statements are one of the most widely used manual debugging methods. In this method, selected variables and/or register contents can be printed on a terminal program for correctness and status checking.

- Status dumps is another manual method which is used for logging the status variables at critical sections of the software program.

Manual debug-segments based on print statements as well as status dumps can be integrated in the user program. These debug-segments can be enabled or disabled either at compile time or at runtime. The compile time enabling is achieved by using conditional compilation statements. On the other hand, encapsulating these debug-segment(s) inside the conditional statements, whose execution is dependent on a test-flag defined as a global variable, allows us to enable/disable their execution by setting or resetting the test-flag. This method includes the copy of the debug-segment(s) in the final application program resulting in an execution as well as memory storage overhead. However, it allows dynamic activation of the debug-segment(s) and many vendors use this approach for on-site customer support.

It is important to recognize that the usefulness of manual methods is limited for real-time systems, since their execution time is quite large. For instance, print statement takes large execution time depending on the baud rate used for serial communication. Furthermore, these methods are highly intrusive and lag sufficient control in terms of selectivity of an event of interest, reproducing an error, etc.

1.8.2   Software-Only Methods

The software only debugging solutions include two main tools; namely, the simulators and monitors. Below we discuss each of them.

Software Simulators

A software simulator is a program running on an independent computer hardware platform (usually a personal computer). The simulator, depending on its complexity, can simulate the execution of the instruction set as well as the behavior of an I/O of the target microcontroller. Simulators offer the lowest-cost development tools for microcontroller-based systems and many microcontroller manufacturers offer their simulators free of cost.

In a simulator environment, the user programs are executed (in non-real-time fashion) allowing the user to insert breakpoints to stop the program execution and perform analysis of the register and memory contents, visualize and modify the program variables, etc. Logical errors can be rectified easily by executing the program in single stepping mode, allowing the user to visualize the status and variable contents after each instruction execution. The key advantages of software simulation tools are listed below.

- One of the biggest advantages of software simulators is their free of cost availability.

- Most of the time they require only a computer for their operation and are supported for many different computer operating systems.

- Simulators support breakpoints allowing code single stepping, visualization as well as modification of registers and memory.

- If any of the peripherals are simulated, then it allows constructing and testing many complex scenarios.

- Timing-related information for performance measurement and analysis can be obtained by enabling tools such as stopwatch, trace time-stamp, etc.

Some of the disadvantages of the simulators are also discussed below.

- The execution speeds may not be fast enough compared to the high speed embedded microcontrollers. In addition, the complexity of the simulation as well as the speed of the computer used to run the simulator will determine the overall speed of simulation.

- The simulation process becomes quite complex, when functionalities including multiple inputs and waveform tracing as well as register content logging to files are enabled.

Monitors

Monitors are also software programs or tools but they are integrated in a microcontroller hardware and can be run for different purposes. A monitor software program usually resides at the top or bottom region of the microcontroller's memory. The monitor program can be used to download to burn the program code in the microcontroller, execute the code, set breakpoints as well as visualize and modify memory contents or registers.

The target microcontroller is usually connected to a terminal program using a serial communication (USB or RS232) interface. The monitor commands implemented on the target microcontroller can be executed by the user. One of the disadvantages of a monitor tool is the requirement to stop the execution of the application program before any memory location or register contents can be visualized or modified. Monitor tools were in common use in the early days of microcontrollers but their use has declined with the introduction of low price hardware-software tools (mainly in-circuit debuggers), which will be discussed next.

1.8.3 Software-Hardware Debugging Tools

Debugging tools, integrating hardware and software, offer many advantages in terms of capability as well as flexibility. These tools have become prevalent mainly due to reduced costs of the high performance microcontrollers. The price gap between a low-end 8-bit microcontroller and a 32-bit microcontroller with reasonably good performance has become almost negligible. This reduced price of high-end microcontrollers has led to the trend of integrating the debugging capability on each microcontroller evaluation board by the vendors. As of today, many microcontroller manufacturers are offering their evaluation platforms with integrated programming and debugging capability while costing less than $20. A few years back this was not the situation. Even today the users of 8-bit microcontrollers are still using the debugging methods of old days. Below we discuss both the conventional as well as the latest debugging methods in the hardware-software category and highlight their pros and cons.

Burn and Learn

In many non-complex embedded system developments, for instance a simple application based on an 8-bit microcontroller, the conventional *burn and learn* method can be used for debugging purposes. Broadly speaking, this method involves the following key steps in the process of debugging.

- Write or modify the program code.
- Burn (program) the executable to the microcontroller chip using a programmer or in circuit programming capability.
- Run the program (by resetting) and learn (observe) for proper functioning.
- If the program does not function as desired, then go back to first step.

- Continue this procedure until the required functionality and performance is achieved.

The learning part can be implemented in multiple ways, including toggling of the I/O pins, dumping the status and variable data through the serial port, sending the test data to an LCD to name a few. For example, the execution result of a task can be sent to an I/O port with LEDs connected to it for observing the behavior of the task under test. Depending on the availability of a digital to analog converter, one can also measure signal levels at the analog pins to check whether the data is as expected. Another possibility is to send the data to the LCD at various points of the program to trace the program execution. The burn and learn method does not require any specialized software tools. This method however is slow, inefficient, and tedious for testing a system's functionality. In general, the following hardware can be useful in burn and learn testing method.

- LEDs: An LED can be blinked on and off for binary type of status indications. One of the disadvantages is that it cannot be used in fast and time-critical applications.

- LCD display: Some of the status variables can be displayed on an LCD for performance analysis. The disadvantage is the tedious interfacing if the LCD uses a parallel interface.

- Serial port (USART or RS232): Using a serial port for data and status display is another possibility and is already discussed in the context of print statements.

- Voltmeter or oscilloscope: Traditionally, the oscilloscope is the most common tool for debugging the hardware-related issues but can also be used for status monitoring on I/O pins. Using a voltmeter or an oscilloscope is one of the simplest ways to analyze the proper functioning of a program. One of the limitations of using an oscilloscope is that it can only monitor a few signals depending on the number of available channels.

- Logic analyzer: A logic analyzer can be a very useful tool for debugging an embedded system where an external memory bus is used. It can also be highly effective and useful in testing different high speed interfaces such as USB, I2C, CAN, or SPI. A logic analyzer can log address, instruction, and some data information as the user program executes. The key advantages of a logic analyzer include high speed tracking capability of the application, tracing program flow in the trace buffer, completely non-intrusive, to name a few. The down side is that it can be highly expensive, does not enable setting breakpoints, and requires moderate effort for setting it up.

In-Circuit Emulators

The traditional debugging method for embedded microcontrollers has been based on in-circuit hardware emulators. In-circuit emulators (ICE) offer many powerful debugging features including real-time execution of the application code, breakpoint capability as well as peripheral implementation. Typically, an ICE uses a specialized hardware from the vendor of the embedded microcontroller. The target microcontroller in an ICE is replaced by a probe. Real-time trace buffers, multi-level conditional program execution, and full internal register and memory access are offered by in-circuit emulators. The key advantages of using the ICEs for debugging are listed below.

- ICE is highly flexible and provides maximum control for application system debugging. In an ICE the program memory is mapped to internal RAM with a trace analyzer connectivity.

- Debugging complex systems with real-time requirements becomes possible using ICEs.

- System development time is reduced significantly and using an ICE enables rectifying elusive problems, which otherwise would have been quite difficult to be localized.

The main disadvantages of ICEs are

- Compared to other competitive debugging tools, for instance in circuit debuggers, ICEs are relatively expansive. This is one of the main reasons for the ICEs to become less common.

- The ICE is a complex tool with highly capable hardware platform. As the embedded platforms are becoming faster and faster, it has become very challenging for an ICE to match the increase in the clock frequency of the microcontrollers.

- Another limitation is attributed to the increase in the speed and the reduced size of the microcontrollers, which requires larger memory and more dense probe connectivity capability from an ICE.

In-Circuit Debuggers

In-circuit debuggers (ICD) are becoming more and more prevalent and have almost replaced the more expensive in-circuit emulators. An ICD is a small hardware device, which is connected between a computer and the microcontroller development system. Many of the user development platforms available in the market today have a built-in ICD circuitry. The key motivating factors leading to the development of the ICDs are: (a) to make a debugging tool available, which is fairly inexpensive compared to ICEs and (b) to match the increasing complexity and speed of the microcontrollers being introduced to the market.

One of the benefits of using ICDs is that it allows the debugging of the complete program code and executes it on the actual microcontroller integrated on the target board. Another important aspect of an ICD is based on the fact that when an application program is debugged and has been deployed, it can be reprogrammed without any hardware changes or modifications.

The ICD utilizes special hardware features as on-chip debugging capability, which are integrated during the microcontroller chip design and fabrication process. The ICD communicates with the microcontroller using the microcontroller's on-chip debugging hardware. The ICD can send commands to the on-chip debugging logic to halt the program execution and can also request for status and application data variables from memory and registers. It also allows single-stepping of the application program as well as enabling of the breakpoints making its functionality quite similar to that of an ICE, but at a much lower price. For proper operation, an ICD requires some memory resources from the microcontroller. In addition, some hardware pins on the microcontroller chip are dedicated for communication with the ICD.

Figure 1.14 illustrates the connectivity of an ICD with a microcontroller platform on one side and the computer on the other side. The comparison among various debugging techniques is summarized in Table 1.1.

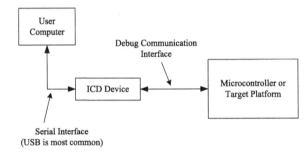

Figure 1.14: Block diagram illustrating the connectivity of an ICD with the target platform.

We conclude this chapter by defining black-box and white-box terms in the context of testing and debugging. *Black-box testing* is simply observing the inputs and outputs without looking inside. It has an important place in debugging a module for its functionality. On the other hand, *white-box testing* allows you to control and observe the internal workings of a system. Effective debugging may require the use of both the approaches. One must always start with black-box testing, subjecting a hardware or software module to appropriate test-cases. Once we document the failed test-cases, we can use them to aid us in effectively developing the insights for performing white-box testing.

Table 1.1: Comparison of various debugging tools.

Parameter	Burn and Learn	Simulator	ICE	ICD
Cost	Free	Very low	High	Low
Single stepping	No	Yes	Yes	Yes
Breakpoints	No	Yes	Yes	Yes
Real time execution	Yes	No	Yes	Yes
Loss of target pins	No	No	No	Yes
Trace	No	Yes	No	Yes

1.9   Summary of Key Concepts

In this chapter, we have introduced the basic concepts related to embedded systems. Specifically, the following key concepts were introduced and discussed.

- The components involved in building a general organization of a computer are central processing unit (CPU) or microprocessor, input and output units, and main and secondary memories. These components together make a computer operational for useful activities.

- The parameters that you need to keep in mind when you want to design an embedded system are power consumption, speed of execution, system size and weight, and performance accuracy.

- The building blocks of a microcontroller are digital and analog input/output units, microprocessor, system buses, and RAM and ROM memory units. All these blocks are integrated in a single chip, which is called a microcontroller.

- Primary memories ROM and RAM can be considered to be divided into logic locations, each with a typical storage size of 8 bits. These locations are uniquely identified with the help of a number called an address.

- Different types of ROMs are available. ROMs can be programmed only once, can be reused by erasing and reprogramming using ultraviolet (UV) light or electric signals. Flash memory is the latest type of ROM which can be erased electrically and can be read and written in large multi-byte blocks.

- Different RAM types are available. Dynamic RAM (DRAM) is the most commonly used type of RAM because of its storage capacity. Static RAM (SRAM) is faster but yields less memory space on a given chip.

- Arithmetic logic unit (ALU), control unit (CU), and buses together make a microprocessor useful. ALU carries out logic and arithmetic operations. CU generates control signals to control the activities inside and outside the processor. Binary signals flow from one component to another over the buses.

- One way of categorizing microprocessors is on the basis of instruction set architecture. In this respect, reduced instruction set architecture (RISC) and complex instruction set architecture (CISC) are two well-known groups. The debate of which one is superior is never-ending.

- Another dimension that is used to group microprocessors is memory interfaces. There are two widely used memory interface architectures. The von Neumann architecture uses a common bus for both data as well as code memory. The Harvard architecture utilizes separate buses for accessing code and data memories.

- There two main steps involved in the conversion of a user program to an executable: (1) compilation and (2) linking. Compilers convert high-level languages to machine code while linkers resolve the dependencies among the codes.

- A debugger can help with identifying both hardware and software problems. Debugging techniques for microcontroller-based systems can be manual methods, software-only methods, and software-hardware methods.

Review Questions

Question 1.1. A memory is accessed for byte, halfword, and word size read/write operations. What memory addresses, when accessed, will result in byte, halfword, and word-aligned accesses simultaneously?

Question 1.2. If there are two or more microcontrollers in an embedded system, then what are possible methods to communicate among them? You can consider one of the microcontrollers, involved in communication, as a peripheral device.

Question 1.3. List different methods that can be used for debugging an embedded system.

Question 1.4. Explain how the Harvard architecture improves the execution speed for memory access operations compared to von Neumann architecture.

Question 1.5. Explain why software (compilation) tool development becomes complex for RISC architecture compared to its CISC counterpart.

Question 1.6. What causes a microcontroller to be designated as 32-bit? Is it the processing width, data bus width, address bus width, or their combination?

Question 1.7. List different attributes that can be used to classify the instruction set architecture for a microprocessor.

Question 1.8. What is the job of an assembler? Can the object code generated by an assembler be executed directly without linking?

Question 1.9. What is the job of a linker?

Exercises

Exercise 1.1. What will be the difference(s) in Cortex-M processor working if its RISC architecture is changed to CISC?

Exercise 1.2. Explain why microprocessors widely use von Neumann architecture, while microcontrollers use Harvard architecture.

Exercise 1.3 An embedded processor developing company is going to build a new processor that is required to perform different functions. Software implementation of those functions costs $0.5/function while hardware implementation costs $3.0/function. Explain whether the company should design the new processor using CISC or RISC processor architecture as an economical choice.

Exercise 1.4. What is the difference between simulator and emulator?

Exercise 1.5. Is it possible for a microcontroller to have an 8-bit address bus and a 32-bit data bus? What about a 32-bit address bus and a 16-bit data bus?

Part I

Architecture

Chapter 2

Cortex-M Architecture

Overview

A microcontroller integrates different subsystems in a single package. Typically, a microcontroller connects a microprocessor with memory and input/output interfaces to perform data exchanges and execute different tasks. In addition, each microcontroller provides one or more special interfaces for programming and debugging purposes. System architecture is responsible for defining the rules to perform this integration of different subsystems.

Different microcontroller architectures have been proposed to meet varying demands of performance, size, efficiency, cost, etc. As introduced in Chapter 1, the instructions set architecture and the bus system architecture are the two key aspects differentiating different microcontrollers in the market. The microcontrollers available in the market today are either completely designed and developed by the chip manufacturers or they are developed by integrating designs from different design vendors. Some examples of the former case are the microcontrollers from Atmel, Microchip, Intel, and Texas Instruments based on their own proprietary designs. In the latter case, the above-mentioned chip manufacturers and many others integrate the subsystem designs from others with their own to develop a complete system. ARM-based microcontrollers always fall in the latter category of microcontroller designs. This is due to the interesting fact that ARM does not manufacture microcontrollers. ARM only designs processors and different other building blocks, which are provided to different silicon chip developers through licensing. Traditionally, these designs are called *intellectual property* (IP) and the associated business model is termed IP licensing.

This chapter first introduces the basic building blocks of an ARM Cortex-M based microcontroller, which is followed by detailed discussion of those building blocks. In particular, the Cortex-M microprocessor architecture is introduced in detail covering its different attributes. This is followed by the bus system architecture, which is responsible for connecting the memory, peripherals as well as debug-related interfaces to the processor. We also provide architectural details related to memory address map, endianness, and bit banding as well as different types of debug interfaces. Interrupt

35

controller is one of the key building blocks of the microprocessor and its architecture-related details are not covered in this chapter; rather, a complete chapter is dedicated for this purpose.

2.1 Introduction to Cortex-M Microcontroller

Typically, in a microcontroller design, the processor is just one subsystem and there are many other subsystems including memories (both volatile and non-volatile), system buses, digital and analog input-output interfaces, communication peripherals, clock and power management system, etc. Although many microcontroller manufacturers have started using ARM Cortex-M processors as the CPU of their choice, the other system blocks can vary from vendor to vendor. As a result, the capabilities of one Cortex-M based microcontroller from one vendor can be significantly different from another microcontroller from a different vendor. This flexibility also provides the microcontroller manufacturers an opportunity to include additional features in their products making them more suitable for a specific market. Based on this background, we now focus our discussion on Cortex-M based microcontroller architecture with a brief history of ARM processor architectures.

The ARM architecture is one of the most widely adopted architectures. One of the early ARM architectures was ARMv4/v4T. ARM7TDMI processor belongs to this family and was one of the widely adopted processors. For example, Nokia6110 cell phone is based on ARM7TDMI processor. The ARMv5/5E architecture was introduced with the ARM9/9E family of processors. This architecture integrated digital signal processing (DSP) instructions aimed for applications involving multimedia content. The introduction of ARMv6 architecture resulted in ARM11 processor family. This architecture included single-instruction multiple-data (SIMD) processing capability as well as memory-related features.

Further extensions to ARM architecture were motivated by the diversification of the microprocessor market and resulted in a multi-profile ARMv7 architecture. Processors based on this architecture have been divided into the following three profiles:

- Cortex-M profile: Processors in this profile are designed for microcontroller-based embedded systems.

- Cortex-A profile: This profile is aimed for addressing the high performance applications mainly covering the cellular market.

- Cortex-R profile: Addressing the demands of real-time applications is the main motive of this profile.

The processor architecture used throughout this text is based on Cortex-M profile of ARMv7 architecture [ARM(2010)]. The evolution of ARM processor architecture is shown in Figure 2.1. The ARM Cortex-M based microcontroller architecture usually integrates the following key building blocks:

1. Microprocessor core

2. Nested vectored interrupt controller

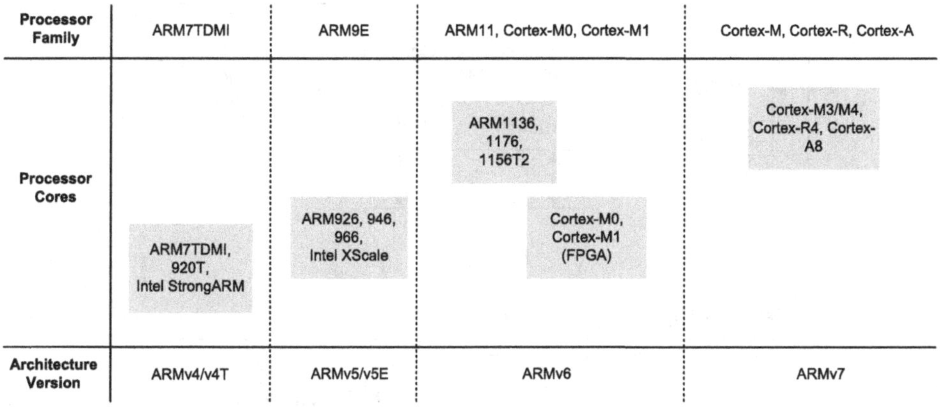

Processor Family	ARM7TDMI	ARM9E	ARM11, Cortex-M0, Cortex-M1	Cortex-M, Cortex-R, Cortex-A
Processor Cores	ARM7TDMI, 920T, Intel StrongARM	ARM926, 946, 966, Intel XScale	ARM1136, 1176, 1156T2 / Cortex-M0, Cortex-M1 (FPGA)	Cortex-M3/M4, Cortex-R4, Cortex-A8
Architecture Version	ARMv4/v4T	ARMv5/v5E	ARMv6	ARMv7

Figure 2.1: Different versions of ARM processor architecture and their evolution.

3. Bus system and bus matrix

4. Memory and peripherals

5. Debug system

Figure 2.2 illustrates the integration of different blocks in a Cortex-M based architecture. We will discuss each of the above-mentioned building blocks in detail. Let us start our discussion with the microprocessor core, which is based on ARM Cortex-M 32-bit architecture.

2.2 Microprocessor Architecture

The Cortex-M profile microcontrollers are based on a 32-bit RISC processor architecture. Currently, Cortex-M processor cores consists of Cortex-M0, Cortex-M1, Cortex-M3 and Cortex-M4 families. Cortex-M0, and Cortex-M1 belong to ARMv6 architecture family while Cortex-M3 and Cortex-M4 are based on ARMv7 architecture. The relationship between Cortex-M3 and Cortex-M4/M4F is shown in Figure 2.3. The key difference between Cortex-M3 and Cortex-M4 is the inclusion of DSP instructions in the latter. Furthermore, in Cortex-M4F, 'F' signifies the presence of an optional floating point unit (FPU), which will be absent in a Cortex-M4 processor. Different features supported by an ARM processor are appended as part of the name. For instance, in ARM7TDMI processor 'T' signifies the Thumb instruction set architecture, 'D' represents Debug interface (JTAG), 'M' shows the integration of fast Multiplier and 'I' indicates the inclusion of ICE (in-circuit emulator). A list of features supported by different ARM processors is tabulated in Table 2.1.

This chapter will mainly focus on Cortex-M3 architecture and any references to Cortex-M4 processor will be made explicitly wherever required. In Cortex-M3 processor register bank, memory interfaces as well as the data-path are all 32-bit. The word *data-path* in a processor core defines the mechanism of data inflow to and outflow from the microprocessor. The Cortex-M3 based microprocessor core includes the following

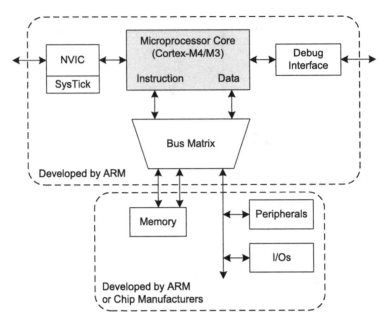

Figure 2.2: Block diagram of ARM Cortex-M based microcontroller architecture. The top dashed box encloses all the blocks developed by ARM, while the bottom dashed box contains the system blocks either developed by ARM or possibly by a chip manufacturer.

key components and the associated features.

- Instruction Set Architecture

- Register Set

- Processor Operating Modes

- Interrupts and Processor Reset Sequence

- Pipelined Architecture and Data Path

- Memory Address Map

Processor-specific information, such as interface details and timing are documented in the corresponding family specific Cortex-M Technical Reference Manuals[1] (TRM) [ARM(2010)]. The microcontroller specific technical reference manuals may also cover those implementation details, which are not covered by the architecture specifications, for instance, the list of supported instructions. Not all of the instructions covered in the ARMv7-M architecture specification manual are present on all ARMv7-M devices.

The Cortex-M3 processor is based on Harvard architecture, where separate instruction and data buses are used. Using this architecture allows instruction and data accesses to take place at the same time and enhances the performance of the processor because data accesses do not affect the instruction pipeline. This feature results in multiple

---

[1]The architecture and reference manuals are freely accessible from ARM website.

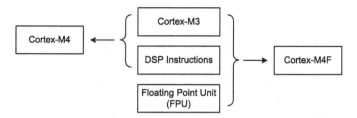

Figure 2.3: Architectural differences between Cortex-M3 and Cortex-M4 microprocessor cores.

Table 2.1: List of features supported by different ARM processors or processor families.

Feature label	Description
T	Thumb
D	On-chip debug support
M	Enhanced multiplier
I	Embedded ICE hardware
T2	Thumb2
S	Synthesizable code
E	Enhanced DSP instruction set
J	JAVA support, Jazelle
Z	TrustZone extension
F	Floating point unit
H	Handshake, clockless design for synchronous or asynchronous design

bus interfaces on Cortex-M3. Each bus has its own optimized usage and can be used
with the other bus simultaneously. But this does not mean that having two separate
bus interfaces allows us to have 8 GB of memory space. Rather, the instruction and
data buses share the same memory space. Recall that the 32-bit interface can access
at most 2^{32} bytes of memory, which is equal to 4 GB.

2.2.1 ARM Instruction Set Architecture

Historically, the ARM processors have supported two different instruction sets: the
ARM instructions that are 32 bits and Thumb instructions that are 16 bits. The
size of an instruction (i.e., assembly instruction) signifies the number of bits required
to store the machine code or opcode of that instruction. From functionality provisioning viewpoint, the Thumb instruction set is a subset of the ARM instructions.
Using Thumb instructions can provide higher code density. It is a preferable choice
for products with tight memory requirements. On the other hand, using ARM instructions can improve the processor execution performance. The higher code density
for Thumb instruction set and better execution performance for ARM instruction set
are explained next.

Consider a simple user program that involves six operations to be performed of which three are simple operations, while the other three are complex operations. The three simple operations are supported by both Thumb as well as ARM instruction set architectures using three assembly instructions. On the other hand, the three complex operations are supported by ARM instruction set and there are three corresponding assembly instructions. However, the Thumb instruction set supports these complex operations by using two assembly instructions for each of these complex operations. We assume that each of the assembly instructions used by the program requires one cycle for execution. Now for the same clock speed, the ARM ISA based processor will require six clock cycles for execution, while it will require nine clock cycles for a Thumb-based processor. On the other hand, it will require 24 bytes of memory space to store the program for ARM ISA based processor, while 18 bytes will be required by the Thumb ISA based processor. This simple example illustrates why ARM ISA has, in general, better execution performance, while Thumb ISA has higher code density.

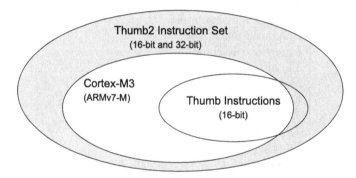

Figure 2.4: The relationship between ARMv7-M Instruction Set Architecture (ISA) with Thumb2 and the traditional Thumb.

Thumb2 is a superset of the Thumb instruction set. Thumb2 introduces 32-bit instructions that are intermixed with the 16-bit instructions. The Thumb2 instruction set covers all the functionality of the Thumb instruction set. Thumb2 has the execution performance close to that of the ARM instruction set and has the code density performance close to the original Thumb Instruction Set Architecture (ISA). The Thumb2 technology extended the Thumb ISA into a highly code density efficient and yet powerful instruction set that delivers significant benefits in terms of ease of use, code size, and performance (see Figure 2.5). According to ARM performance evaluation results, Thumb2 instruction set gives approximately 26% improvement in code density compared to ARM, while it also provides approximately 25% performance gain over the conventional 16-bit only Thumb architecture.

It is important to realize that Thumb2 instruction set is not constructed by simple union of Thumb and ARM instruction sets. Rather, the 32-bit instructions in Thumb2 are different from ARM 32-bit instructions. In Thumb2 ISA, the 32-bit instructions are constructed in such a way that its leading 16-bits are different from all 16-bit Thumb instructions to avoid any ambiguity that may otherwise arise while decoding that instruction.

Cortex-M3 is primarily focused on small memory system devices such as microcontrollers and also aiming to reduce the size of the processor, supports only the Thumb2 instruction set. Supporting Thumb2 instructions automatically ensures the support for Thumb instructions. The details of the instruction set are provided in a document called The ARM Architecture Reference Manual (also known as the ARM ARM). For the Cortex-M3 instruction set, the complete details are specified in the ARM v7-M Architecture Reference Manual [ARM(2010)].

In traditional ARM processors, the processor can use ARM instructions for some operations and Thumb instructions for others. However, doing so requires the processor to change from ARM state to Thumb state, which incurs additional overhead due to the execution of state change instructions. However, this is no more an issue with Thumb2 architecture, since the processor uses Thumb2 instructions for all the operations.

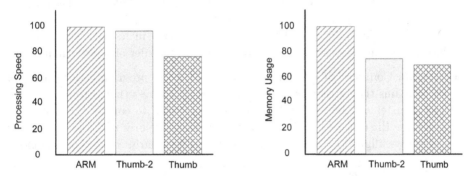

Figure 2.5: Performance comparison of Thumb2 instruction set with the conventional Thumb and ARM architectures.

2.2.2 Register Set

Processor registers are one of the most important components of a microprocessor core. The data processing related registers are responsible for temporary storage of data elements before and after the execution of instructions. In addition, control registers are responsible for configuring the desired mode of operation of the processor, while the status registers provide information regrading the current state as well as mode of operation. The Cortex-M3 processor has 32-bit registers, which are shown in Figure 2.6. The registers can be differentiated based on their functionality.

- General-purpose registers, R0-R12

    1. Registers R0-R7 are called low registers and are accessible by all instructions that specify a general-purpose register.

    2. Registers R8-R12 are called high registers and are accessible by all 32-bit instructions that specify a general-purpose register. Registers R8-R12 are not accessible by any Thumb (16-bit) instructions.

- Stack Pointer (SP): Register R13 is used as the Stack Pointer (SP). Because the SP ignores any writes to bits [1:0], this makes it aligned to a word (4 byte) boundary automatically. In addition, it is important to remember that stack pointer is a banked register with two copies, namely Main Stack Pointer (MSP) and Process Stack Pointer (PSP). Only one copy of the stack pointer (R13) is visible and active at a given time. This means that stack pointer logically has one copy at any arbitrary time instant, while physically it has always two copies. Handler mode always uses MSP, but you can configure Thread mode to use either MSP or PSP. Processor modes of operation (handler and thread) will be introduced in Section 2.2.3.

- Link Register (LR): Register R14 is the subroutine Link Register (LR). The LR receives the return address from the program counter register, discussed next. The link register contains the return address to be used by the processor, when returning from a function or service routine. When the link register is not used for holding a return address, it can be treated as a general-purpose register. Despite the dependency of LR (R14) on PC (R15) in its definition, we have introduced this register before PC to maintain the order of register numbering.

- Program Counter (PC): Register R15 is called the program counter register. PC contains the current program or instruction address that is to be executed. This register can be modified by the program itself to control the flow of the program. Bit 0 of this register is always 0, which ensures that the instructions are always aligned to either word or halfword boundaries in the code memory.

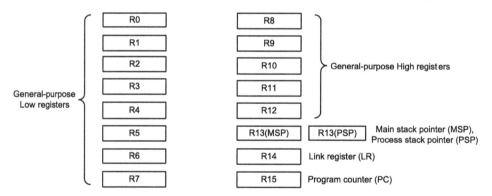

Figure 2.6: General purpose along with some special purpose registers. All of the registers are 32-bit.

The usage and allocation of general-purpose registers in the execution of a specific task can be performed either automatically by the compiler or manually by writing an assembly program. In addition to the above-mentioned registers, Cortex-M3 also contains special purpose registers, which are used to provide the microprocessor current status and to control the mode of its operation. The key status and control registers are listed below:

- Program Status Registers (PSRs): This set of registers consists of the following three status registers:

1. Application Program Status Register (APSR)

2. Interrupt Program Status Register (IPSR)

3. Execution Program Status Register (EPSR)

These registers have special functions and can be accessed only by special instructions. They cannot be used for normal data processing. Figure 2.7 shows different bit fields used by program status register and the definitions of these bit fields are explained in Table 2.2. The negative flag denoted by N is set to 1 if the result of the previous operation carried out by the ALU was negative. In case the previous operation result was positive, or zero, then N becomes 0. If the zero flag denoted by Z is equal to 1, then it indicates that the previous operation result was zero. Otherwise, result of the previous operation was non-zero. If the previous addition operation resulted in a carry bit or the previous subtraction did not result in a borrow bit, then the carry flag becomes 1. Otherwise, C = 0. The overflow flag represented by V becomes a 1 when the previous operation resulted in an overflow. Otherwise, V flag is cleared. The DSP overflow and saturation flag denoted by Q will be set to 1 when DSP overflow or saturation has occurred. Otherwise, it is 0. From Figure 2.7, it can be observed that the bit field allocations in the three program status registers are done in such a way that they occupy non-overlapping bit locations. This permits us to combine these three registers into one register without affecting the functionality of each bit field.

- Control register (CONTROL): The control register is used to configure the privilege level as well as to perform the selection of stack pointer register. The processor has two modes of operation, namely the handler mode and thread mode. Associated with these modes are two levels of operation called privileged level and unprivileged (user) level. The handler mode always operates at privileged level. However, the thread mode can either operate at privileged or unprivileged level. The control register has 2 bits, namely CONTROL-Bit1 and CONTROL-Bit0. The CONTROL-Bit0 is writable only at privileged level. Writing '1' to this bit makes the processor state change to unprivileged level. However, it is not possible to directly change to a privileged level from an unprivileged level. For that purpose, the processor at unprivileged (user) level can switch back to privileged level by triggering an interrupt or exception. The CONTROL-Bit0 of the control register is always 0 in handler mode. However, in the thread mode, it can be either 0 or 1. The CONTROL-Bit1 is used to select which stack main or process is to be used in thread mode as described in Table 2.3. The CONTROL register Bit0 and Bit1 are writable only when the processor has the privileged access level. At the unprivileged (user) level, writing to the CONTROL register is not allowed. The functionality of these bits is described in Table 2.3. The use of control register is further illustrated later in this section in the context of processor operating modes.

	31	30	29	28	27	26:25	24	23:20	19:16	15:10	9	8	7	6	5	4:0
APSR	N	Z	C	V	Q											
IPSR										Exception number						
EPSR					ICI/IT		T			ICI/IT						

Figure 2.7: Program status registers and their bit field allocations.

Table 2.2: Bit field definitions for different program status registers.

Bit field	Description
N	Negative flag.
Z	Zero flag.
C	Carry flag.
V	Overflow flag.
Q	Saturation flag.
ICI/IT	Interrupt continuable instruction (ICI) bits. IF-THEN instruction block status bit.
T	Thumb state (always 1 in case of Cortex-M processor using Thumb2 ISA).
Exception number	Shows which exception number is being handled by the processor.

The Concept of Overflow

Why microprocessors have separate carry and overflow flags? The answer to this question is that overflow occurs differently for signed and unsigned numbers. For a 32-bit register, 2^{32} different values can be stored. The range is 0 through 4,294,967,295 (2^{32} - 1) for representing an unsigned number, and -2,147,483,648 (-2^{31}) through 2,147,483,647 (2^{31} - 1) for a signed number. The carry flag represents overflow for unsigned numbers while signed number overflow is indicated by the overflow flag. Signed and unsigned overflows occur independently. There can be four possible outcomes when an arithmetic operation such as addition is performed. (1) no overflow, (2) unsigned overflow only, (3) signed overflow only, and (4) both signed and unsigned overflows.

Let's first look at an example of an unsigned overflow but not signed overflow (i.e., case 2). Assume that we want to perform addition of R0 = 0xFFFFFFFF and R1= 0x00000001. The 32-bit answer comes out to be 0x00000000 with a carry, i.e., 1 comes out of the most significant bit (MSB). If these numbers are interpreted to be unsigned, then our expected answer is 4,294,967,295 + 1 = 4,294,967,296. As the correct answer cannot be accommodated in a 32-bit register, we got an incorrect answer (i.e., 0). This situation is reflected by a 1 carried out of the MSB and we say that an unsigned overflow has occurred. In case the numbers are taken to be signed numbers, an answer of 0 (-1 + 1 = 0) is as per our expectation. As the answer is within the range for signed numbers, no signed overflow occurred and V = 0.

Table 2.3: Control register bit field definitions.

Bit field	Description
CONTROL-Bit1	When this bit is set to 0, main stack is used (through main stack pointer (MSP) register) and is the default setting. Setting this bit to 1 results in the use of process stack (through process stack pointer (PSP) register).
CONTROL-Bit0	When this bit is 0 thread mode is privileged and setting it to 1 makes thread mode switch to unprivileged (user) level.

Now, we look at a signed but not unsigned overflow scenario (i.e., case 3). Let's assume that R0= 0x7FFFFFFF and R1= 0x7FFFFFFF. Either interpreted as signed or unsigned, 0x7FFFFFFF is 2,147,483,647. The answer is 0xFFFF FFFE. There is no unsigned overflow as the answer (i.e., 4,294,967,294) is within the range for unsigned numbers. Thus, making the C flag equal to 0. However, the answer is out of the range for signed numbers and we say that a signed overflow has occurred. It will be indicated by V = 1. The signed interpretation of the answer is -2.

An example of the forth case is when we wish to add R0= 0x80000000 and R1= 0x80000000. In case the numbers are interpreted as unsigned numbers, we are trying to add 2,147,483,648 with 2,147,483,648. The answer is 4,294,967,296, which cannot be accommodated in a 32-bit register. The carry coming out of the MSB indicates that an unsigned overflow has occurred. When we take the numbers to be signed numbers (i.e., -2,147,483,648 + -2,147,483,648), the answer also exceeds the signed number's range. Therefore, both C and V flags are turned on.

The processor directly uses the carry coming out of the MSB for determining the unsigned overflow. On the other hand, it uses the carry into and out of the MSB to update the V flag. Actually, if the carry into and out of the MSB do not match (i.e.,there is a carry into the MSB but no carry out, or if there is a carry out but no carry in), then the signed overflow has occurred.

2.2.3 Processor Operating Modes

The processor supports two modes of operation, Thread mode and Handler mode. These modes of operation have an associated pair of access levels namely Privileged and Unprivileged (user) access levels (see Figure 2.8).

- Thread Mode: The processor enters Thread mode on reset or as a result of an exception return. Code execution in thread mode can have either privileged or unprivileged (user) access levels.

- Handler Mode: The processor enters Handler mode as a result of an exception. All code is privileged in handler mode.

In thread mode with unprivileged (user) access level, permission to modify the contents in system control space (SCS), which is a part of the memory region for configuration registers and debugging components, is blocked. Furthermore, instructions that access special registers cannot be used. If a program running at the user access level tries to access SCS or special registers, a fault exception will occur. When an exception takes place or an interrupt occurs, the processor will always switch to a privileged access level and returns to the previous state when exiting the exception handler. A program with user access level cannot change to the privileged level directly by writing to the control register. As mentioned previously, it has to go through an exception handler which takes it to privileged access level. However, the program running at the privileged access level can always switch the processor execution level to the unprivileged access level by writing to the control register. The possible state transitions from one operating mode to the other as well as the corresponding privileged/unprivileged access levels at each state are shown in Figure 2.8.

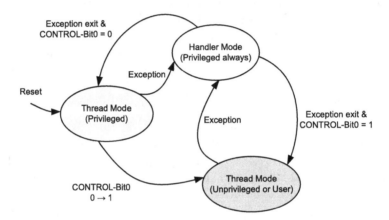

Figure 2.8: Possible transitions for operating modes and privilege levels.

Apart from the processor modes of operation and the privileged/unprivileged access levels, the processor has dual stack capability as well. The utilization of two privilege levels as well as the dual stack provisioning can be best understood in the context of an operating system based user application development. When an operating system is being used as part of the system design, then the user can run different applications on top of the operating system. Now suppose the user is running an application using an operating system and has the permission to run it at the privileged level. This will grant access to the critical SCS memory region as well as the CONTROL register enabling it to modify their content.

Now consider the scenario where the user is trying to run a third-party application, which is not bug free. In that case, the application crash might lead to the entire system crash due to the capability of modifying critical memory and CONTROL register contents. However, granting an unprivileged access level will only lead to the application crash and rest of the system and other application processes will continue functioning normally. Based on these facts, the operating system is granted privileged access and uses the main stack, while the applications run at unprivileged level and use the process stack. This interface between operating system and user application is provided through system calls and application program interface as shown in Figure 2.9.

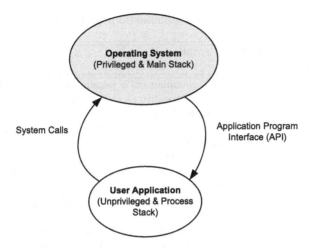

Figure 2.9: Selection of privilege level and stack between operating system and user application program.

In the context of the above discussion we can explain the possible choices for mode of operation and the associated stack pointer used. These possible combinations based on the CONTROL-Bit0 and CONTROL-Bit1 values are given in Table 2.4.

2.2.4 Interrupts and Processor Reset Sequence

Under normal circumstances, a processor executes the program in a predefined sequence. However, due to the interaction with real world physical phenomenon, we might be interested in knowing if an event of interest has occurred. One possible solution is to keep on checking an appropriate indicator for the occurrence of the event. This solution is inefficient because we use processing resources in checking the event even if it has not occurred. An alternative, more efficient, solution is based on interrupts, where the processor is informed by the dedicated interrupt related hardware, configured appropriately, about the occurrence of an event. Based on the above discussion, the interrupt can be thought of as an indicator about the occurrence of an event, which can be generated by either hardware or software. An interrupt causes a deviation from the normal sequence of program execution. In case of an event or exception, it is natural to think that a response is required from the processor. The processor response generally involves a sequence of operations to be performed and is implemented as a dedicated response function and is called service routine. The sequence of operations involved from the occurrence of an event to the resumption of normal program execution again has the following main steps.

- When an event occurs, an interrupt request is generated to the processor.

- The processor suspends current task execution in response to the interrupt and starts executing an interrupt service routine to generate the response to the event.

Table 2.4: Control register bit combinations and their possible usage.

CONTROL-Bit0	CONTROL-Bit1	Description
0	0	This combination is possible for simple applications, where the entire program runs at *privileged* level and only the Main Stack Pointer (MSP) is used by both the main program as well as interrupt handlers.
0	1	This is the scenario when the user application program runs, in *privileged* thread mode, on top of an embedded operating system. The user application uses Process Stack Pointer (PSP), while the MSP is used by the operating system kernel and exception or interrupt handlers.
1	0	The user application is running with *unprivileged* access level and still uses MSP. It is highly unlikely that this scenario will be used by the user applications, where the user application and the operating system share the stack memory.
1	1	In this case the user application program runs, in *unprivileged* thread mode, on top of an embedded operating system. The user application uses Process Stack Pointer (PSP), while the MSP is used by the operating system kernel and exception or interrupt handlers.

- Once the response to the event is completed, the processor resumes execution of the suspended task.

When a microcontroller is powered or if it reset by pressing the reset button, an interrupt called *reset interrupt* is generated. Every processor has an associated sequence of operations, which it performs to reset event. The first operation performed by the Cortex-M processor after a reset or power-on involves reading of two words from instruction memory.

- Address 0x00000000: This address in the code memory (most likely a Flash memory) contains the starting value that is loaded to the main stack pointer, MSP (R13).

- Address 0x00000004: Reset vector is contained at this address and the program counter is loaded with this value to jump to the reset interrupt service routine. The *main* function is called from within reset interrupt service routine.

The reset sequence will be further discussed in Chapter 4.

2.2.5   Pipelined Architecture and Data Path

The three basic steps involved in processing of an instruction are fetch, decode, and execute. Performance of a microprocessor that follows these three steps in a sequential manner is depicted in Figure 2.10(a). It is clear from Figure 2.10(a) that the microprocessor requires six clock cycles to complete the execution of two instructions. We also observe that while the microprocessor is decoding the first instruction the buses remain idle. This time slot could have been utilized in pre-fetching the second instruction because decoding of the first instruction and fetching of the second instruction can be performed simultaneously without affecting each other. This would thus result in an improved throughput of the instructions executed by the microprocessor. Such a microprocessor is said to be based on the pipelined architecture and is shown in Figure 2.10(b). We see that a microprocessor with pipelined architecture decodes the first instruction and pre-fetches the next one. Similarly when the first instruction is being executed the second instruction will be decoded, while the third instruction is being fetched and all these three operations are performed simultaneously. This can be observed from Figure 2.10(b). The non-pipelined architecture takes six clock cycles to complete two instructions while the pipelined architecture based microprocessor takes four clock cycles to complete the execution of two instructions.

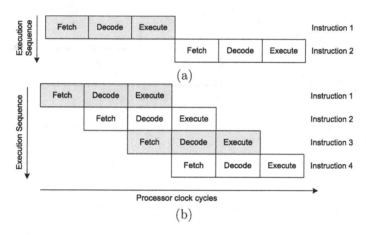

Figure 2.10: Comparison between non-pipelined and pipelined architectures. (a) Non-pipelined architecture and (b) three stage pipelined architecture.

Here we have made the assumption that saving the result after instruction execution back to the memory is part of the instruction execute step. Having a closer look at Figure 2.10(b), we observe that when the first instruction result is being written back to memory, the third instruction is being fetched at the same time. This simultaneous writing of the result (data) to memory and reading of an instruction (code) is possible due to the fact that we have two physically separate buses for the instruction memory and data memory. So when the data from the first instruction execution is being written to memory on the data bus the third instruction will be fetched using the instruction bus. Recall that both instruction and data buses have separate address, control, and data lines.

During the instruction fetch phase, the microprocessor generates the address on the instruction address bus along with the read control signal. The memory location that contains the instruction is selected and the corresponding instruction encoded information is sent on the instruction bus. The microprocessor retrieves this information from the instruction bus and stores it in a register. During the second stage, decoding of the instruction is carried out, while the third stage is responsible for execution of the instruction and saving the result to either register or memory. In the execution phase if the instruction happens to be a memory load or store operation, then the data is read (in case of load instruction) or written back (if the instruction is a store operation) using the data bus. The sequence of operations as discussed above can be easily understood by looking at the data path block diagram shown in Figure 2.11.

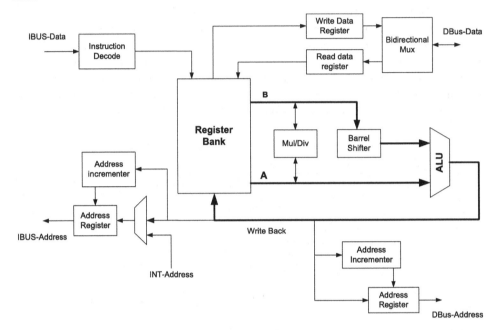

Figure 2.11: Processor data path.

The Cortex-M3 processor has a three-stage pipeline. The pipeline stages are

1. Instruction Fetch.

2. Instruction Decode.

3. Instruction Execute.

When running programs with mostly 16-bit instructions, you may observe that the processor is not fetching instructions in each clock cycle. This is because the processor fetches up to two instructions (32 bits) simultaneously. When 32-bit information is fetched from the code memory, we get two 16-bit instructions fetched in one cycle. In this case, the processor bus interface may need to fetch the instructions after each alternate cycle. In case the instruction holding buffer inside the processor is full, the instruction bus interface could remain idle. If some of the instructions take multiple cycles to execute, which is possible due to multiple reasons, then in these circumstances, the pipeline will be stalled. While executing a branch instruction, first of all the pipeline has to be flushed. The processor will then fetch instructions from the branch destination to fill up the pipeline again and start executing this new piece of code.

More complex microprocessor designs divide three above-mentioned broad phases of processing into multiple sub-phases thus making the pipeline have a larger number of stages and as a result requiring more number of phases per instruction execution. For instance, ARM's Cortex-A8 has 13 stage pipeline while Intel's Penyrn microprocessor has 12 to 14 pipeline stages.

The data path block diagram also shows the key functional unit called *Barrel Shifter*. The barrel shifter is a functional unit which can be used in a number of different circumstances. The key advantage of barrel shifter is its capability to perform the shift and rotate operations in one cycle independent of the number of bit positions required by the operation. It provides five types of shift and rotate operations, which can be applied to second operand of the assembly instructions as can be observed from Figure 2.11. The functionality of barrel shifter will be elaborated further in Chapter 5.

2.2.6 Memory Address Map

Cortex-M3 has a predefined memory and device address map. This makes the built-in peripherals, such as the debug components and the interrupt controller, to be accessed by simple data access instructions. Additionally, the predefined address map also allows us to optimize the Cortex-M3 processor for speed and ease of integration for designing a system-on-chip (SoC). Overall, the 4 GB memory space is divided into different ranges and is shown in Figure 2.12. From the figure we observe that different memory regions have a pre-assigned primary usage. For instance, the first 500 MB of memory space is preferred to be used as code (instruction) memory. This is due to the fact that Cortex-M3 microprocessor has an internal bus infrastructure optimized for this memory usage. However, it is not a mandatory requirement and the design allows these regions to be used differently. For example, data memory can still be put into the code region. Similarly, the program code can be located in the code region, the static random access memory (SRAM) region, or the external RAM region. It is however best to put the program code in the code region because this allows to carry out the instruction fetches and data accesses simultaneously on two separate bus interfaces.

Based on the preferred memory address space usage, we can categorize the memory space in the following regions:

- Code region for storing program or code. In general, a Flash or EEPROM memory is located in this address space.

- Data region for application data. Mostly SRAM is placed in this region.

- Peripheral region hosts different input and output peripherals.

- External memory and peripheral region, through a bus interface, can have access to external memory or peripherals.

- Private Peripheral Bus (PPB) (both internal and external) region is meant for debug interfaces. This address space also hosts the registers for the Nested Vectored Interrupt Controller, processor's configuration registers, as well as registers for debug components. Note that the Private Peripheral Bus (PPB) is not used for normal peripherals. This will be further explained in this chapter later.

- Vendor specific region contains information about device ID, versions, release, etc.

The Cortex-M3 devices have fixed addresses for many of the peripherals in the address space. This leads to a simpler porting of applications between different Cortex-M3 products. Porting is the process of modifying the existing software code developed for one platform to be used for another platform and can have different levels of difficulty.

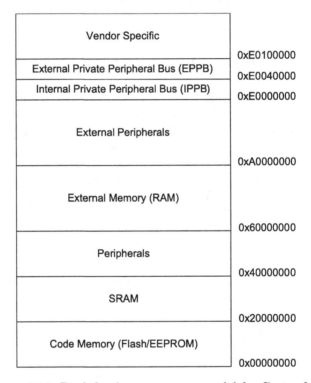

Figure 2.12: Predefined memory map model for Cortex-M3.

2.3   Nested Interrupt Vector Controller

The Cortex-M3 processor also includes an interrupt controller called the Nested Vectored Interrupt Controller (NVIC). It is closely coupled to the processor core and provides a number of features as listed below:

- Nested interrupt support,

- Vectored interrupt support,

- Dynamic priority change support,

- Reduction of interrupt latency,

- Interrupt masking.

Further details regarding nested vectored interrupt controller will be discussed in Chapter 3.

2.4   Bus System and Bus Matrix

The Cortex-M3 processor contains not only the processor core but also a number of components for system management, as well as debugging support. These components are connected together through an Advanced High-Performance Bus (AHB) and an Advanced Peripheral Bus (APB). The AHB and APB bus protocols are part of the Advanced Microcontroller Bus Architecture (AMBA) standard. AMBA standard consists of a set of multiple bus protocol specifications. The AMBA specifications are freely downloadable from the ARM website. Under normal circumstances, the chip manufacturer will hook up all the bus signals to memory blocks and peripherals, while in some cases, it is possible that the designer connects the bus to a bus bridge and allows external bus systems to be connected off-chip as well. It is unlikely to have a provision for an external direct access to the bus interface signals.

A *bus matrix* is used at the heart of the Cortex-M3 internal bus system as can be seen from Figure 2.13. The bus matrix effectively is an AHB interconnection network, allowing data and instruction code transfers to take place on different buses simultaneously unless both buses are trying to access the same memory region. Most processor implementations contain multiple external AHB interfaces and one APB interface. An internal AHB-to-APB bus bridge is used to connect a number of APB devices, such as debugging components, which follow the private peripheral bus interface. Different bus interfaces used by the Cortex-M3 as well as Cortex-M4 processors for memory and peripheral connectivity are discussed below.

- ICode Bus: The ICode bus is a 32-bit bus based on the AHB-Lite bus protocol. It allows to perform instruction fetches in memory regions from 0x00000000 to 0x1FFFFFFF. Instruction fetches are performed in word size, even for 16-bit Thumb instructions. Therefore, the CPU core can fetch up to two Thumb instructions at a time.

- DCode Bus: The DCode bus is a 32-bit bus based on the AHB-Lite bus protocol. Data access in memory regions from 0x00000000 to 0x1FFFFFFF can be performed via this bus. Although the Cortex-M3 processor supports unaligned transfers, you would not get any unaligned transfer on this bus because the bus interface on the processor core converts the unaligned transfers into aligned transfers. As a result, the devices (such as memory) that attach to this bus need only support AHB based aligned transfers.

- System Bus: The system bus is a 32-bit bus based on the AHB-Lite bus protocol. It is used for instruction fetch and data access in memory regions from 0x20000000 to 0xDFFFFFFF and from 0xE0100000 to 0xFFFFFFFF. Similar to the D-Code bus, all the transfers on the system bus are aligned.

- AHB-AP Bus: The processor contains an AHB-AP (access port) interface for debug accesses. An external Debug Port (DP) component accesses this interface. The Cortex-M3 system supports three possible DP implementations:

  1. The Serial Wire JTAG Debug Port (SWJ-DP): The SWJ-DP is a standard CoreSight debug port that combines JTAG-DP and Serial Wire Debug Port (SW-DP).

  2. The Serial Wire DP: This provides a two-pin interface to the AHB-AP for debugging purposes and is widely used in today's microcontrollers for debugging interface.

  3. No DP present: If no debug functionality is present within the processor, a DP is not required.

  The DP and AP together are referred to as the Debug Access Port (DAP).

- Private Peripheral Bus (PPB): The private peripheral bus is a 32-bit bus based on AMBA-based APB protocol. This is intended for private peripheral accesses in memory regions 0xE0040000 to 0xE00FFFFF. However, since some part of this APB memory is already used for different debug interfaces, the memory region that can be used for attaching extra peripherals on this bus is only 0xE0042000 to 0xE00FF000. Transfers on this bus are word aligned.

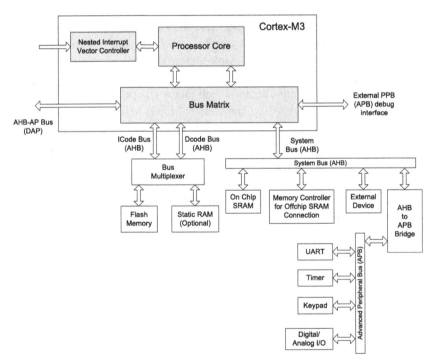

Figure 2.13: Bus system for memory, peripherals, and debug interconnection.

2.5 Memory and Peripherals

Cortex-M3 architecture uses a simple linear 4GB memory map. The bus matrix allows memory and other peripherals to be accessed using the AHB and APB buses as discussed in the previous section. As mentioned earlier, an AHB-to-APB bus bridge is used to connect a number of APB devices to the Cortex-M3 processor, which operates at relatively slower speeds, such as I/O devices and some of the debugging components. The APB protocol is used for connecting devices following PPB interface. In addition, the Cortex-M3 allows chip manufacturers to attach additional devices through the external private peripheral bus using this APB bus protocol.

Using the AHB bus matrix, if the instruction bus and the data bus are accessing different memory devices at the same time (from different memory regions as well as accessible from different bus interfaces), the transfers can be carried out simultaneously. In case both the data and instruction buses are trying to access the same memory device at the same time, it is possible to assign higher priority to the data bus for better performance.

Common Cortex-M3 microcontroller designs use system bus for SRAM connection. The main SRAM block should be connected through the system bus interface, using the SRAM memory address region. This allows data accesses to be carried out at the same time as instruction accesses. Some microcontrollers might allow an interface for external memory connection. This requires an additional memory controller to provide external memory interface, since it is not possible to connect off chip memory devices directly to the system bus based on AHB protocol. In addition, direct memory access (DMA) controller capability can lead to the data transfers among different peripherals without requiring processor intervention.

Simple peripherals can be connected to the Cortex-M3 processor through an AHB-to-APB bridge. This allows the use of simpler APB bus protocol for connecting peripherals to the processor core. The block diagram shown in Figure 2.13 is one possible illustration and chip manufacturers might choose different bus connection designs.

From software design and development viewpoint, the application programmer only needs to know the memory map. As can be seen from the Figure 2.13, an AHB-to-APB bus bridge is used to connect I/O interfaces, timer, keypad, universal asynchronous receiver/transmitter (UART), etc. to the processor core through system bus (AHB) and bus matrix. The microcontroller development may involve different vendors providing the required peripherals and interfaces, which are connected with the processor core from ARM. In such a situation, the application programmer may also need to refer to specific microcontroller's data sheet for further details in the process of developing software for Cortex-M3 based systems. This will be further elaborated in Chapters 8 to 11, when we will learn how to perform basic input-output interfacing and how to use available peripherals when using an actual microcontroller from a specific vendor. Different memory regions can be accessed using multiple buses. In addition, each bus is capable of addressing only a specific region of memory. The mapping between different memory regions and the corresponding buses, which have an access to those memory regions, is shown in Figure 2.14.

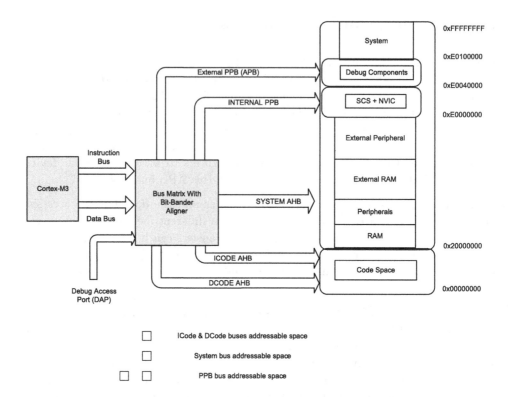

Figure 2.14: Interfacing of different buses with corresponding memory areas.

2.5.1   Memory Endianness

Different processors, irrespective of their processing capability (i.e., 8-bit, 16-bit, or 32-bit processor) and the associated bus width (in terms of number of data lines) support commonly used data types. When either the processor or the associated bus or even both, due to their size limit, are not capable of directly handling larger data types, software libraries are provided as part of the tools to make the user applications run on those resource limited platforms without any problem. However, ARM processors are 32-bit and their memory interface is also 32-bit and are not limited in that aspect. ARM Cortex-M processors support the following common data types when performing operations or transferring data to or from memory.

- *Byte* data type of size 8-bits

- *Halfword* data type of size 16-bits

- *Word* data type of size 32-bits

The processor views memory as a linear collection of bytes numbered in ascending order from zero. For example, bytes 0-3 hold the first stored word, and bytes 4-7 hold the second stored word. Similarly, bytes 0-1 and 2-3 hold first and second halfwords, respectively. Now let us consider the case of a halfword. If we are interested in finding the numeric value of a halfword, we need to put its corresponding two bytes in an order to obtain that value. This is what the processor will do when asked to compare the value of this halfword against say some constant. Constructing the halfword value by taking byte 0 as high byte and byte 1 as low byte will lead to a different result, compared to the case when the order is reversed. Similarly, when we are constructing a word from bytes or halfwords, we need to know in which order to put the four bytes or two halfwords to construct a word. This byte order in the construction of halfwords and byte or halfword order for the construction of words is called the *memory endianness*. The are two possible types of memory endianness, namely *little endian* and *big endian* and they are explained next.

- *Little Endian*: In little-endian format, the processor stores the least significant byte (halfword) of a word at the lowest-numbered byte (halfword) address, and the most significant byte (halfword) at the highest-numbered byte (halfword) address. Similarly, the least significant byte in a halfword occupies the lower address. See Figure 2.15 for illustration.

- *Byte Invariant Big Endian*: In byte-invariant big-endian format, the processor stores the most significant byte of a word at the lowest byte address, and the least significant byte at the highest byte address. See Figure 2.15 for illustration.

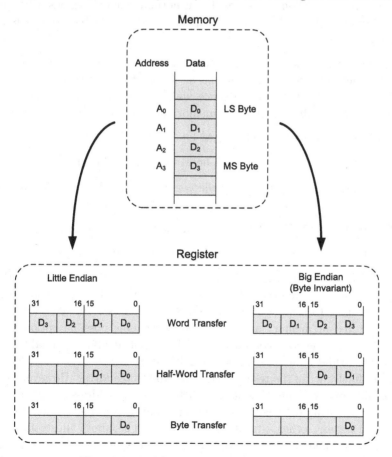

Figure 2.15: Memory endanness formats.

The Cortex-M architecture supports both little-endian and byte invariant big-endian modes, however most of the Cortex-M based microcontrollers use little-endian format. The memory endianness used is implementation specific.

2.5.2   Bit Banding

Normally memories are byte addressable and any read or write operation involves either a single or multiple bytes. However, integrating *bit banding* support in ARM architecture allows single bit accesses to perform read and write memory operations. In the Cortex-M architecture, bit-band operations are supported in two separate and predefined memory regions, which are termed *bit-band regions*. One of these regions is located in the first 1 MB of the SRAM, while the second is placed in the first 1 MB of the peripheral region. Access to these two regions as bit-band regions is not direct. Rather, we need to access a separate memory region called the *bit-band alias*, to perform bit-band operations in the two predefined bit-band memory regions. In other words, normal read/write operations performed in the bit-band alias memory regions result in single bit read/write operations in the actual bit-band region. The two bit-band memory regions are accessible as normal memory regions, when accessed using their actual address. Now we will explain in detail how bit-banding is actually performed.

Let us first locate the two bit-band memory regions in the memory address map. Based on the fact that one of the regions is at the start of SRAM space and the other is located at the start of peripheral space and each of them is 1 MB in size, their starting and ending addresses in linear memory address space are

- 0x20000000 − 0x200FFFFF (SRAM bit-band region, 1 MB)
- 0x40000000 − 0x400FFFFF (peripheral bit-band region, 1 MB)

The starting and ending addresses for the corresponding bit-band alias memory regions are

- 0x22000000 − 0x23FFFFFF (SRAM bit-band alias region, 32 MB)
- 0x42000000 − 0x43FFFFFF (peripheral bit-band alias region, 32 MB)

The mapping between bit-band region and its corresponding alias is illustrated in Figure 2.16. Using the above address information, we will next illustrate the bit-banding operation. Consider the case where we want to set the least significant bit of the data stored at memory address 0x20000000. One possible option to achieve this is to read the memory contents from the specified address, modify the least significant bit, while maintaining the integrity of all other data bits using proper masking and finally write the result back to 0x20000000 address. However, using bit-band capability, we will simply write 0x00000001 to the address 0x22000000, which is equivalent to writing to the least significant bit of the data at address 0x20000000 without affecting any other bits at that address. The above illustration explains how the bit-band operations can simplify the memory accesses involving single bit operations. The comparison of key steps involved using bit-band operation with that of conventional load-modify-store procedure is illustrated in Example 2.1.

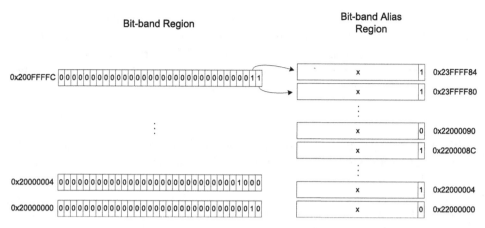

Figure 2.16: Access to bit-banding region. Each individual bit in the bit-band region is accessed separately in the least significant bit (LSB) of 32-bit contents at the word-aligned bit-band alias address.

Example 2.1 (Writing to bit-band region for bit setting.). Let's first see how bit setting using conventional load-modify-store procedure looks like. The pseudo-code listed below outlines the key steps involved in this process.

```
Step 1: Setup address in bit-band region
Step 2: Read data from the address to register
Step 3: Set the selected bit
Step 4: Write the result back to same address
```

Now the same activity will be performed using bit band alias region, then the following steps are required.

```
Step 1: Setup address in bit-band alias region
Step 2: Setup data for setting bit
Step 3: Write to the bit-band alias region
```

Since the memory addresses assigned to bit-band regions as well as their bit-band alias counterparts are fixed, we can define a general mapping between the two addresses. For example, when we want to set the fifth bit at address 0x20000400, what would be the corresponding bit-band alias address to perform this operation? Since the MSB of 0x20000000 in the bit-band region is mapped to 0x2200007F in bit-band alias, correspondingly the LSB of 0x20000001 is mapped to 0x22000080 in bit-band alias. Let x represent the memory address and c is the bit location in the bit-band region that we want to operate on. Then the corresponding bit-band alias address y is given by

$$y = ((0xF0000000 \ \& \ x) \ | \ 0x02000000) + ((0x000FFFFF \ \& \ x) \ll 5) + (c \ll 2). \quad (2.1)$$

Similar to write operation for modifying the bit contents of memory, we can use bit-band capability to simplify the memory bit read operations in the application code as well. For instance, if we need to determine what is the value of bit 2 at the address 0x20000000, we will rather read the bit-band alias region at address 0x22000004. The value obtained from this read operation can be used directly by any conditional statement without requiring masking or shifting operations. The comparison of bit-band based read operation with the conventional load-store mechanism is illustrated in Example 2.2.

Example 2.2 (Reading single bit from memory.). The pseudo-code listing below provides the steps involved to read a bit from memory using a conventional approach.

```
Step 1: Setup address in bit-band region
Step 2: Read the entire word from the address to register
Step 3: Extract the selected bit
```

The same activity when performed using bit band alias region, we require the following steps.

```
Step 1: Setup address in bit-band alias region
Step 2: Read the bit from the address
```

Bit-banding in its original essence is not an entirely new idea. Rather, a similar capability has been supported by the 8-bit microcontrollers, such as the 8051, for many decades. Unlike the 8051 microcontroller, the Cortex-M processor does not support any special instructions to perform memory and I/O bit operations, rather special memory regions (bit-band alias) are defined, so that any data accesses to these special regions are automatically converted to their equivalent bit-band operations using another memory address space.

Advantages of Bit-Band Operations

We have already seen that bit-band operations can reduce the number of instructions required when bit read or write operations are performed in a specific memory region. In addition, bit-band operation can also help in simplifying the branch decisions. This is explained using Example 2.3.

Example 2.3 (Performing branch operation using bit banding.). Let us consider that a branch operation is to be performed after testing a status bit in one of the registers associated with a peripheral. This bit can be set or reset based on the presence or absence of a certain condition related to that peripheral. Under these situations the normal sequence of operations to perform the above-mentioned task involves the following steps.

```
Step 1: Read the complete status register.
Step 2: Mask unwanted bits and perform any arbitrary bit shifting
        if required.
Step 3: Compare with the test value and then perform branch
        operation if the test condition is true.
```

However, this very same activity can be efficiently performed based on bit-banding by reading the status bit using band alias region and involves the following steps.

```
Step 1: Read the status bit using bit-band alias region.
Step 2: Compare and perform the branch operation if the test
        condition is valid.
```

It is important to remember that any read or write operation performed in the bit-band alias region should be aligned. If an operation using an unaligned address is performed in the bit-band alias memory region, the operation result will be unpredictable. In addition, bit-band feature is optional on Cortex-M3/M4 and as a result it is quite possible that the Cortex-M3/M4 microcontroller being used does not support bit-banding. Further details can be found in the specific documentation of the microcontroller provided by its vendor.

2.5.3 System Stack Architecture

System stack architecture defines the mechanism of using a special memory region, called stack memory region. The stack memory is accessed using stack pointers (MSP or PSP). Stack memory read and write operations always follow Last-In-First-Out (LIFO) data buffer format. Cortex-M processors allocate a small region from the main system memory (RAM) as the stack memory. Data is written to stack memory using PUSH instruction and the POP instruction is used to retrieve data from the stack.

The selected stack pointer is adjusted automatically when PUSH and POP instructions are executed and as a result multiple data PUSH operations will not cause old stacked data to be overwritten. For normal usage of the stack, each write (PUSH) to the stack, must be paired with a read (POP) and the address of the POP operation should match that of the PUSH operation. Stack memory is used in the following situations:

- Some of the registers (holding application data currently) may need to be freed for some other use as required by the task execution. This will result in pushing the register contents onto the stack memory and will be retrieved later when the task has been performed. Doing so guarantees that the application program data integrity is maintained.

- Stack memory can also be used to pass parameter or argument values when a function or subroutine is called.

- Stack is also utilized for declaring any local variables used by a software function or subroutine. As a result local variables use the memory temporarily. In contrast, the global variables are declared using main memory, i.e., RAM, and permanently occupy the memory space.

- In case of exceptions and/or interrupts, the processor status and a set of general purpose registers are stored before the corresponding interrupt service routine is executed.

Stack Implementation in Cortex-M

The Cortex-M processor uses a full-descending stack operation model, where the SP points to the largest address (also called the stack starting address) when stack memory is empty, i.e., no PUSH operation has been performed. In normal operation, SP points to the last data pushed onto the stack memory and gets decremented before a new PUSH operation. During a POP operation, data is read from the memory location pointed to by SP and then SP is incremented. In this case, the contents of the memory location remain unchanged and will be overwritten when the next PUSH operation takes place on this location.

Because each PUSH/POP operation transfers 4 bytes of data (each register contains 1 word, or 4 bytes), the SP decrements/increments by 4 at a time. In case of more than one register are pushed or popped, SP decrements/increments in multiples of 4 depending on the number of the registers. In the Cortex-M processor, R13 is defined as the SP. In case of interrupts, a number of registers will be pushed automatically. Similarly, the pushed registers will be popped automatically when exiting an interrupt handler and the SP will also be adjusted accordingly.

The Two-Stack Model

As mentioned before, the Cortex-M processor has two SPs: the MSPS and the PSP. Which of the two stack pointers should be used is controlled by the control register bit 1 (CONTROL-Bit1). When CONTROL-Bit1 is 0, the MSP is used for both thread mode and handler mode as shown in Figure 2.17. In this arrangement, the main program and the exception handlers share the same stack memory region. This is the default setting after power-up. From Figure 2.17(b), we can observe that MSP points to the top of the stack when the stack is empty.

When the CONTROL-Bit1 is 1, the PSP is used in thread mode and MSP is used in handler mode. In this arrangement, the application program and the operating system (including exception handlers) have separate stack memory regions (see Figure 2.18). This can prevent a system crash, when a stack error is caused due to user application, but the main system stack (accessed by MSP) used by the operating system remains intact. This is based on the assumption that the user application runs in thread mode and the OS kernel executes in handler mode. From Figure 2.18(b), we can observe that when two stacks are used then two different non-overlapping memory regions are allocated separately for each stack. However, it is the responsibility of the software developer to ensure that the two stack pointers are properly initialized before using the stack memory.

2.6 Debug System

With the advancement of processor architectures and integration of more and more complex functionalities, the corresponding system firmware/software is also getting more complex. To ensure the proper functionality of these complex system software as

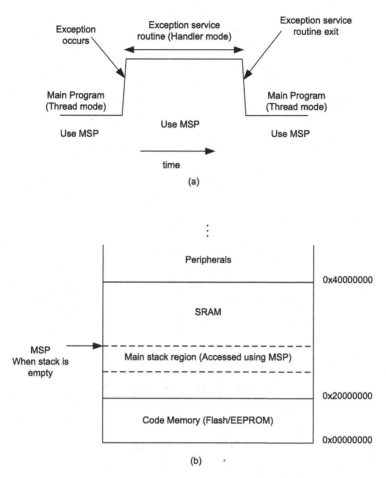

Figure 2.17: When CONTROL-Bit1 is 0 both thread and handler modes use main stack.

well as applications, the debugging features are becoming more and more important in the modern processor architectures.

The Cortex-M3 processor includes a number of fixed internal debugging components. These components provide debugging operation supports and features, such as breakpoints and watchpoints. Additional debugging features, such as instruction trace and various types of debugging interfaces can be provided by optional components.

Single-step, processor halt, CPU core register access, vector catch, software breakpoints and full system memory access are the basic debug functionalities. The processor implementation determines whether the debug configuration, including debug, is implemented. In case the processor does not support debug, no ROM table is present and the halt, watch point, and breakpoint functionalities are not present. Figure 2.19 gives a detailed processor block diagram detailing different debug and trace interfaces. The following are the optional debug and trace interfaces that can be implemented by the chip manufacturers.

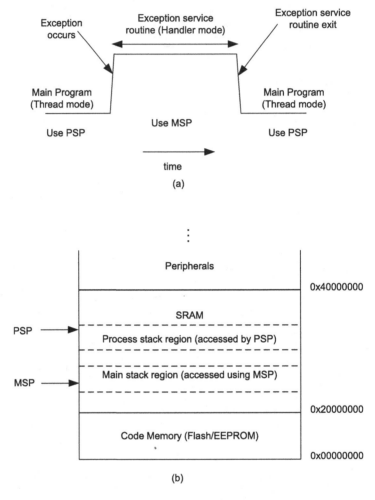

Figure 2.18: When CONTROL-Bit1=1, thread mode uses process stack and handler mode uses main stack.

2.6.1   AHB Access Port

The AHB-AP is a *Memory Access Port* as defined in the ARM Debug Interface v5 Architecture Specification. The AHB-AP is an optional debug access port in the Cortex-M3 system and provides access to all memory and registers in the system, including processor registers. System access is independent of the processor status. Either Serial Wire Debug Port (SW-DP) or Serial Wire JTAG (SWJ-DP) is used to access the AHB-AP. The AHB-AP is a master into the bus matrix.

2.6.2   Flash Patch and Breakpoint Unit (FPB)

The FPB unit is responsible for implementing both hardware breakpoints as well as code and data patching in the code memory region. A fully capable FPB unit

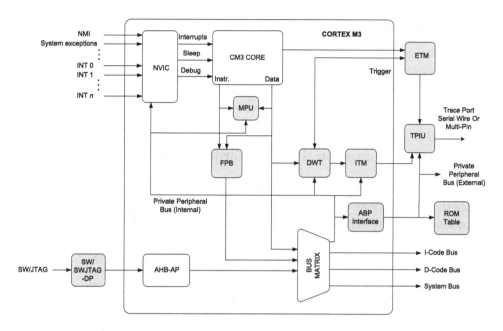

Figure 2.19: Detailed processor block diagram from the perspective of different debugging interfaces that can be implemented on ARMv7M-based microcontrollers. PPB is private peripheral bus, FPB is flash patch and breakpoint, DWT is data watchpoint and trace, ITM is instrumentation trace macrocell, ETM is embedded trace macrocell, and TPIU is trace port interface unit.

contains six instruction comparators to match against instruction fetches and two literal comparators. The instruction address comparators can be used to generate breakpoint events and as a result a halt mode or debug exception can be triggered. A reduced size FPB unit has only two instruction address comparators.

The FPB unit can be used to patch instructions or literal data from code memory space to the system SRAM. For instance an example user subroutine placed in SRAM can be tested by patching the appropriate region in flash memory or code region, to allow a branch to the test program.

2.6.3 Data Watchpoint and Trace (DWT)

The DWT is an optional debug unit that provides watchpoints, data tracing, and system profiling for the processor. A brief description of those features is provided below:

- Comparators that support watchpoints, which can cause the processor to enter debug state or take a debug monitor exception, data tracing, signaling for use with an external resource, for example an ETM, PC value tracing, cycle count matching.

- PC sample trace output as a result of a cycle count event and external PC sampling using a PC sample register.

- Exception trace.

- Performance profiling counters.

Which DWT features are supported is implementation specific. The DWT data is encapsulated in the form of packets and these packets are sent to the instrumentation trace macrocell (ITM) for transmission.

2.6.4   Instrumentation Trace Macrocell (ITM)

The Instrumentation Trace Macrocell (ITM) provides a memory-mapped register interface that applications can use to write logging or event words to a trace sink, for example to the optional external Trace Port Interface Unit (TPIU). The ITM forms event words and timestamp information into packets and multiplexes them with hardware event packets from the Data Watchpoint and Trace (DWT) block.

2.6.5   Embedded Trace Macrocell (ETM)

An Embedded Trace Macrocell (ETM) is an optional feature of an ARMv7-M implementation. Where it is implemented, the device must implement a trace port interface unit that can format the combined output packet stream from the ETM, the DWT, and ITM.

2.6.6   Trace Port Interface Unit (TPIU)

The TPIU is used to output trace packets from the ITM, DWT, and ETM to the external capture device. This packet stream might:

- Terminate in the processor

- Be visible externally through a trace buffer

- Be visible externally through a trace interface connection

The Cortex-M3 TPIU supports two output interfaces. One of them is a parallel trace port interface which has a clock pin and up to 4-bit parallel data output pins. The other one, a serial wire asynchronous port, can also be provided. Either one or both of these interfaces can be available simultaneously.

2.6.7   Memory Protection Unit

In addition to debugging capabilities, we also mention some other special features provided by the ARM architecture. For example, in complex applications, which require additional memory related features, the Cortex-M3 processor has an optional Memory Protection Unit (MPU). In addition, it is also possible to use an external cache memory if required by the system designer.

2.7 Summary of Key Concepts

In this chapter, the ARM Cortex-M architecture has been introduced, with a focus on ARM Cortex-M3/M4 profiles. Below is the summary of key concepts introduced in this chapter.

- The ARM processors based on ARM architecture version 7 are classified in three different profiles, namely, Cortex-M, Cortex-R, and Cortex-A and are considered RISC processors.

- Cortex-M based processor architecture is aimed for microcontroller-based embedded applications.

- The fundamental building blocks of Cortex-M processor include a processor core (M3/M4), a bus matrix, a nested vectored interrupt controller, and multiple debug interfaces.

- Cortex-M profile is based on Thumb2 instruction set architecture (ISA), which includes both 16-bit as well as 32-bit instructions.

- The key advantage of mixing 16-bit and 32-bit instruction encoding is reduced memory requirement (with respect to 32-bit ARM ISA) as well as improved execution performance (with respect to 16-bit Thumb ISA).

- There are 13 general purpose registers, R0-R12, which are further grouped into low registers (R0-R7) and high registers (R8-R12).

- In general, high registers are only accessible to 32-bit instructions.

- The registers R13-R15 are special purpose registers, with R13 used as stack pointer (SP), R14 is link register, and R15 is program counter (PC).

- The register R13 has two physical registers, main stack pointer (MSP) and program stack pointer (PSP) associated with it.

- Additionally, there are three status registers, one control register, and three interrupt masking registers.

- The Cortex-M based processor can operate in two modes, the thread mode and handler mode. The thread mode can be either privileged or unprivileged, while handler mode is always privileged.

- The Cortex-M processors have a very simple reset sequence. On reset the hardware copies word size variables from address 0x00000000 to SP and from 0x00000004 to PC.

- The Cortex-M processor has a stage pipeline, with fetch, decode, and execute as the pipeline stages.

- Cortex-M processors follow a predefined and fixed memory and device address map.

- The bus matrix in Cortex-M processor is responsible for interconnectivity among the processor, memories, and other interfaces using Harvard bus architecture.

- Cortex-M processor supports both little-endian as well as byte invariant big-endian formats, however most of the processors follow little-endian format.

- The memory addressing in Cortex-M processor supports bit banding, which provides bit addressing capability for selected memory regions.

- The Cortex-M processor follows a full descending stack model.

- The Cortex-M architecture supports multiple debug interfaces. However, JTAG and serial wire debug are more widely used.

Review Questions

Question 2.1. What are the possible instruction sizes in Thumb and ARM modes?

Question 2.2. How does the instruction size in Thumb2 mode differ from that of Thumb and ARM modes?

Question 2.3. How does the performance (both processing speed and memory usage) of Thumb2 instruction set architecture (ISA) compare with that of Thumb and ARM ISAs?

Question 2.4. Is it possible for a 16-bit instruction to use high registers, i.e., R8-R12?

Question 2.5. Which of the three special registers, SP (R13), LR (R14), and PC (R15) has two physical copies?

Question 2.6. What happens if the contents of register LR are modified inside a subroutine?

Question 2.7. Let the processor be in thread mode with unprivileged access level. Can the processor modify the contents of PC or LR in this operating mode?

Question 2.8. What are the very first two operations performed after reset?

Question 2.9. List the three phases of pipelined execution of an instruction.

Question 2.10. What are the default starting addresses and sizes of code and data memory regions allocated in the memory map?

Question 2.11. Does little-endianness and big-endianness make any difference if all the memory operations are of size byte?

Question 2.12. The processor uses multiple buses including ICode, DCode, and System bus to access different regions in the memory address space. Which of these buses is used to access data memory region?

Question 2.13. What are the key uses of stack memory region?

Question 2.14. If the user application is required to use process stack, then what configurations should be performed by the system before the user application program running in thread mode can perform PUSH/POP operations on the process stack?

Question 2.15. List the key advantages of bit band aliasing. What is the key disadvantage of using bit banding?

Question 2.16. Name different debug interfaces that are available on Cortex-M processor. What are the main uses of Serial Wire Debug (SWD) interface?

Exercises

Exercise 2.1. What are the values of the T, F, and I bits in the Program Status register on the processor reset?

Exercise 2.2. Which of the following ARM processor grouping implements the Thumb2 technology?

1. All ARM processors

2. All ARMv7 processors

3. ARMv7-A processors only

4. ARMv7-A and ARMv7-R but not ARMv7-M

Exercise 2.3. Derive an expression that maps the addresses in bit-band alias region to the corresponding addresses in the bit-band region.

Exercise 2.4. Is it possible for main stack (accessed by MSP) and process stack (accessed by PSP) to be nonempty simultaneously? If the answer is YES, then explain the scenario and if the answer is NO, then explain why.

Exercise 2.5. What are the possible solutions to increase the execution speed for Thumb instruction set architecture?

Exercise 2.6. Consider a microprocessor system where the processor has 16-bit data bus and 20-bit address bus. What is the maximum size of the byte addressable memory that can be connected with this processor?

Exercise 2.7. In continuation to the previous exercise, now consider a bit addressable memory with 32-bit bus. What is the maximum memory size in (bytes) that can be accessed?

Exercise 2.8. A microprocessor executes single instruction in 2 clock cycles. Its non-pipelined architecture version takes 200 clock cycles to execute a test program with 100 instructions. How many clock cycles will be required to execute the same test program on its pipelined version? The pipelined version of the processor is implemented by dividing a single instruction to four stages (phases) and each stage takes half clock cycle. Moreover the test program does not include branch instructions that can cause pipeline flushing and refilling.

Exercise 2.9. What are different possibilities to switch the Cortex-M processor from user (unprivileged) mode to handler (privileged) mode?

Exercise 2.10. We want the processor to operate in thread mode with unprivileged access level and use main stack pointer. What value should be configured to the CONTROL register for this purpose?

Exercise 2.11. Pipelined architecture helps to improve the execution speed. What type of program can completely nullify the advantage of using pipelined architecture?

Exercise 2.12. Four bytes 0x12, 0x34, 0x56, and 0x78 are stored at contiguous memory locations in order of increasing address. Assume that these bytes constitute one 32-bit word in little-endian format. Now what will be the 32-bit value read as word variable, if the starting address of the word variable is same as that of lowest address byte, i.e., 0x12?

Exercise 2.13. We want to modify the 13th bit of the memory word stored at address 0x20000FF0 using bit banding. What memory address, from the bit band alias region, should be used to modify this bit?

Chapter 3

Exceptions and Interrupts Architecture

Overview

Today almost every microcontroller integrates interrupt capability, which can range from simple interrupts to multi-level prioritized interrupts. Both hardware (e.g., external inputs or peripherals) as well as software events can generate interrupts. An interrupt causes deviation to the normal program execution flow. Whenever an event or exception happens, the corresponding peripheral or hardware requires a response from the processor. The response from the processor is implemented as a function call and is also called *interrupt service routine* (ISR). The sequence of operations involved from the occurrence of the exception/interrupt to the resumption of normal program execution again often involves the following key steps.

- Step 1: One of the interrupt or exception sources generates a request. For instance, the peripheral asserts an interrupt request to the processor.

- Step 2: In response to the interrupt, the processor suspends the currently executing task.

- Step 3: The processor executes an interrupt service routine to generate the response and service the source of interrupt. In addition, sometimes it is required to clear the interrupt request by software.

- Step 4: Finally, the processor resumes the execution of previously suspended task from the same state.

This chapter describes the concepts related to Cortex-M exceptions and interrupts in a programming-free fashion. It looks at the role of the nested vectored interrupt controller (NVIC) in handling interrupts on Cortex-M. We explain how to configure interrupts. How many interrupts does Cortex-M support? Why do interrupts need to be prioritized? How does Cortex-M make the invocation of interrupt service routines more efficient? At how many levels does NVIC allow to mask interrupts?

3.1 The Cortex-M Exceptions and Interrupts

All Cortex-M processors include a Nested Vectored Interrupt Controller (NVIC), which is responsible for handling exceptions and interrupts. According to ARM nomenclature, an interrupt is simply a specific type of an exception. Exceptions numbered 1 to 15 are also called system exceptions. Exceptions numbered from 16 and above are used for peripheral interrupts. Following ARM terminology, we will call the system exceptions (i.e., exceptions 1-15) simply *exceptions*, while exceptions numbered 16-255 will be termed *interrupts*.

3.1.1 Nested Vectored Interrupt Controller

The NVIC is an integral part of the Cortex-M based microprocessor architecture. It is tightly integrated with Cortex-M processor core. NVIC can be configured for the desired functionality using its memory-mapped control registers. Most of interrupt status and control registers can only be accessed in the privileged mode, except the software trigger interrupt register, which can be configured for user mode accessibility. Further details regarding NVIC registers will be provided later.

In addition to interrupt processing, the NVIC also contains a timing module called SYSTICK timer and its associated control registers. The SYSTICK timer is responsible for generating a timing reference that is used by the system software to mange and schedule its activities. The NVIC supports 1-240 peripheral interrupts (correspondingly exceptions 16-255), which are also known as interrupt requests (IRQs). Almost all the Cortex-M based microcontrollers support exceptions 1-15, however, the actual number of interrupts that are supported by the microcontroller is determined by the hardware manufacturers. Block diagram showing the interfacing of different interrupt sources with NVIC as well as the connectivity between ARM core and NVIC is illustrated in Figure 3.1.

3.1.2 Exception Types

The Cortex-M provides an exception architecture for both system as well as for external interrupts. Exceptions numbered 1-15 are assigned to system exceptions while exceptions numbered from 16 onward are allocated for external interrupts. Some of the system exceptions are assigned fixed priority, while the external interrupts have programmable priority. The number of external interrupt inputs in Cortex-M3 chips can be between 1 and 240, and can be assigned different priority levels. This allows the system developer to configure Cortex-M based microcontrollers for a wide variety of applications.

Exceptions numbered 1 to 15 are specifically called system exceptions. The label of each system exception and its brief description is outlined in Table 3.1. Exceptions numbered from 16 and above are used for external or peripheral interrupt sources, which are also included in Table 3.1.

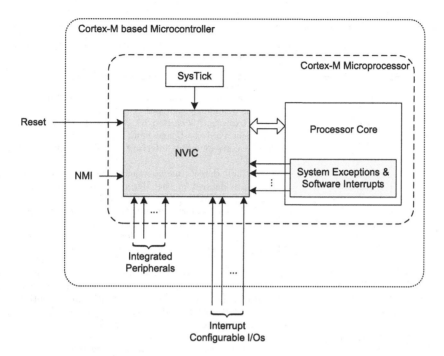

Figure 3.1: Block diagram illustrating the interfacing of different sources of interrupts with nested vectored interrupt controller.

3.2 Exception and Interrupt Priority

In its simplest implementation, we can have all the interrupts and exceptions of the same priority. In this situation, the execution of an ISR corresponding to a newly arrived interrupt request only begins after the completion of the previously executing ISR corresponding to an earlier interrupt request. However, it is quite natural to have multiple tasks to be executed by the microcontroller and these tasks have different priorities. For the microcontroller to be able to implement a prioritized response to the events corresponding to these tasks, a priority assignment capability for different interrupts is required.

The NVIC allows multiple interrupt priority levels. The total number of priority levels supported by a microcontroller can vary depending on the architectural requirements of that microcontroller. Interrupt handling of an exception is affected by the priority assigned to that interrupt. In Cortex-M architecture, a higher-priority corresponds to a smaller number assigned for priority level. When the exception priorities are enabled, a higher priority exception can preempt a lower priority (correspondingly a larger value in priority level) exception. Table 3.2 shows the priority assignment to different exceptions. From Table 3.2, we can see that some of the exceptions have fixed priorities assigned to them, e.g., reset, NMI, and hard fault have fixed priorities. The priority of reset exception or interrupt is −3 while that of NMI and hard fault are −2 and −1, respectively. Since higher priority corresponds to lower priority value, reset interrupt is the highest priority interrupt. In addition, the priorities assigned to other

Table 3.1: System exceptions and interrupts.

Exception No.	Label	Description
1	Reset	System reset exception.
2	NMI	Nonmaskable interrupt. The use of this system exception is defined by the user.
3	Hard fault	This exception is caused by Bus Fault, Memory Management Fault or Usage Fault. The usage fault occurs if the corresponding interrupt handler cannot be executed.
4	Memory management fault	This fault detects memory access violations to regions that are defined in the Memory Protection (Management) Unit (MPU). One possibility can be code execution from a memory region with read/write access only.
5	Bus fault	This system exception occurs when memory access errors are detected when performing instruction fetch, data read or write, interrupt vector fetch or register stacking.
6	Usage fault	Can occur due to the execution of undefined instructions, unaligned memory access (in case of multiple data word load/store instructions). When this exception is enabled it can detect, divide-by-zero as well as unaligned memory accesses.
7-10	Reserved	-
11	SVC	SuperVisor Call used by operating system.
12	Debug Monitor	Debug exception due to events including breakpoints, watchpoints, etc.
13	Reserved	-
14	PendSV	An OS based software exception for scenarios like context switching.
15	Systick	Exception generates by system tick timer. This timer can be used by the OS for system timing reference generation.
16	Interrupt 0	Peripheral interrupt 0 also called IRQ0. Can be connected to on chip peripherals or interrupt I/O lines. This argument is valid for all IRQs.
17	Interrupt 1	Peripheral interrupt 1 also called IRQ1.
⋮	⋮	⋮
255	Interrupt 239	Peripheral interrupt also called IRQ239.

exceptions are programmable and can take values between 0 and 255. As a result, they are of lower priority compared to reset, NMI, and hard fault exceptions.

The Cortex-M3/M4 based microcontrollers support three fixed highest-priority levels and up to 256 programmable priorities, while the maximum number of priority preemption levels is 128. The reason for 128 preemption priority levels will be explained in Section 3.6. Though 128 programmable priority levels are available, most Cortex-M3/M4 chips have fewer supported priority levels, for example, 8, 16, 32, etc. When a Cortex-M3/M4 architecture based chip is designed, the number of priority levels required can be customized by the chip designer.

Interrupt-priority level configuration registers are used to assign the required priority-level to a specific interrupt. The minimum number of priority levels is 8 corresponding to 3-bit priority configuration, while using an 8 bit priority configuration, we can have

Table 3.2: System exceptions and interrupts with associated priorities.

Exception No.	Label	Priority
1	Reset	-3
2	NMI	-2
3	Hard fault	-1
4	Memory management fault	Programmable
5	Bus fault	Programmable
⋮	⋮	⋮
16	Interrupt 0	Programmable
17	Interrupt 1	Programmable
⋮	⋮	⋮
255	Interrupt 239	Programmable

256 priority levels and is the maximum number of priority levels available with Cortex-M architecture. To reduce the number of priority levels, fewer bits of the priority configuration registers are used by ignoring the least significant bits. For instance, when 3-bit priority is implemented in the design, a priority-level configuration register will look like as shown in Figure 3.2. Since bit-0 to bit-4 are not used when implementing 3-bit priority, these bits are read as zero, while any write operation to these bits is ignored. With this setup, we have possible priority levels of 0x00 (corresponding to high priority), 0x20, 0x40, ⋯ 0xE0 (correspondingly lowest priority), which are also shown in Figure 3.2. Similarly, if 4-bit priority is implemented by the processor, a priority-level configuration register will look like as shown in Figure 3.3.

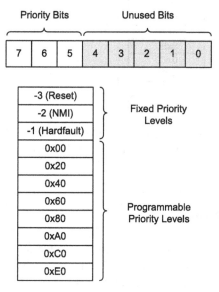

Figure 3.2: Bit allocation in the priority configuration register for configuring 3-bit priority. The priority values corresponding to programmable priorities are also tabulated.

Using most significant bits of the priority configuration registers instead of the least significant bits makes it simpler to reuse the existing software of one Cortex-M3/M4 based microcontroller for another device. For example, if the number of priority levels on the current platform were implemented using 4-bits, while the newer platform has only 3-bit priority levels, then the software program will be able to execute with its interrupt priorities intact. On the other hand, if least significant bits are used for interrupt priority configuration, then going from 4-bit priority to 3-bit priority will not be possible in a straightforward manner. This might lead to an inversion of priority. For example, if an application uses priority level 0x05 for IRQ 0 and level 0x03 for IRQ 1, then IRQ 1 has higher priority. However, when the most significant bit, i.e., bit2 is eliminated, IRQ 0 will become level 0x01 and have a higher priority than IRQ 1, which still will have the same priority 0x03.

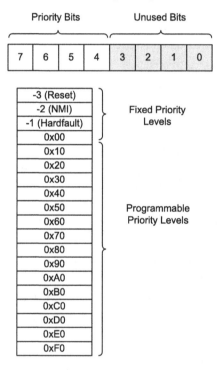

Figure 3.3: Bit allocation in the priority configuration register for configuring 4-bit priority. The priority values corresponding to programmable priorities are also tabulated.

3.2.1   Interrupt States

When implementing multiple priority levels for different interrupts, the execution of an ISR corresponding to a low priority interrupt might be suspended due to the occurrence of a high priority interrupt. This leads to different possible operating states and correspondingly the interrupt can be in one of the following states.

- Active: The interrupt is in active state when it is being serviced by the processor but servicing has not been completed yet. An exception handler of higher priority can interrupt the execution of another lower priority exception handler. In this case, both exceptions are in the active state.

- Inactive: The interrupt is neither active nor pending in this state. In other words, an inactive state of an interrupt corresponds to the situation when no interrupt condition has been generated from the corresponding interrupt source.

- Pending: The interrupt is waiting to be serviced by the processor. An interrupt request generated from a peripheral or from software changes the interrupt state for that interrupt source to pending. The pending state changes to active state when the servicing corresponding to that interrupt starts.

- Active and pending: An interrupt is being serviced by the processor and there is a pending interrupt from the same source.

3.3 Interrupt Configuration

Interrupt configuration process has multiple components. Global interrupt configuration is applicable to all the interrupts, while local interrupt configuration is specific to that particular source of interrupt. Both local as well as global interrupt configurations are required for proper functioning of interrupts. In addition to interrupt configuration, we need to set up a table of interrupt vectors holding the information related to interrupt service routines. This table is called interrupt vector table and is used at the time of occurrence of an interrupt to determine where the corresponding service routine is located in the instruction memory. It is also possible to relocate the vector table in data memory.

3.3.1 Basic Interrupt Configuration

There are two aspects related to interrupt configuration. The first aspect is the global configuration for the desired interrupt processing behavior. The following configurations are normally involved in setting up the interrupt behavior globally.

- Interrupt/exception masking registers configuration
- Setting up interrupt vector table
- Configuring interrupt priority groups

Configuration of masking registers and interrupt vector table are discussed in this section, while the discussion regarding the configuration of interrupt priority groups is postponed until Section 3.6.

The second aspect is the device or peripheral specific interrupt configuration of the source of interrupt and can be configured using the associated registers. Some of the important local device specific configurations, based on the functionality, are listed below.

- Enabling and disabling of interrupts locally

- Interrupt pending control and status

- Priority level configuration

- Active status indication

Only the global configurations are discussed in this chapter. The device specific local interrupt configurations will be discussed in later chapters.

3.3.2   Interrupt Masking

There are three different interrupt/exception masking special registers in a Cortex-M processor. They are priority masking register (PRIMASK), fault mask register (FAULTMASK), and the base priority masking register (BASEPRI). These registers are useful for interrupt enabling or disabling and mask the interrupts based on the assigned priority levels. These registers can only be accessed when the processor is operating at privileged access level. In case of unprivileged access level, any write operation to these registers is ignored, while a read operation returns zero. On reset these registers are cleared to zero resulting in no interrupt masking.

The PRIMASK interrupt masking register is a 1-bit (only least significant bit is used) register. When it is set, all the exceptions/interrupts are blocked except the Reset Interrupt, the Non-Maskable Interrupt (NMI), and the HardFault exception. Setting PRIMASK to 1 is equivalent to raising the priority level to 0. One of the common usages of PRIMASK is to disable all of the interrupts when executing a critical code section that should not be interrupted once its execution starts.

The FAULTMASK register is quite similar to PRIMASK as this is a 1-bit register too. Setting FAULTMASK to 1 only allows reset and NMI but masks the HardFault exception. This is effectively equivalent to the scenario that the priority level of current exception has been set to -1. The interrupt service routine corresponding to FAULTMASK can efficiently avoid any further triggering of fault interrupts. For instance, FAULTMASK can be used to suppress any bus faults. In contrast to PRIMASK, the FAULTMASK is cleared automatically when returning from an exception.

The flexibility in masking the interrupts is introduced by the BASEPRI register. The interrupt masking by the BASEPRI is performed depending on the current priority level configuration. Since the number of priority levels in ARM Cortex-M architecture can be between 8 and 256, it correspondingly requires 3 to 8 bits to be used by the BASEPRI register. The actual number of bits used by the BASEPRI are determined by the microcontroller manufacturer. Most Cortex-M3 and Cortex-M4 based microcontrollers have 8 or 16 programmable interrupt priority levels. When BASEPRI is set to a non-zero value, it blocks all the interrupts of either the same or lower priority, while it allows the processor to accept the interrupts of higher priority for processing. When BASEPRI is set to 0, it is disabled.

The BASEPRI and the PRIMASK system registers can be used for disabling interrupts temporarily when dealing with time critical applications. The FAULTMASK register can be used by the operating system for disabling temporarily the fault handling, in case a task has crashed. For this case, the crashing of a task might have been the result of different faults. When the operating system is busy in system recovery, it might be desirable to not allow some other faults to interrupt the system. Therefore, the FAULTMASK allows the operating system kernel to operate uninterrupted while dealing with fault condition. Figure 3.4 shows the definitions for the three interrupt mask resisters.

These interrupt masking registers are accessed through special register access instructions. Move to special register from general-purpose register (MSR) and move special register to general-purpose register (MRS) assembly programming instructions are used for this purpose and will be explained in Chapter 6.

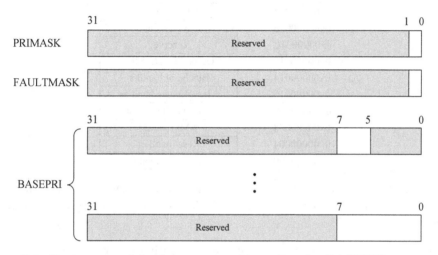

Figure 3.4: Registers used for interrupt masking. For the BASEPRI register two extreme scenarios corresponding to 3-bit and 8-bit implementation are shown.

3.3.3 Setting Up Interrupt Vector Table

The response to an interrupt by the Cortex-M processor is implemented in the form of a service routine. When an interrupt has occurred and is accepted for processing based on the masking registers configurations as well as priority settings, the next step for the processor is to obtain the starting address of the corresponding interrupt service routine or the exception handler. The starting addresses of the interrupt service routines, for all the interrupts used by the application or the system, are stored in the memory in the form of an interrupt vector table. In Cortex-M processor architecture, the vector table starts at memory address 0 by default. For 32-bit address bus the starting address of an arbitrary interrupt service routine can be stored in 4 bytes. Due to this fact, the vector addresses are stored sequentially in ascending order of exception numbers as depicted in Figure 3.5. It is important to remember that the order as well as the assigned location to the entries in the vector table is fixed and they cannot be rearranged.

The vector table is usually provided as part of the startup code made available by the microcontroller chip manufacturers. Since the vector table has 256 entries, corre-

Memory Address   Memory Contents

Memory Address	Memory Contents
0x00000048	Interrupt #2 handler
0x00000044	Interrupt #1 handler
0x00000040	Interrupt #0 handler
0x0000003C	Systick handler
0x00000038	PendSV handler
0x00000034	Reserved
0x00000030	Debug Monitor handler
0x0000002C	SVC handler
0x00000028	Reserved
0x00000024	Reserved
0x00000020	Reserved
0x0000001C	Reserved
0x00000018	Usage Fault handler
0x00000014	Bus Fault handler
0x00000010	MemManage handler
0x0000000C	Hard Fault handler
0x00000008	NMI handler
0x00000004	Reset handler
0x00000000	MSP initial value

Figure 3.5: Interrupt vector table at the default location. Observe that some of the unused entries are also included to maintain the location of each entry in the vector table.

spondingly it requires as many interrupt service routines. However, it is possible to have only a few interrupt service routines implemented based on the interrupts used by the system and application. The first 16 entries of the interrupt vector table as shown in Figure 3.5, correspond to system exceptions and are always the same across all the Cortex-M based microcontrollers independent of the chip manufacturer. However, the vector table entries from 17 onward are used for peripheral interrupts and can be assigned to different peripheral modules in arbitrary order by the microcontroller chip manufacturer.

The first entry in the vector table contains the initial value of the main stack pointer (MSP). It is worth mentioning that considering the main stack pointer initial value as part of the vector table is optional and can be excluded from it. In Cortex-M architecture, the default starting address of the interrupt vector table is 0x00000000, when the main stack pointer initial value is considered as part of the vector table. This address is located in the code memory region, which usually is based on either flash memory or an EEPROM. Since the vector table is located in non-volatile permanent memory region, it cannot be modified or updated at runtime.

Based on the application requirements, sometimes it is quite useful to have the ability to modify interrupt vectors or correspondingly use a different service routine for the same interrupt source, at run-time. This can be implemented by first relocating the interrupt vector table in the RAM area. For that purpose, Cortex-M processor architecture supports *vector table relocation* capability.

3.3.4 Configuring an Interrupt

For simple applications, the program is stored in flash memory or EEPROM and there is no need to change the exception handlers at run time. In this situation we can have the interrupt vector table located at the beginning of the code memory region. As a result, the vector table offset will be zero and the sequence of steps required for setting up the interrupts is listed below.

- Step 1: This step is related to global enabling of the interrupts. It involves configuring PRIMASK, FAULTMASK, and BASEPRI masking registers. Specifically, bit 0 of PRIMASK and FAULTMASK registers should be set to 0 to enable all the interrupts. In addition, BASEPRI is configured for the base priority level to disable the interrupts below a certain priority level. The default value is zero for all of these interrupt masking registers.

- Step 2: This optional step is related to setting the priority of the interrupt. Priority setting may require two configurations, one for priority level and second for priority group. This is applicable when the priority register is decomposed into two bit fields. In case the priority groups are not implemented, the entire priority configuration register is used for assigning priority levels. On reset, the priority configuration register is 0 and correspondingly all the interrupts are at priority level 0 which is the highest configurable priority level.

- Step 3: In addition to global configuration of the interrupts, each potential interrupt source should be configured to enable the specific interrupt.

Priority configuration is implemented by NVIC for each interrupt source separately. The general ideas related to priority configuration will be discussed in this chapter. The hardware platform dependent details regarding the priority configuration will be discussed in Chapter 9. In addition the local interrupt configuration for a peripheral source will also be discussed in Chapter 9.

When dealing with interrupts, allocating enough size of the stack memory is quite important. The stack memory size is of significance when dealing with a large number of nested interrupt levels. Since the exception handlers always use the main stack pointer, it is the size of the main stack memory that should have enough space.

3.4 Handling of Exceptions or Interrupts

When an exception or an interrupt occurs, the processor uses either handlers for exceptions or service routines for interrupts. This differentiation can also be understood in the context that user application programs are mostly responsible for implementing interrupt service routines, while an operating system deals with the exception handlers. Based on this terminology we have defined the following handlers and service routines.

- Interrupt Service Routines: The external or peripheral interrupts (i.e., exceptions 16 to 255) are handled by interrupt service routines.

- Fault Handlers: These handlers are used for HardFault, MemManage fault, UsageFault, and BusFault.

- System Handlers: NMI, PendSV, SVCall SysTick as well as the above-mentioned fault exceptions are all considered system exceptions. These exceptions are handled by system handlers. Usually these handlers are implemented as part of the operating system.

The Cortex-M3 processor uses a vector table that contains the addresses of the service routines. When an interrupt from a particular interrupt source occurs, then the corresponding interrupt service routine is executed. When the processor has accepted an interrupt, it fetches the service routine address from the vector table using instruction bus interface. On reset the vector table is located at address zero. However, it can be relocated to any arbitrary address by using the associated configuration registers. When an exception takes place, a sequence of operations is performed for proper handling of the exception. The following basic steps are involved in that sequence.

- Execution of the current instruction is either completed or its execution is terminated without completing, depending on the value of interrupt continuable instruction (ICI) field in the xPSR register.

- The current status is preserved by pushing registers R0-R3, R12, LR, PC, and PSR to the stack.

- The processor reads the exception number field from the updated value of xPSR register and performs vector fetch by reading the exception handler starting address from the interrupt vector table.

- Before the execution of the interrupt service routine begins, the processor updates the stack pointer, link register (LR), and program counter (PC). The LR is loaded with a specific value to signify the type of interrupt return as will be explained in this section later.

- Specific interrupt service routine corresponding to the interrupt source is performed.

- The final step is to exit or return from the interrupt service routine, which also involves popping the registers from the stack.

3.4.1   Register Stacking in Response to Interrupt Occurrence

When an exception takes place, the instruction currently being executed is either finished or terminated (based on the value of ICI field in xPSR register) and then registers R0-R3, R12, LR, PC, and program status register (xPSR) are pushed on the stack. If the application program executing at the time of occurrence of interrupt or exception is using the process stack pointer, then the process stack will be used for storing the current status. On the other hand, if the application program was using the main stack pointer, then the registers are stacked to the main stack.

From this point onward (i.e., after stacking of the registers as described above), only the main stack will be used during the handler mode. For nested interrupts, where a higher priority interrupt occurs during the execution of a lower priority ISR, the main stack will be used during the execution of both ISRs. The set of eight registers pushed to the stack is commonly termed *stack frame*.

The order followed to stack the registers in the stack memory region is shown in Figure 3.6(a). The time sequence of stacking the registers is shown in Figure 3.6(b). The PC and xPSR registers are stacked first. This allows updating the exception number field in xPSR register by the NVIC to let the processor know about the source of interrupt. Once the processor knows about the interrupt source and has stacked the PC register, it can start fetching the interrupt service routine, by updating the PC, in parallel to stacking of other registers. This is possible due to the fact that register stacking is done using the data bus, while fetching of ISR is performed using the instruction bus.

It is important to realize that the processor automatically stacks its state before entering the interrupt service routine and unstacks the system state while exiting from the ISR. In other words, no PUSH and POP instructions are executed while stacking and unstacking the registers, which provides low latency interrupt handling. The built in hardware capability of register stacking and unstacking on one hand simplifies the interrupt programming task, while on the other hand it makes the latency independent of user implementation.

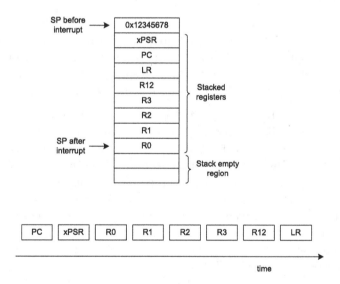

Figure 3.6: Stacking of the registers in case of interrupts. (a) The order in which eight registers are stacked in memory and (b) time sequence of stacking the registers.

After stacking, SP will be updated and points to the new stack top in the stack memory as can be seen from Figure 3.6(a). According to the C/C++ procedure call standard for the ARM architecture, the registers R0-R3, R12, LR, PC, and xPSR are caller saved registers and are required to be stacked when making a function call [ARM(2012)]. This arrangement allows the interrupt handler to be a normal C function. Any other general purpose register that can be modified by the exception handler should be saved on the stack by the user. The general registers (R0-R3 and

R12) are located at the end of the stacked region, allowing an easier access to these registers using SP relative addressing. In addition, these general purpose registers allow simpler parameter passing in case of software interrupts.

3.4.2   Updating Registers

After stacking of the registers is completed, few of the stacked registers are updated to their new values. Specifically, xPSR, SP, and PC registers are updated during the process of stacking and vector fetching, while the LR register is updated before the execution of ISR. A brief description of this updating of registers is provided below.

- SP: The stack pointer (either the MSP or the PSP) will be updated to a new value during register stacking. During execution of the interrupt service routine, the MSP will be used if any further stacking of some registers is required.

- xPSR: The least significant nine bits of xPSR (corresponding to the interrupt program status register, IPSR) will be updated to the new exception number.

- PC: The program counter will be updated to the exception handler and starts fetching the service routine instructions from the location of the exception handler.

- LR: The LR will be updated to a special value called EXC_RETURN. This special value controls the exception or interrupt return type, when returning from the service routine. The least significant 4 bits (5 bits in case of Cortex-M4F processors) of LR are used to provide exception return information. Our discussion below, regarding the exception return, is applicable to Cortex-M3/M4.

The bit description of LR register for determining the type of EXC_RETURN is provided in Table 3.3. Bits 31 to 4 are all set to 1 and do not have any effect on the exception return process. Bit 3 of LR register determines which mode, either handler or thread, will be used by the processor on returning from the exception handler. Bit 2 determines whether the main stack or process stack will be used by the system, after returning from the exception or interrupt service routine. Bit 1 is reserved and is not used. The value of bit 0 determines the processor state, being ARM or Thumb, after the exception return. Since the Cortex-M3/M4 based processors support only the Thumb state, bit 0 must be 1. The valid values for register LR that can be used for exception return are listed in Table 3.4.

At the time of exception occurrence of low priority, if the processor was operating in thread mode and was using MSP (main stack), then the value of LR is updated to 0xFFFFFFF9 before entering the interrupt service routine. On the other hand, if the processor was operating in thread mode and was using PSP (process stack), the value of LR register will be updated to 0xFFFFFFFD. This can be observed from Figures 3.7 and 3.8.

If multiple interrupts are enabled with different priorities, then there is always a possibility that a higher priority interrupt occurs while the processor is busy in executing a low priority interrupt service routine. Consider the scenario where a low priority

Table 3.3: Description of bit fields on LR register for exception return.

Bits	Value	Description
31-4	0xFFFFFFF	No effect on exception return behavior.
3	0	Return to handler mode.
	1	Return to thread mode.
2	0	Use main stack on exception return.
	1	Use process stack on exception return.
1	0	This bit is reserved and must be zero.
0	0	Processor returns to ARM mode of operation.
	1	Processor returns to Thumb mode of operation.

Table 3.4: Possible exception returns based on LR register.

LR value	Description
0xFFFFFFF1	Return from exception handler to handler mode.
0xFFFFFFF9	Return from exception handler to thread mode and use main stack on return.
0xFFFFFFFD	Return from exception handler to thread mode and use process stack on return.

interrupt has occurred at time t_1 and the processor is in handler mode executing the ISR corresponding to low priority interrupt. Now at time t_2, a higher priority interrupt occurs as can be seen from Figure 3.7. At this time, the value of LR register is updated (always) to 0xFFFFFFF1. This scenario corresponds to nested interrupts. From time t_2 to time t_3 the ISR corresponding to higher priority interrupt is executed and completed, assuming no further higher priority interrupts occur during the interval t_2 to t_3. At time t_3, the processor returns from higher priority ISR and resumes lower priority ISR. The lower priority ISR is completed in the time interval from t_3 to t_4, under the same assumption that no higher priority interrupts have occurred during this time.

If the nested interrupt occurs with LR value of 0xFFFFFFF9, then the corresponding processor states and its stack usage are pictorially depicted in Figure 3.7. For the case when the LR register value is 0xFFFFFFFD at the time of high priority exception occurrence, the corresponding processor states and its stack usage are shown in Figure 3.8. When returning from the exception or interrupt service routine, the registers are unstacked and the return behavior can also be observed from Figures 3.7 and 3.8.

Due to the special EXC_RETURN value format, one cannot perform an interrupt return to a memory address in the range 0xFFFFFFF0-0xFFFFFFFF. However, this requirement does not lead to any restrictions, since this address does not fall in the code memory region.

Some other NVIC related registers are also updated during the process of interrupt handling. For instance, when the processor is busy in handling a high priority interrupt, while a low priority interrupt occurs, then the status corresponding to low priority interrupt is set to pending in the corresponding NVIC register. When the

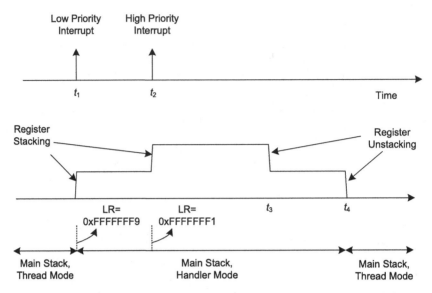

Figure 3.7: LR is set to 0xFFFFFFF9 on the occurrence of low priority interrupt and is updated to 0xFFFFFFF1 on the higher priority interrupt. In this case, main stack is used in thread mode.

processor is done with the high priority interrupt and initiates the handling of low priority interrupt, its status is automatically updated from pending to active in the corresponding NVIC register. We refer the reader to see either the reference manual or microcontroller specific data sheet for further details.

3.4.3   Exception Exit or Return

After completing the execution of exception or interrupt handler, an exception or interrupt return is performed to restore the system status, by unstacking the registers to resume normal execution of the interrupted program. In assembly programming, different instructions can be used to generate exception return. Some of the instructions, which can be used for returning from ISR, are listed below for reference purpose only and will be explained in later chapters.

- POP instruction

- Load instruction with PC as the destination register

- Branch instruction with any register holding the label

Some of the microprocessors support special instructions for returning from interrupts (for instance, 8051 has RETI instruction). In the case of Cortex-M3, return from interrupt can be performed using normal return instruction, allowing to implement the entire interrupt service routine as a normal C function. During the execution of interrupt return process, the register unstacking as well as updating of NVIC registers are also carried out.

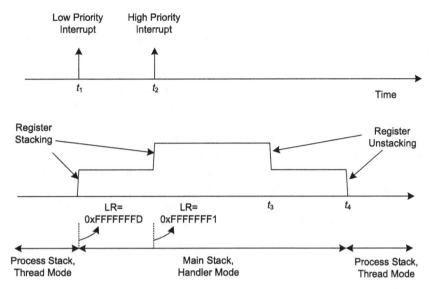

Figure 3.8: LR is set to 0xFFFFFFFD on the occurrence of low priority interrupt and is updated to 0xFFFFFFF1 on higher priority interrupt. In this case, process stack is used in thread mode.

- Unstacking: All the registers that have been pushed to the stack are restored during this process. The sequence of POP operation will match the one followed during register stacking. The stack pointer register is also updated during this process.

- NVIC register update: The active bit corresponding to current exception is cleared. If the interrupt input, for external interrupts, is still asserted, the pending bit is set again. As a result the program execution leads to reentering the interrupt service routine.

3.4.4 Interrupt Latency

The interrupt latency has two components. The first component is the interrupt entry latency and is quantified as the number of processor clock cycles from the arrival of an interrupt signal at the processor until the execution of the first instruction in the interrupt service routine starts. The second component corresponds to the interrupt exit latency and is equal to the number of clock cycles required from the execution of the interrupt return instruction until the execution of the next instruction of the interrupted task. Interrupt latency for the same hardware platform can be different under different operating conditions as discussed below. What is of importance is the minimum achievable interrupt latency for a given processor architecture. The interrupt latency is a crucial parameter for real time applications.

The Cortex-M3/M4 processor has the minimum interrupt entry latency of 12-cycles and interrupt exit latency of 10-cycles. Achieving these minimum interrupt entry and exit latencies requires a number of conditions to be fulfilled as discussed next.

The Cortex-M processor has three main memory interfaces. The I-Code and D-Code buses are used to access memory in the address space 0x00000000-0x1FFFFFFF while system bus is responsible for accessing memory addresses 0x20000000 and higher. To achieve 12-cycle interrupt entry latency it is required that register stacking (requiring 9 cycles to PUSH 8 registers) takes place on one memory interface (usually system bus) in parallel with vector fetch and interrupt handler fetch (requiring 6 cycles) on the other memory interfaces (usually I-Code). When these two tasks cannot be performed in parallel, the resulting interrupt entry latency will be higher. For instance, this might happen if the interrupt vector table is relocated to RAM area then both register stacking as well as interrupt handler fetch will be performed using system bus. Performing these two operations one after the other, rather than in parallel, will increase the latency. In addition, 12-cycle interrupt entry latency also requires that there are no memory wait-states. Memory access without wait-states is also required to achieve 10-cycle interrupt exit latency.

3.5   Interrupts Tail Chaining

Consider the scenario where interrupts from multiple sources with different priorities are configured simultaneously. In this situation, it is always possible that when an ISR corresponding to an interrupt is being executed a second interrupt occurs. Assuming the second interrupt is of the same or lower priority, it will be in pending state until the completion of the first ISR. On exiting from the first ISR the processor is required to enter the ISR corresponding to the second interrupt. Traditional interrupt controllers under this situation will perform register unstacking operation on exiting from the first ISR, immediately followed by register stacking to enter the second ISR.

Cortex-M3/M4 architecture avoids this unnecessary register unstacking/stacking cycle to improve both interrupt latency as well as its power efficiency (due to reduced number of memory read/write operations). This performance improvement in Cortex-M3/M4 processor is achieved by implementing tail-chaining in the NVIC. Tail-chining allows NVIC to switch from pending interrupt to active interrupt without unstacking/stacking of the registers. Tail-chaining reduces the latency between two interrupt handlers to 6 cycles compared to 22 (10+12) cycles. The 6-cycle latency in tail-chaining is due to the vector fetch and interrupt handler fetch corresponding to the second ISR. Comparison between tail-chaining and conventional interrupt handling is illustrated using Figure 3.9.

In the case of multiple nested interrupts, the processor will tail-chain a pending interrupt that has higher priority than all other interrupts in pending state. The reader is referred to [Yiu(2013)] for further details as well as discussions related to late-arrivals and unstacking preemption.

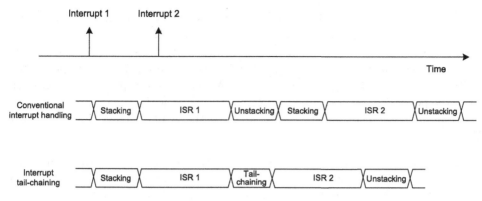

Figure 3.9: Illustration of tail-chaining implementation in nested vectored interrupt controller.

3.6 Interrupt Nesting with Multi-Level Priority

When multiple different priorities can be assigned to interrupts, the occurrence of nested interrupts is a natural consequence. The NVIC in Cortex-M3/M4 processor core provides nested interrupt support, allowing higher priority interrupts to preempt lower priority interrupts. The NVIC is responsible for decoding interrupt priorities and assign appropriate interrupt status to each interrupt event. When an interrupt occurs and has priority higher than any active interrupt already being processed by the processor, its state is changed to pending and newly arriving interrupt is processed immediately by the processor. If this is not the case, then newly arriving interrupt is assigned pending state and the processor continues executing the ISR corresponding to currently active interrupt.

When a large number of different priorities are assigned to multiple interrupts, then the size of the main stack should be assigned by taking into account the possibility of multiple nested interrupts. At each interrupt nesting level, at least 8 registers are pushed onto the stack assuming the ISR does not require extra stack space. For n nested interrupt levels, the minimum stack size required for worst case sequence of nested interrupt occurrences will be $32n$ bytes. These $32n$ bytes will be used just for register stacking by the hardware and any stack operations inside ISRs will require additional stack space.

When an interrupt is enabled with an appropriate priority level and processor is executing its ISR, all the interrupts with the same or lower priority will be blocked by pending them. This leads to the fact that a second interrupt event from the same source will not be serviced until the processor has returned from its ISR, making them non-reentrant interrupts.

3.6.1 Multi-Level Interrupt Priority

The 8-bit interrupt priority register is divided into two sub-fields. The most significant bit field determines the *group priority*, while the least significant bit field determines

the *sub-priority* of an interrupt within the same priority group. Appropriate configuration of the application interrupt and reset control (AIRC) register in the system
control block splits the programmable interrupt priority register into group priority
and sub-priority fields with required bit field lengths. Recall that interrupt priority
register can be from 3 bits to 8 bits in size. Assuming interrupt priority register is
8-bit, the possible number of group priority levels along with the permissible number
of sub-priorities, which can be configured using the AIRC register, is tabulated in
Table 3.5.

Table 3.5: Relation between number of group priorities and sub-priorities for different configurations of PRIGROUP bit field of AIRC register. The interrupt priority
register of size 8-bit is considered.

PRIGROUP	Group priorities (associated bits)	Sub-priorities (associated bits)
0	128 (b7-b1)	2 (b0)
1	64 (b7-b2)	4 (b1-b0)
2	32 (b7-b3)	8 (b2-b0)
3	16 (b7-b4)	16 (b3-b0)
4	8 (b7-b5)	32 (b4-b0)
5	4 (b7-b6)	64 (b5-b0)
6	2 (b7)	128 (b6-b0)
7	1 (none)	256 (b7-b0)

From Table 3.5 it can be observed that the maximum number of priority groups is
128 for default configuration with 7 bits (b7-b1) of interrupt priority register assigned
to priority group bit field. For this configuration, the PRIGROUP bit field of AIRC
register is set to 0. Since group priority is responsible for determining the interrupt
preemption, the maximum number of preemptable priority levels is also 128. When
the processor is executing an ISR, then another interrupt of the same group priority
cannot preempt the processor. For multiple interrupts in the pending state with the
same group priority, the sub-priority field is responsible for determining the order of
their processing. Setting PRIGROUP bit field of AIRC register to 7 configures one
(same) priority group for all the interrupts and corresponds to the last row in Table
3.5. In this case no preemption is permitted among any interrupts with programmable
priority levels. The above discussion is equally applicable for the case when interrupt
priority register size is less than 8 bits.

3.6.2   Interrupt Pending Behavior

The NVIC has dedicated programmable registers for storing pending status of interrupts. When an interrupt signal is asserted on an arbitrary input of NVIC, it results
in setting the pending bit of the corresponding pending status register associated
with that interrupt. The pending status remains active even after interrupt signal is
deasserted, allowing NVIC to function properly for edge (pulsed) interrupts.

The pending state represents that the interrupt is waiting to be served by the processor. When no other higher priority interrupts are being served, the processor

immediately starts serving this interrupt changing the pending status to become active. On the other hand, when the processor is busy serving an interrupt of higher or equal priority that occurred earlier, the current interrupt request is kept in pending state until the other interrupt handler has finished.

As soon as the processor begins processing an interrupt, its pending state is changed to active automatically as shown in Figure 3.10. The interrupt pending status is stored in dedicated interrupt pending status registers of NVIC, which can be accessed from user program. This leads to the possibility of manually setting or clearing interrupt pending status. When an interrupt source continuously asserts an interrupt signal, its interrupt request will be put in the pending state again and its ISR will be executed again by the processor as soon as the first ISR execution has finished and no new interrupt of higher priority has occurred during this time. For edge triggered (pulsed) interrupts, if the interrupt signal has occurred multiple times before the processor begins executing ISR, the interrupt request will be considered as a single interrupt.

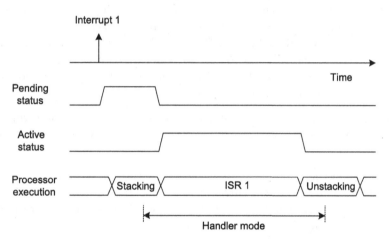

Figure 3.10: Relationship of interrupt pending and active states with the processor execution.

3.7 Summary of Key Concepts

In this chapter, we have introduced the basic concepts of interrupts and exceptions. Specifically, the following key concepts were introduced and discussed.

- The Cortex-M processor includes a nested vectored interrupt controller (NVIC); a unit responsible for handling exceptions and interrupts. It allows us to configure the interrupts according to the requirement of our application.

- Cortex-M can support up to 256 interrupts. 1-15 are system exceptions and the remaining are termed interrupts. The number of remaining interrupts can vary from one board manufacturer to another.

- Interrupts and exceptions are assigned priorities. For example, the RESET button when pressed should be entertained immediately irrespective of the activity being carried out by the microprocessor. So, assign it the highest priority. A value of -3 is assigned to it. For Cortex-M, higher priority corresponds to lower priority value. Some of the exceptions have a fixed priority and cannot be programmed by the user.

- There are various registers associated with interrupts to configure them. For example, we would like to mask interrupts and assign priority to them. In this regard, the interrupt configurations can be applied globally and locally. By global settings, we mean those settings which when programmed are applicable on all interrupts while local settings allow us to control interrupts independently.

- When an interrupt has occurred and is accepted for processing, the next step for the processor is to serve it. The service is a set of instructions collectively called an interrupt service routine (ISR) or the exception handler. The starting address of an ISR is stored in a table called interrupt vector table.

- When an interrupt occurs, a sequence of activities has to be carried out. For example, to preserve the current status of the Cortex-M processor, we need to push registers R0-R3, R12, LR, PC, and PSR onto the stack for temporary storage. Exiting an ISR will require to perform another sequence of activities. For example, unstacking all the registers that have been pushed onto the stack earlier at the entry of the ISR.

- When interrupts occur back to back, Cortex-M improves the interrupt latency by avoiding unnecessary register unstacking/stacking cycle. This is termed tail chaining.

- Interrupt nesting. For example, when an interrupt occurs and is of higher priority as compared to the currently executing interrupt, the state of the latter is changed to pending so that the processor can pay attention to the higher priority task immediately.

Review Questions

Question 3.1. How does the Harvard architecture help to speed up interrupt processing?

Question 3.2. What are the similarities and differences between system exceptions and external interrupts?

Question 3.3. What are different interrupt states?

Question 3.4. Which system exceptions have fixed priority?

Question 3.5. Is it possible to assign the same priority, to an external interrupt, as that of Hard Fault or NMI?

Question 3.6. What are the minimum and maximum number of priority levels supported by Cortex-M3/M4 based processor?

Question 3.7. For less than 8-bit priority, why do we use the most significant bits of priority register?

Question 3.8. If the BASEPRI register is configured with a value of 3, then what are the priorities of possible interrupt sources?

Question 3.9. What is the default memory location of the interrupt vector table (IVT)? Can IVT be moved to data memory region?

Question 3.10. What will happen if the two interrupt service routines, in the vector table, are swapped?

Question 3.11. When an interrupt occurs, which registers are pushed on to the stack? Are those registers pushed automatically or is the application programmer responsible for that?

Question 3.12. On the occurrence of an interrupt, certain registers are pushed onto the stack. Which stack (process or main) is used for that purpose?

Question 3.13. If the interrupt occurs during the execution of an instruction, then did the execution of that instruction complete or terminate before execution of the ISR?

Question 3.14. What is the special use of register LR in handling the interrupts?

Question 3.15. Under what conditions is the interrupt latency minimized?

Question 3.16. How does tail-chaining reduce interrupt latency?

Question 3.17. If there are eight group priorities, then what is the maximum number of sub-priority levels in each group?

Exercises

Exercise 3.1. When an interrupt occurs, it is expected that a corresponding ISR will be executed to generate system response. List at least four interrupt configuration steps that should be performed to ensure the execution of ISR.

Exercise 3.2. While executing an ISR, is it possible to tell that this ISR has put on hold the execution of the user application program or another ISR of lower priority?

Exercise 3.3. Based on the steps performed by the hardware in response to an interrupt, estimate the delay (in terms of clock cycles) from the occurrence of an interrupt to the execution of the first instruction inside the corresponding ISR. It can be assumed that no other interrupt (of higher priority) has occurred during this time.

Exercise 3.4. Why are there so many interrupts supported by a microcontroller? Look into two real embedded systems around you and list different interrupts that each system requires.

Exercise 3.5. Why do some of the interrupts have fixed priority while priority of others can be programmed? Justify your answer with real-life examples. Can something go wrong if we allow a fixed priority interrupt to become a programmable priority interrupt?

Exercise 3.6. A Cortex-M microcontroller supports negative values for the priority of some system exceptions. Justify the need to use negative priorities.

Exercise 3.7. Discuss the differences between local and global mechanisms available on Cortex-M for masking interrupts.

Exercise 3.8. What is interrupt latency? What are its two main components? How does Cortex-M improve the interrupt handling time by reducing interrupt latency?

Exercise 3.9. A Cortex-M based microcontroller is configured to enable three interrupts with associated priorities. Let source A be assigned priority 1 (highest), while sources B and C be assigned, respectively, priority 3 and 6. Now consider the scenario where interrupt from source C suspends the execution of application program and its ISR is being executed. While its ISR is not finished yet, the interrupts from sources A, B, and C occur and require servicing. Describe the sequence in which these interrupts will be handled by the processor until the application program is resumed. You can assume that no further interrupts occur during this phase.

Exercise 3.10. Consider a scenario where n multiple interrupts are configured with m different priority levels. What will be minimum stack size required, keeping in view the interrupt nesting possibility, for the following two cases:

1. $n > m$,

2. $n < m$.

Part II

Programming

Chapter 4

Basics of Assembly Programming

Overview

This chapter is an introduction to assembly language programming based on Thumb2 instruction set architecture, where we first present the format of assembly language instructions. After introducing a selected group of instructions, we discuss the other necessary details (including assembler directives), which are required to transform a set of assembly instructions to a standalone program that can be compiled and built to an executable. In addition, the processor reset sequence is discussed. Finally, both 16-bit and 32-bit instruction encodings, used by Cortex-M, are explained.

4.1 Introduction to ARM Instruction Sets

Before the introduction of Thumb2 architecture, the ARM processors used to operate mainly in two states; ARM state and Thumb state. The ARM state allowed only the execution of 32-bit word aligned instructions providing higher performance, while Thumb state allowed only 16-bit half-word aligned instruction execution providing higher instruction code density. In Thumb state all of the functionality provided by ARM instructions was not possible and it required multiple Thumb instructions to complete certain single cycle operations of ARM state. The processors that supported both 32-bit as well as 16-bit instructions required to switch between ARM state and Thumb state incurring an additional overhead of state switching. A third operating state called Jazelle state was introduced for byte-aligned Java byte codes with variable word length. However, Cortex-M profile does not support Jazelle state. It is important to mention that when we refer to Cortex-M processor it implies any of the Cortex-M3, Cortex-M4, or Cortex-M4F processor.

With Thumb2 instruction set, it has now become possible to use one operating state while meeting different processing requirements. The Cortex-M processor is based on ARMv7-M architecture and as a result uses Thumb2 instruction set. One limitation of Thumb2 instruction set architecture is that the older ARM code is no longer reusable on Cortex-M processor, which only supports Thumb2 architecture. However, the processors from Cortex-A profile support both Thumb2 and ARM states. The absence of the requirement to switch between states gives Cortex-M processor some key advantages over the conventional ARM processors as outlined below.

- Since no state switching is required, this results in an improved execution performance and saves instruction memory space as well.

- In conventional ARM processors separate ARM and Thumb code source files were maintained. However, this is not required for Cortex-M processors, providing simpler software development as well as maintenance.

Thumb2 architecture introduced a large number of 32-bit instructions in addition to the existing 16-bit Thumb instruction set. As a result, Thumb2 instruction set includes both 16-bit and 32-bit instructions, which can be intermixed freely as shown in Figure 4.1. This is one of the most important features of the Cortex-M processor, since it allows 32-bit instructions and 16-bit instructions to be used together for high code density as well as better execution efficiency.

It should be noted that Cortex-M3 does not implement all of the instructions provided by the Thumb2 instruction set. The ARMv7-M Architecture Reference Manual [ARM(2010)] only requires a subset of Thumb2 instructions to be implemented. For instance, co-processor instructions as well as Single-Instruction-Multiple-Data (SIMD) are not implemented on Cortex-M3. Similarly Cortex-M4 processor does not support the floating point instructions.

Instruction Execution in Thumb2 Architecture

Figure 4.1: The 16-bit and 32-bit mixed instruction execution in Thumb2 ISA.

Since Cortex-M based processors support only Thumb2 instruction set, reusing of the ARM instruction set based existing software requires porting it to the new Thumb2 instruction set. However, it should be realized many C programming based applications will only require to be recompiled using newer software development tools, which provide support for Cortex-M. The assembly codes on the other hand might need modifications and should be ported to the new architecture.

4.2   Cortex-M Assembly Programming Basics

Assembly language, in general, is considered a low-level programming language, which is specific for a processor architecture. This is in contrast to many of the high-level languages, e.g., C, Java, etc., which are easily portable across different architectures or platforms. Based on this fact one might argue that what is the point in studying the assembly programming at all. However, there are many advantages of learning assembly language, which are summarized below.

- An assembly language program provides complete as well as precise control of the available hardware resources.

- When writing a program in a high level language, the user is mostly unaware of the underlying hardware capabilities, which can lead to highly inefficient implementation.

- Programming in assembly language also gives the user complete understanding of the system architecture.

In an assembly language program for Cortex-M, each instruction might contain four different fields as illustrated in Listing 4.1. The left-most field called *label* is an optional field with unique symbolic representation in an assembly instruction. If an instruction has a label in front of it, then it can be used to determine the address of that instruction and correspondingly of that memory location uniquely. Every instruction is not required to have a label as it will become clear later in this chapter. The next field is called the *opcode* field, which contains the instruction or the opcode. The opcode field is a unique symbolic representation of the instruction's machine code, which is used by the processor for execution and is a mandatory field. Next to the opcode field is the *operand* field that might contain a different number of operands. Some of the Cortex-M assembly instructions will have no operand, while others might have as many as four operands, which makes this also an optional field. The number of operands in an instruction depends on the type of instruction. In addition, the syntax format of the operand can also be different for different instructions. Normally, the first operand represents the destination of the operation. The last field in an assembly instruction is also an optional field and is called *comment* field. This field comprises the text after each semicolon. These comments do not affect the program operation but make the programs easier to understand by humans.

```
label opcode operand_1, operand_2, ..., operand_n ; Comments
```

Listing 4.1: Syntax of an assembly program.

The assembly instructions for Thumb2 instruction set architecture can be grouped based on the functionality as shown in Figure 4.2. The instruction grouping in Figure 4.2 also shows their dependency on the Cortex-M processor families. In learning assembly programming language, we will discuss some of the instructions in more detail, while the remaining instructions will be introduced briefly. Our coverage of assembly language is confined to Cortex-M3 processor. For further details, we refer the reader to ARMv7-M Architecture Reference Manual [ARM(2010)].

To introduce Cortex-M assembly programming in this chapter, we have selected multiple assembly instructions from three different groups, namely data processing instructions, memory access instructions and branch and control instructions. In particular, we have selected instructions from data processing/data movement, memory access, and branch/control classes. This will be an informal introduction to the assembly

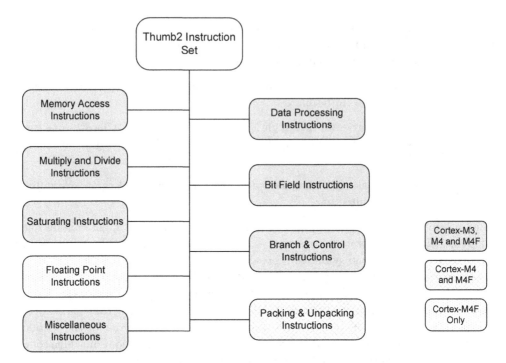

Figure 4.2: Instruction grouping based on functionality along with supported processor architecture information. For the data processing, multiply and divide, and saturating instruction groups, Cortex-M3 only supports a subset of instructions from these groups.

instructions, while a more formal coverage with additional details will be presented in the subsequent chapters.

4.2.1   ADD and MOV Data Processing Instructions

We start the assembly programming introduction using ADD (addition) and MOV (move) instructions, which belong to the group of data processing instructions. Here the use of these instructions is illustrated directly, without discussing the instruction syntax, which is postponed until the next chapter. Listing 4.2 illustrates the use of ADD instruction. In Listing 4.2, ADD and ADDS are called instruction mnemonics and represent the opcode field. The registers R1 - R6 and immediate value #0x123 are the operands. The # symbol is used to specify an immediate value or number. The integer value following the # symbol can be specified in different formats. For instance, we can use hexadecimal, decimal, or ASCII character format. The comments provide useful description of the instructions. The ADD instruction only performs addition of operands and the result is stored in the destination operand. However, the optional suffix S in ADDS instruction is responsible for updating the condition code flags in the application program status register in addition to performing addition.

```
ADD   R1, #0x6      ; R1 = R1 + 0x6
ADD   R4, R2        ; R4 = R4 + R2
```

```
ADD  R6, R5, R3     ; R6 = R5 + R3
ADDS R6, R5, R3     ; R6 = R5 + R3, flags are updated
```

Listing 4.2: Simple addition instruction illustration.

From Listing 4.2 we observe that the first two instructions use two operands, while the third instruction uses three operands. In the first two instructions the destination register is the same as one of the source operands. However, the third instruction uses a destination operand that is different from the source operands. Based on this fact, it can be concluded that the number of operands for an instruction may vary, depending on the instruction format used. Further explanation and illustrations regarding this fact will be provided in the subsequent chapters.

The second assembly instruction is MOV instruction, which is used for simple data transfers within the processor. This is illustrated with an immediate data value transferred to a register in Listing 4.3. Another data movement operation may require to transfer the contents of one register to another register. The use of MOV assembly instruction to perform these data exchanges, involving an immediate data transfer to a register and data transfer between two registers, is illustrated in Listing 4.3. The second instruction in Listing 4.3 illustrates how the ASCII value can be used as an immediate data operand. The MOV instruction cannot be used for data transfers between processor registers and memory.

```
MOV R2, #0x123      ; move immediate value of 0x123 to R2
MOV R4, #'A'        ; move ASCII value of A (i.e., 0x41) to R4
MOV R7, R3          ; move the contents of R3 to R7
```

Listing 4.3: Data movement instructions.

4.2.2 LDR and STR Memory Access Instructions

As discussed above, the MOV instruction is used to transfer data within the processor, i.e., between different registers or an immediate value to a register. On the other hand, LDR and STR instructions allow data to be transferred between processor and memory. The load register, LDR, instruction is used to transfer data from memory to the processor register, while store register, STR, instruction is used to transfer data from processor register to memory.

```
LDR R2, [R1]        ; load R2 with data from memory pointed to by R1
LDR R0, =NUM1       ; load R0 with the value of constant NUM1 (this
                    ; constant might represent the memory address)
STR R4, [R3]        ; store R4 to a memory location addressed by R3
```

Listing 4.4: Memory access instructions.

Listing 4.4 illustrates the use of LDR and STR instructions. The first LDR instruction in Listing 4.4 retrieves data from memory address specified by the value of register R1 and transfers it to register R2. In other words, assume variable x has been assigned an integer value, then $[x]$ represents accessing memory contents from an address specified by the value of x. The second LDR instruction in Listing 4.4 loads the immediate value to the destination register and in that context it is functionally equivalent to the MOV instruction used for loading immediate value. However, these instructions differ in the maximum immediate value that can be used. The maximum permissible immediate value for MOV instruction is either 8-bit for 16-bit encoding or 12-bit for 32-bit encoding. On the other hand, LDR instruction allows any arbitrary 32-bit immediate value that can be loaded to the specified register. This will be explained in detail in Chapter 6. Similarly, the last instruction in Listing 4.4 stores the contents of R4 to the memory location addressed by the value in R3. It should be realized that here we only introduced the simplest load and store instruction formats, however, there are many different possible syntaxes for LDR and STR instructions and will be discussed in detail later in Chapter 6.

4.2.3   Unconditional- and Combined Compare-Branch Instructions

The last set of assembly instructions in this introduction includes an unconditional branch and a conditional branch that performs comparison in addition to branching. The instruction mnemonic B represents unconditional branch, while compare and branch on zero instruction CBZ performs conditional branch if the result of comparison is zero. On the other hand, if the result of comparison is not zero, then CBZ behaves as failed conditional branch instruction. Listing 4.5 illustrates the use of unconditional branch instruction, where 'loop1' is the label used by the program. The label 'loop1' effectively is the address of the instruction LDR R2, [R1]. The same program also uses CBZ, which compares the value in register R5 with zero and if the result of comparison is true a branch to address 'label1' occurs. The use of instruction CBZ R5, label1 also implies that there is an instruction inside 'loop1' block that decrements the value of register R5 and eventually a branch to 'label1' occurs.

```
loop1   LDR R2, [R1]      ; load R2 with data from memory pointed
                          ;  to by register R1

    ...
        CBZ R5, label1    ; Compare R5 with zero. If comparison
                          ; result is true then branch to label1

    ...
        B   loop1         ; jump to the memory location labeled as
                          ; loop
label1
        MOV R3, #0x034
```

Listing 4.5: Simple memory access instructions.

4.3   Our First Assembly Program

Having gained some familiarity with Cortex-M assembly instructions, we are now in a position to write our first assembly language program. Let's take a simple task. We want to write a program that determines the sum of the five even numbers. The first five even numbers are 2, 4, 6, 8, 10. When we add them up, the sum is 30. To accomplish this task, we need to implement a loop so that we can avoid writing the same instructions again and again. Below is the code to implement this functionality, which uses the assembly language instructions that have already been introduced in the previous section.

```
        MOV R0, #0       ; R0 will accumulate the sum
        MOV R1, #2       ; R1 will have the updated even number
        MOV R2, #5       ; the counter for the loop

lbegin
        CBZ R2, lend     ; If R2 != 0 continue with the next
                         ; instruction
        ADD R0, R1
        ADD R1, #2
        SUB R2, #1
        B lbegin         ; branch unconditionally to lbegin
lend
```

Listing 4.6: A simple program calculating the sum of the first five even numbers.

Let's analyze the working of this program before we learn how to execute it on Cortex-M processor. The register R0 will accumulate the sum while R1 register is initialized to 2 and will get incremented by 2 during each iteration of the loop in order to generate the next even number. The R2 register is initialized to five as it will act as a counter for the loop. Its value will be decremented by 1 within the loop. The label 'lbegin' indicates the beginning of the loop. Each label can be an arbitrary string consisting of alphanumeric values. Recall that the label is an optional field and its name should be selected carefully. Specifically, we should refrain from using the label names the same as mnemonics for assembly language instructions; otherwise, it might result in compilation errors.

The 'CBZ' instruction following the label 'lbegin' is responsible for testing whether the counter (i.e., register R2) has decremented to zero. In case the counter is equal to zero, the loop terminates and branches to the label 'lend'. For the first iteration, the value of R2 is not equal to zero, therefore the loop condition is false and the processor will execute the next instruction. The next three instructions update the sum that is accumulated in register R0, generate the next even number and decrement the loop counter, respectively. The last instruction is an unconditional branch instruction, which takes the control back to the label 'lbegin' without checking any condition. Understanding the difference between the functioning of instructions 'B' and 'CBZ' clarifies the difference between the working of unconditional and conditional branch instructions.

Next is the question that most of us have in our mind. Can we run this program straightaway? In other words, what else do we need to do before we can make this program execute on an ARM Cortex-M processor? Let us explore the issues related to this question. In doing so, we will also learn the corresponding solutions. Once these basic concepts are properly understood, they can be easily extended to other processors from the Cortex-M family with minimal changes. As a first step, the following set of basic issues needs to be addressed.

- First of all we need a software package that will allow us to write the code, compile (assemble) it to generate the machine code, and link it to generate the executable.

- We must be aware of the reset sequence and the memory map of Cortex-M processor so that we know how the microcontroller is initialized, where our code and data will reside, and where the interrupts/exceptions handlers are loaded in the memory.

- Next, we need to decide whether we want to download and run the program on an actual hardware using a microcontroller or are we interested in executing it using a simulator. Recall that simulator is a software tool that mimics the hardware (i.e., the microcontroller) when executing user program and does not require the actual hardware platform for its functioning.

We will use the KEIL μVision5 Tools from ARM [Keil(2014)] for compilation as well as linking of assembly and C programs throughout this book. We recommend the reader to see [Keil(2014)] for gaining familiarity and usage understanding of the KEIL μVision environment for ARM processors. The KEIL tools provide two types of development environments. One environment configuration allows the user to simulate the ARM Cortex-M microcontroller at the software level, while the second environment enables the user to download, run, and debug the program on an actual hardware platform. Throughout this chapter, we will focus only on the simulator environment pending the discussion of running programs on the target hardware for later chapters. Our discussion will be simple so that the reader can easily grasp the basic concepts.

The program in Listing 4.6 cannot be assembled to build the executable for testing its proper functionality using either simulator or hardware platform. Our next step is to introduce all the missing components that will make Listing 4.6 become a standalone program. The first instruction of the user program is not the first instruction that is executed when the program is run. There are certain system level initializations that need to be performed before the user program execution is started. Since our objective at this point is to keep the discussion simple for easier understanding, we will only be concerned with minimum system initialization that allows us to run the user program. In this context, we first discuss the following two key components as the basic building blocks.

- Assembler directives

- Reset sequence

4.3.1   Assembler Directives

An assembly source file is a collection of assembly language instructions, assembler directives and, if required, may also include macro directives. But before those instructions can be executed we need to first assemble them, which is the process of converting mnemonic representation of instructions to the equivalent 16-bit or 32-bit machine codes. In other words, the assembler translates assembly language source files to machine understandable object file. The file generated by the assembler containing machine codes (of the assembly language instructions) is usually called an *object* file. This file is used by the linker along with linker command file (based on linker directives) and any libraries (if required) to generate the executable.

It should be clear that the assembler and linker syntax depends on which software tools are being used. Since we are using Keil μVision tools, the assembler syntax for these tools will be introduced next. To get familiarity with the syntax of other assemblers, the best place to start with is the code examples that are provided along with the tools by the vendor. The assembler syntax is based on the use of *assembler directives*, which are pseudo-opcodes or pseudo-operations performed by an assembler and are not part of the assembly instruction set. These assembler directives are executed during the assembling process at compilation time, in contrast to assembly instructions that are executed at runtime.

One batch of these directives is used to control the placement of data and code in the memory. For example, the AREA directive is used to instruct the assembler for a new code or data section. Sections are independent, named individually, and comprise indivisible chunks of code or data that are manipulated by the linker. The general syntax for the AREA directive is given below.

$$\text{AREA} \quad sectionname \ \{, attr\}\{, attr\}...$$

In the above expression *sectionname* is the name assigned by the user to a particular data or code section. Some of the section names are conventional. For instance, |.text| is used to mention code sections generated by the C compiler, or it can also be used for code sections associated with the C library. The *attr* is one or more comma-delimited optional section attributes. For example, READONLY and READWRITE are two possible section attributes and only one of these two can be used for a given section. A list of some useful section attributes is given in Table 4.1.

The AREA assembler directive syntax is explained in Listing 4.7 using a few examples. The section attribute DATA in Listing 4.7 indicates the location for application data, which also includes the global variables and is located in RAM. Similarly the section name STACK indicates the stack area, which is also in the RAM. The ALIGN attribute is explained in Table 4.1.

```
AREA    Mydata, DATA, READWRITE  ; Define a data section
                                 ; with section name Mydata

AREA    STACK, NOINIT, READWRITE, ALIGN=3
AREA    |.text|, CODE, READONLY, ALIGN=2
```

Listing 4.7: Examples of AREA section directive.

The second batch of directives is used for defining variables or constants of different sizes. For example, SPACE is used to allocate a zero initialized block of memory for one or more variables. Similarly, DCB, DCW, and DCD, respectively, are used to allocate byte, half-word, or word size of memory and specify the initial content. The DCB, DCW, and DCD assembler directives can be used in both CODE as well as DATA memory sections. The EQU directive gives a symbolic name to a numeric constant. In addition, EQU can also be used to generate register-relative or PC-

Table 4.1: Some useful attributes for AREA directive.

Attribute	Description
READONLY	Indicates that this section must not be written to. This is the default for code areas.
NOINIT	Use of this attribute indicates that the data section is uninitialized, or initialized to zero. A section with this attribute contains only space reservation directives SPACE or DCB, DCD, DCDU, DCQ, DCQU, DCW, or DCWU with initialized values of zero. You can decide at link time whether an area is uninitialized or zero initialized.
READWRITE	Indicates that this section can be read from and written to. This is the default for Data areas.
DATA	Contains data, not instructions. READWRITE is the default.
CODE	Contains machine instructions. READONLY is the default.
ALIGN=*number*	By default, sections are aligned on a four-byte boundary. The *number* can have any integer value from 0 to 31. The section is aligned on a 2^{number} byte boundary. For example, if expression is 10, the section is aligned on a 1 KB boundary.

relative addresses. One can define constants using EQU, and then use them within the user program code.

Another set of useful directives include ENTRY, END, IMPORT, and EXPORT. The ENTRY directive declares an entry point to a program. The END directive informs the assembler that it has reached the end of a source file. The EXPORT directive declares a symbol that can be used by the linker to resolve symbol references in two different object files. GLOBAL is a synonym for EXPORT. GLOBAL and EXPORT directives tell the assembler that the associated labels are not defined in the current assembly program file and need to be looked into some other object files (generated from other source files) or libraries.

The THUMB directive instructs the assembler to interpret subsequent instructions as Thumb or Thumb2 instructions. The PRESERVE8 directive specifies that the current file preserves 8-byte alignment of the stack. We will use these assembler directives to complete our first assembly program, so that it can be compiled and executed. However, before doing this, we first revisit the reset sequence that was introduced in Chapter 2.

4.3.2 Reset Sequence

Next, we look at the reset sequence for Cortex-M processor. As discussed in Chapter 2, Cortex-M after a reset enters the thread mode running at a privileged level and uses the MSP stack pointer to access the main stack area. Figure 4.3 shows the timing diagram of the reset sequence. From this figure, we observe that after a reset Cortex-M processor reads two words from code memory which are described below.

- First, it fetches a 32-bit value from address 0x00000000 that contains the address of the stack area in memory (RAM) and initializes the main stack pointer (MSP) register with that value. It is important to remember that only MSP is initialized on the reset and the value of PSP is undefined on reset.

- Second, it fetches another 32-bit value from address 0x00000004 that contains the address where the boot code or reset handler resides and initializes its program counter (PC) register with this value.

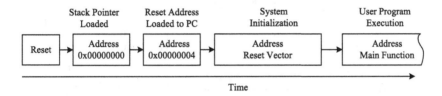

Figure 4.3: Reset sequence of Cortex-M processor.

In addition, the bit field T in the execution program status register (EPSR) is set to '1' to mark the Thumb state. The link register LR is reset to 0xFFFFFFFF. Once the PC is initialized, Cortex-M processor will start executing the reset handler instructions and from this point onward, it is ready to fetch and execute the user application program.

The understanding of the reset sequence and the memory map loaded with the object file will help us in writing assembly programs. Therefore, we next illustrate the memory map in the context of reset sequence to complement our understanding developed so far. After writing a program, we will assemble and link it, load it into the memory so that Cortex-M processor can execute it. When we are using a simulator such as KEIL μVision, the memory is also simulated, in addition to the functioning of the Cortex-M processor.

Figure 4.4 provides an illustration of the memory map, when the memory is loaded with a program. We observe that address 0x00000000 is initialized with a 32-bit value (0x20008000 in this example). The linker is responsible for assigning absolute addresses and generates this value. Beside other factors, these tools will take the stack size defined by the programmer for this program into account. For this typical example, the programmer is required to have a stack of size 1 KB. So, the linker reserved a memory region starting from 0x20007C00 to 0x20007FFF (0x20007FFF - 0x20007C00 = 0x400 ≡ 1024 or 1 KB) for the stack. The stack grows in the downward direction. We will talk more about the stack and its associated operations in Chapter 6. At this point, we need to remember that all stack accesses are word aligned. Therefore, the least significant two bits of SP register must be 0.

Following the contents of 0x00000000, we find that the memory contains a table called an interrupt/exception vector table. It contains vectors or addresses of those functions which are executed whenever an interrupt/exception occurs so that the processor can serve the interrupt/exception by executing its corresponding service routine. Let us look at the first vector, which is located at address 0x00000004 in the memory. This vector is called the reset vector, which contains the address of the reset handler. The instructions written as part of the reset handler are also sometimes referred to as boot code. From Figure 4.4, we can see that the code of the reset handler is located at

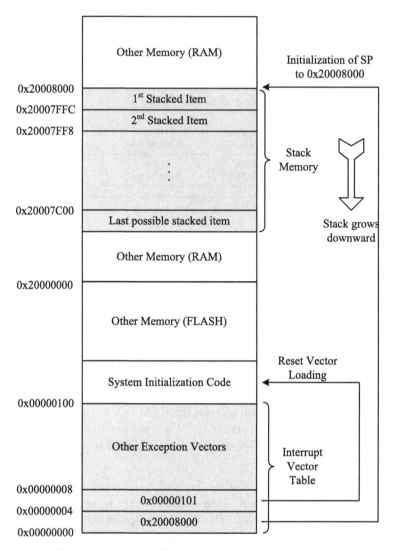

Figure 4.4: Reset sequence of Cortex-M mapped to memory address map.

address 0x00000100. The remaining interrupt/exception vectors, if any, are loaded to the memory after the reset vector in the exception table. It is important to realize that the vectors are placed in a pre-specified order in the interrupt vector table. We only discuss the reset vector here. For the discussion regarding other exceptions and interrupts, the reader is referred to Chapters 3 and 9.

Having looked at the reset sequence and memory map, let's see how our assembly program will run on a Cortex-M microprocessor. On reset or when powered up for the first time, the processor will fetch the 32-bit value from 0x00000000 and initialize its SP register. This will provide a stack region for our program. A program needs a stack to store information temporarily. The processor will then fetch the next 32-bit value from 0x00000004 and initialize its PC register. But since all instructions are half-word aligned, therefore, the least significant bit of PC must be 0. Then, why do we find 0x00000101 at location 0x00000004 instead of 0x00000100? It is assembler (or linker) who has set the least significant bit in the reset vector so that Cortex-M processor will properly initialize the Thumb (T) bit in the program status register (PSR) of the processor. For Cortex-M processor, the T bit should always be set to 1 so that it runs Thumb instructions. We will shortly learn how we can provide the assembler this information about the Thumb instructions.

The PC when initialized to 0x00000100 will allow the processor to fetch and execute instructions from the reset handler or boot code. The reset handler first performs some initializations before it makes a call to the user application function. In summary, an assembly program needs to perform the following sequence of activities to ensure that the user application program is executed properly.

- Define the stack size and reserve appropriate memory space for stack.

- Define the reset vector.

- Write the reset handler code to perform any system related initializations and then make a jump to the user application program.

4.3.3 Complete First Assembly Program

Listing 4.6 of our first assembly program can now be converted to a complete assembly program that can be complied to build the executable. Example 4.1 provides the complete assembly program and also illustrates the use of assembler directives. It can be seen that a stack area of size 256 bytes is reserved in this example program. The label __Vectors marks the beginning of the interrupt vector table and is initialized inside the code memory region. This memory segment is placed at the starting address of 0x00000000 in the code memory region. The first entry is defined as top of the stack region address. This is implemented by first reserving a memory region of size 'StackSize' with starting address given by label 'MyStackMem' in the RAM memory. The top of the stack address is then defined as 'MyStackMem + StackSize' and being the first entry of the code memory region it is stored at address 0x00000000, from where it is loaded to the SP (R13) after reset.

Example 4.1 (The first assembly program.). The complete assembly program for summing first five even integers is given below, which can be compiled and executed. This assembly program does not require 'startup.s' file.

```
    THUMB                   ; Marks the THUMB mode of operation

StackSize   EQU   0x00000100 ; Define stack size of 256 byes

    AREA STACK, NOINIT, READWRITE, ALIGN=3
MyStackMem   SPACE StackSize

    AREA     RESET, READONLY
    EXPORT   __Vectors
__Vectors
    DCD MyStackMem + StackSize ; stack pointer for empty stack
    DCD Reset_Handler          ; reset vector
```

```
    AREA     MYCODE, CODE, READONLY
    ENTRY
    EXPORT Reset_Handler
Reset_Handler
    MOV RO, #0              ; Initial value of sum
    MOV R1, #2              ; First even number
    MOV R2, #5              ; Counter for the loop iterations

lbegin
    CBZ R2, lend           ; Terminate loop if counter is zero
    ADD RO, R1             ; Build the sum
    ADD R1, #2             ; Generate next even number
    SUB R2, #1             ; Decrement the counter
    B lbegin
lend
    END
```

It is important to mention that SP initialization value is not an interrupt vector entry but we have made it the first entry of the vector table to simplify the implementation. The second entry of the vector table is 'Reset_Handler', which is the label of the reset interrupt service routine. The first instruction in the reset handler is indeed the first instruction that is fetched and executed by the processor after reset. The reset is the first interrupt entry in the interrupt vector table and is of highest priority as well.

In Example 4.1, three different memory sections are defined. The STACK section is defined in the RAM memory area using READWRITE attribute, while RESET and MYCODE sections are defined in the Flash memory area using READONLY attribute. We observe that the RESET section defines two word size entries, of which the first entry holds the stack top address while the second entry is the reset handler address. In this example, the user program is implemented as part of the reset interrupt service routine and is made part of the MYCODE section. When the executable for this program is constructed, it is important that the RESET section is placed at address 0x00000000 in the code memory. This is achieved by instructing the linker using linker directives. Managing different memory areas and placement of code and data sections to appropriate memory areas is implemented using a linker script or linker scatter file. For further details related to linker script file, see Appendix A.2.

4.3.4   Assembly Program for Multiplication

Let us look at another assembly program where we implement integer multiplication using addition operation. In Listing 4.8, complete program assembly code is provided that can be compiled to build the executable and test it for its functionality. This program also introduces the use of read-only (Flash or EEPROM) and read-write (RAM) memory segments for data storage purpose. In this program, two variables X and Y are multiplied and the result is stored in PRODUCT variable. Precisely speaking X and Y are constants rather than variables because they are stored in read-only memory, while PRODUCT is indeed a variable since the associated memory space is allocated in RAM. The multiplication is implemented using repeated addition

inside a conditional loop. The loop terminates when the parameter Y is decremented
to zero.

```
; This ARM Assembly language program multiplies two positive
; numbers by repeated addition.

        THUMB                   ; Marks the THUMB mode of operation
StackSize   EQU 0x00000100      ; Define stack size to be 256 byes

; Allocate space for the stack.
    AREA    STACK, NOINIT, READWRITE, ALIGN=3
StackMem    SPACE   StackSize

; Initialize the two entries of vector table.
    AREA    RESET, DATA, READONLY
    EXPORT  __Vectors
__Vectors
    DCD     StackMem + StackSize ; SP value when stack is empty
    DCD     Reset_Handler        ; reset vector
    ALIGN

; Data Variables are declared in DATA AREA
    AREA        ROMdata, DATA, READONLY
X       DCD     10
Y       DCD     3

    AREA        RAMdata, DATA, READWRITE
PRODUCT DCD     0               ; accumulates  product of X and Y

; The user code (program) is placed in CODE AREA
        AREA    |.text|, CODE, READONLY, ALIGN=2
        ENTRY                   ; ENTRY marks the starting point
                                ; of the code execution
        EXPORT  Reset_Handler
Reset_Handler                   ; User Code Starts from next line
        LDR R0, =X              ; load the address of X in R0
        LDR R1, =Y              ; load the address of Y in R1
        LDR R2, =PRODUCT        ; load the address of PRODUCT in R2

        LDR R3, [R0]            ; load the value of X in R3
        LDR R4, [R1]            ; load the value of Y in R4
        LDR R5, [R2]            ; load the value of PRODUCT in R5

Lbegin
        CBZ R4, Lend
        ADD R5, R3
        SUB R4, #1
        B Lbegin
Lend
        STR R5, [R2]            ; store the product back to PRODUCT
        ALIGN
        END                     ; End of program, matched with ENTRY
```

Listing 4.8: Assembly program example for multiplication.

4.4   Instruction Encoding

Instruction encoding is the process of assigning a unique binary codeword to each assembly instruction. One of the main jobs of an assembler is to translate each assembly instruction to an equivalent codeword. Some of the assembly instructions are encoded to 16-bit codewords, while some other instructions are encoded using 32-bit codewords. However, there are many assembly instructions that can be encoded to either 16-bit or 32-bit codewords. In the following subsections, we will explain 16-bit and 32-bit instruction encodings. If an instruction can be encoded using either of the two encoding formats, then the factors that affect the choice between 16-bit and 32-bit encoding formats will also be discussed in the following subsections.

4.4.1   16-bit Instruction Encoding

Consider ADDS instruction for register and immediate data as well as register and register addition. The first two instructions in Listing 4.9 illustrate two possibilities resulting in 16-bit encodings, primarily due to the appropriate choice of immediate values. The third instruction, illustrating register to register addition, is also encoded using 16-bit encoding as it will become clear shortly. Figure 4.5 shows different possible instruction encodings for ADDS instruction. From Figure 4.5 we observe that bit field of size 3 bits is allocated for each register operand. That allows the user to select an arbitrary register from R0-R7, i.e., from the general purpose low registers. In other words, general purpose high registers (R8-R12) cannot be used as operands in 16-bit instruction encoding.

```
ADDS R2, #16        ; R2 = R2 + 16, Instruction is 16-bit encoded
ADDS R4, R2, #6     ; R4 = R2 + 6, Instruction is 16-bit encoded
ADDS R6, R5, R3     ; R6 = R5 + R3, Instruction is 16-bit encoded
```

Listing 4.9: ADD instruction illustration for 16-bit encoding.

Figure 4.5 also illustrates that the immediate value operand field size depends on the choice of the destination register operand. If the same register is used as source and destination operand, the immediate value can have an 8-bit size. On the other hand, choosing different destination and source registers allows only a 3-bit immediate value operand.

4.4.2   32-bit Instruction Encoding

Now let us start with 32-bit instruction encoding format. The advantages of extending instruction encoding to 32-bit format will also become clear. However, we emphasize that the advantages of 32-bit encoding discussed in this section are only a subset of the possible advantages.

Starting with the ADDS instructions in Listing 4.9, we show that what minimum changes to these instructions will make it mandatory to use 32-bit encoding. For the first instruction in Listing 4.9 we change the value of immediate operand from 16 to

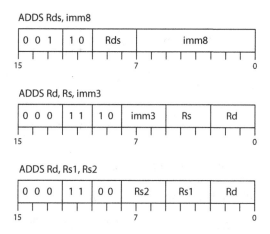

Figure 4.5: 16-bit encoding of ADD instruction.

356 as can be seen from Listing 4.10. This new value cannot be encoded in 8-bit field allocated, for immediate value of 16-bit encoding format and forces the assembler to rather use 32-bit instruction encoding. Figure 4.6 illustrates this case and shows the 32-bit encoding format. The other two ADDS instructions are also modified in the value of immediate operand and the choice of operand register as can be observed from Listing 4.10. It should be noted that even if any of the operand registers belongs to the general purpose high register group, it will become mandatory to use 32-bit instruction encoding.

```
ADDS R2, #356       ; R2 = R2 + 356, Instruction is 32-bit encoded
ADDS R4, R2, #14    ; R4 = R2 + 14, Instruction is 32-bit encoded
ADDS R6, R5, R9     ; R6 = R5 + R9, Instruction is 32-bit encoded
```

Listing 4.10: Illustration of 32-bit encoding requirement for ADD instruction.

Table 4.2 summarizes the scenarios requiring the use of 16-bit encoding along with the possibility, when it becomes necessary, to use 32-bit encoding. It is important to realize that for different instructions, 32-bit encoding becomes mandatory for different immediate values of Operand 2. In some cases, an immediate value larger than 3-bit is sufficient for an instruction to be encoded as 32-bit, while in some other cases an immediate value greater than either 8-bit or 12-bit is required for mandating 32-bit encoding of an instruction. This can be seen from the 16-bit encoding of ADDS instruction, where in one case we require 32-bit encoding if the immediate value is larger than 3-bit and in the other case the immediate value required for 32-bit encoding has to be larger than 8-bit number.

4.4.3   Visualizing Instruction Encoding

Based on the instruction encoding discussed above we illustrate actual instruction encoding performed by the assembler tools giving a few examples. Let us start with ADDS R2, #16 instruction which is encoded using 16-bit encoding. The encoding format that is applicable from Figure 4.5 is ADDS Rds, imm8. For the instruction

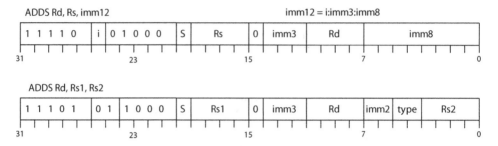

Figure 4.6: ADDS instruction encoding using 32-bit encoding format.

Table 4.2: Instruction encoding dependency.

Operand 1	Operand 2	Encoding
Register R0 to R7	Register R0 to R7	16 bit
Register R0 to R7	Immediate value is limited to 3-bit (for different) or 8/12-bit (for same) source-destination registers	16 bit
Register R8 to R12	Register R0 to R7	32 bit
Register R0 to R7	Register R8 to R12	32 bit
Register R0 to R7	Immediate value exceeds 3-bit or 8/12-bit	32 bit

under consideration Rds = R2 and imm8 = 16 and the corresponding 16-bit sequence becomes |001|10|010|00010000| which is equal to 0x3210. The 16- and 32-bit illustrations for ADDS instruction are assembled by using KEIL tools and the result is shown in Figure 4.7. The lower half of Figure 4.7 shows different instructions compiled, while the upper half shows the disassembly of these instructions generated by the tools. Now observing the second column in the upper half of Figure 4.7 shows that this is indeed the encoded value generated by the assembler.

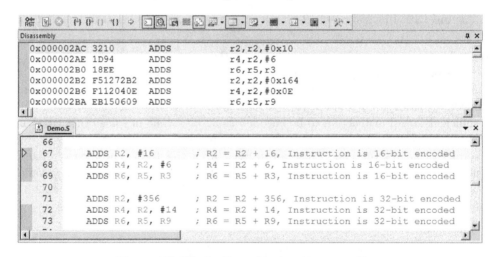

Figure 4.7: Illustration of instruction encoding.

Let us consider another example of 32-bit encoding. We consider ADDS R6, R5, R9 for this case and the corresponding encoding format is ADDS Rd, Rs1, Rs2. From Figure 4.6, the imm2, imm3, and type fields are not used by the instruction under consideration and are zero, while the field S is set to 1. Using these facts we have the 32-bit sequence as |11101|01|1000|1|0101|0|000|0110|00|00|1001| and is equal to 0xEB150609. The 32-bit codeword generated by the tools can be observed from Figure 4.7 and is the same as calculated above.

4.5 Instruction Set

It should be kept in mind that only the instruction set for Cortex-M3 processor will be covered in this book. We refer the reader to *Cortex-M4 Devices Generic User Guide* and *ARMv7-M Architecture Reference Manual* [ARM(2010)] for details regarding additional instructions supported by Cortex-M4 and Cortex-M4F processor families. In addition it is worth mentioning that Cortex-M0 and M1 only support a subset of the instructions supported by Cortex-M3.

Only a subset of Thumb2 instruction set is supported by the Cortex-M3 processor. Figure 4.8 provides an overview of the instructions supported by the Cortex-M3 architecture, where different instructions are divided into groups and subgroups based on functionality. A formal treatment of assembly language programming will be provided, in the subsequent chapters, based on this grouping. Specifically, Chapter 5 will discuss the data processing instruction group. The memory access instruction group will be discussed in Chapter 6, while branch and control instructions along with some other instructions will be covered in Chapter 7.

4.6 Summary of Key Concepts

In this chapter, we have introduced the basics of Cortex-M assembly programming. Specifically, the following key concepts were introduced and discussed.

- ARM processors can operate in two different states; ARM state and Thumb state. In the ARM state, only 32-bit word aligned instructions are allowed to execute while Thumb state allows the execution of only 16-bit half-word aligned instructions. The former provides higher performance and the latter is attractive for higher instruction code density.

- An ARM processor supporting both 32-bit and 16-bit instructions has to switch between ARM state and Thumb state. This results in an additional overhead associated with the state switching. Introduction of Thumb2 instruction set has now made it possible to use one operating state while meeting different processing requirements. Being based on ARMv7-M architecture, the Cortex-M processor employs Thumb2 instruction set.

- Fundamentals of Cortex-M assembly language. The syntax is: label opcode $operand_1, operand_2, ..., operand_n$; Comments. The label indicates a location in an assembly program and is typically used with flow control instructions. The opcode fields contains a mnemonic, while the operation specified by the opcode is carried out on the operands. Finally, we can provide a comment on

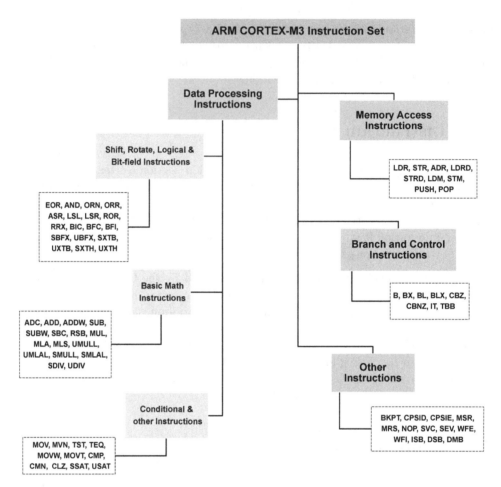

Figure 4.8: Cortex-M3 instruction set summary based on functionality.

an instruction with the help of a semicolon. Comments help readers who have not written the code to easily review it.

- Several basic instructions of Cortex-M were introduced. ADD adds two numbers. MOV moves data from one place to another. LDR allows to load data from memory while STR stores data back to the memory. B and CBZ allow to jump unconditionally and conditionally respectively. We learned the usage of each instruction with the help of examples.

- Assembler directives were presented. An assembler directive is an instruction that directs/assists the assembler to carry out activities at assembly time. These are not executable instructions. For example, the AREA directive is used to instruct the assembler about the beginning of a new code or data section.

- We learned how to write our first assembly program. The above-mentioned Cortex-M's assembly language instructions were used to write a program that adds the first five even numbers. We explained the working of the program when it was the only piece of code run by the processor and the RESET button is pressed. In this context, the memory map was explained.

- Instruction encoding. Assembly instructions are mnemonics such as ADD, MOV. These mnemonics have to be translated into machine code that is actually executed by the hardware, i.e., processor. For this purpose, the assembler employs some kind of coding scheme. The details of this aspect are covered under the term instruction set architecture. We have briefly touched on it here to make the reader understand the relationship between assembly and machine code. The 16-bit and 32-bit encoding schemes supported by Cortex-M were discussed with examples.

Review Questions

Question 4.1. How do we specify the start of execution after reset?

Question 4.2. Which interrupt is the first one to occur on Cortex-M based microcontroller?

Question 4.3. What are the possible sizes (in bits) of an encoded instruction in Thumb2 instruction set architecture?

Question 4.4. List the possible fields of an assembly instruction.

Question 4.5. Which assembly instruction fields are mandatory? Also name the optional fields.

Question 4.6. What is the minimum and maximum number of operands that can be used in an assembly instruction?

Question 4.7. Is it possible to transfer data from memory to a register using MOV instruction? If yes, then provide an example.

Question 4.8. Which two operations are combined in CBZ assembly instruction? How is the flexibility compromised in CBZ instruction?

Question 4.9. What are pseudo assembly instruction? Are those executed at compile time or run time?

Question 4.10. How is the AREA assembler directive used to create code or data sections in the memory?

Question 4.11. Is it possible to create code section in data memory region? If yes, give an example.

Question 4.12. If an assembly instruction uses one of the high registers (R8-R12) as its operand, then which encoding (16-bit or 32-bit) can be used for this instruction?

Question 4.13. List the factors that affect the instruction encoding size.

Exercises

Exercise 4.1. What are assembler directives? How are they different from assembly instructions?

Exercise 4.2. Explain the sequence of steps that are followed after reset.

Exercise 4.3. How is the stack initialized in assembly language? In assembly language, how will you reserve a stack space of 2048 bytes?

Exercise 4.4. What happens if an assembly instruction starts from the first column? Additionally, is it possible to write two assembly instructions on one line?

Exercise 4.5. Modify the program given in Example 4.1 so that we can find the sum of the first 10 odd numbers.

Exercise 4.6. Write a program in Assembly language so that it determines the sum of the five even numbers starting from 20.

Exercise 4.7. Execute the program given in Listing 4.8 and confirm that it works correctly. Modify it by assigning the Data AREA of X and Y the READWRITE attribute and see its impact on the working of the program. In case it does not work properly, try to find the reason for it.

Exercise 4.8. Explain the functionality of the following assembly program. What will be the contents of the 'dst_data' string?

```
STACKSIZE     EQU     0x100
SLOPE         EQU     5
OFFSET        EQU     10
ARRAYSIZE     EQU     4

; Stack area initialization in RAM memory
      AREA     STACK, NOINIT,   READWRITE,   ALIGN=3

StackMem  SPACE    STACKSIZE

; Vector table initialization in ROM memory
    AREA     RESET,  DATA, READONLY
    EXPORT   Vectors
Vectors
  DCD   StackMem + STACKSIZE   ; stack pointer for empty stack
  DCD   Reset_Handler          ; reset vector
  ALIGN

; An array of variables in RAM
    AREA     MyDSTdata, DATA, READWRITE
dst_data      DCW       0, 0, 0, 0

; An array of constants in ROM
    AREA     MySRCdata, DATA, READONLY
src_data      DCW       1, 2, 3, 4

; User program
    AREA     |.text|,  CODE,  READONLY,  ALIGN=2
    ENTRY
    EXPORT Reset_Handler
```

```
Reset_Handler
    MOV       r2, #ARRAYSIZE
    MOV       r3, #OFFSET
    MOV       r4, #SLOPE
    LDR       r5, =src_data
    LDR       r6, =dst_data

Loop
    LDRH    r0, [r5], #2
    MLA     r0, r0, r4, r3
    STRH    r0, [r6], #2
    SUBS    r2, r2, #1
    BNE     Loop

Loop_Forever
    B         Loop_Forever
    END
```

Exercise 4.9. Figure 4.9 shows the memory contents of an ARM processor. Write the contents of PC, SP, R0, and R1 after every instruction cycle, starting from the processor reset. Show contents for PC, SP, R0, and R1 registers for first five instruction cycles.

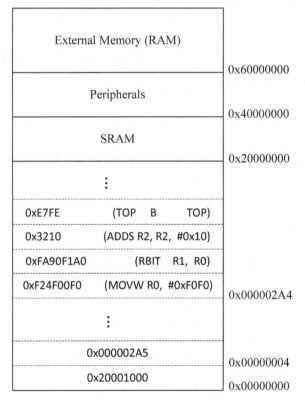

Figure 4.9: Memory contents for Exercise 4.9.

Exercise 4.10. Whether the following instructions are encoded as 16-bit or 32-bit, if 32-bit encoding is used, give the reason why 16-bit encoding is not possible for that case.

```
SUBS  R2,  R2,  #8        ; R2 = R2 - 8
ADDS  R8,  R6,  #4        ; R8 = R6 + 4
MOV   R2,  R5             ; R2 = R5
MOV   R2,  #0x8C30        ; R2 = 0x8C30
```

Exercise 4.11. An assembler translates the assembly instructions to the corresponding machine codes. In Figure 4.10 are a few instructions with their encoding formats.

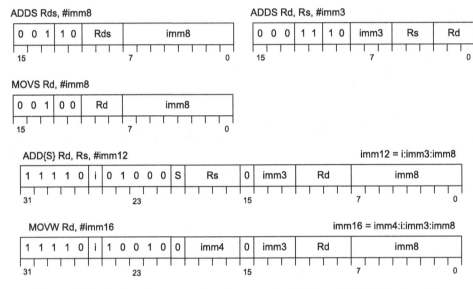

Figure 4.10: Instruction encoding formats for ADD and MOV instructions.

Using the above-mentioned encoding formats; decode the following machine codes to the corresponding assembly instructions.

1. 0x3213

2. 0xF24F00FE

Exercise 4.12. If an instruction is encoded using 16-bit encoding, is it possible to encode the same instruction using 32-bit encoding? If 0x3213 is the 16-bit machine code of an instruction, then if possible, what will be the 32-bit machine code for the same instruction?

Chapter 5

Data Processing Instructions

Overview

In the previous chapter, we have segregated all the Cortex-M3 assembly instructions in three groups. One instruction grouping belongs to the data processing instructions, which are the focus of our discussion in this chapter. The data processing instructions are further divided into multiple subgroups based on either their functionality or if they follow similar generic instruction syntax. Specifically, the data processing instructions have been divided into the following subgroups.

- Shift, rotate, and logical instructions

- Basic math instructions

- Data movement instructions

- Bitfield instructions

- Test and compare instructions

- Saturating instructions

The bitwise operations implemented by shift, rotate, and logical instructions are grouped together. Elementary arithmetic operations of addition, subtraction, multiplication, and division are grouped into basic math instructions. The bitfield instructions operate on a group of adjacent bits and differ from the instructions involving bitwise operations, which can operate on individual bits and do not have any adjacency requirement. The data movement instructions are capable of performing an operation (from a set of operations) on the data before moving it to the register. However, the operations before data movement are not mandatory and when skipped result in simple data transfer operation. The test and compare instructions are used to affect the application program status register to control the flow of program execution. Finally, we will discuss the saturating instructions that allow us to reduce distortion caused by the saturation phenomenon.

Instruction grouping permits us to explain, in detail, the general syntax for many different instructions simultaneously. For each instruction group, in addition to instruction syntax explanation, we provide details about the possibility of using 16-bit and/or 32-bit instruction encoding as well as the set of flags that are updated due to an instruction execution. It should be remembered that we only cover the assembly programming instructions for Cortex-M3 processor. The Cortex-M4 supports all the instructions of Cortex-M3 architecture. In addition, it supports DSP related instructions and optionally the SIMD (single instruction multiple data) arithmetic instructions. In case of Cortex-M4F the optional floating point instructions are also supported by the processor. The reader is referred to the reference manual [ARM(2010)] for any details about the additional instructions supported by Cortex-M4/M4F.

5.1   Shift, Rotate, and Logical Instructions

Bit operations are considered one of the simplest processing operations. The simplicity of the actual operation being performed allows us to explain in detail some of the important aspects related to these assembly instructions. In particular, the instruction syntax and the flexibility available in the instruction operands is discussed. We illustrate the usage of different instructions with examples.

5.1.1   Shift and Rotate Instructions

Shift operations move the bits in a register left or right by a specified number of bits, termed the shift length. Register shift operation can be performed directly by using the appropriate shift instructions and the result is written to a destination register. Another usage of shift operation is by some of the assembly instructions during the evaluation of their second operand in the form of a register with shift. This scenario will be explained in detail in Section 5.1.2, where logical instructions are discussed.

The shift lengths that are permitted depend not only on the type of shift operation but also on the specific instruction being used. When the specified shift length is 0 then no shift occurs. Register shift operations can optionally update the flag(s) except when the specified shift length is 0. In case of rotate instruction the bits in a register are circulated in the specified direction by the given number of bit locations.

In the following, we first describe general syntax for shift and rotate instructions followed by the explanation of each of these instructions. In these descriptions, Rm is the register containing the value to be shifted, and n is the shift length. The Cortex-M3 supports *logical shift left* (LSL), *logical shift right* (LSR), *arithmetic shift right* (ASR), *rotate right* (ROR), and *rotate right extended* (RRX). The general syntax for

Table 5.1: Instruction encoding and flag(s) affected by shift and rotate instructions.

Mnemonic	Brief description	Encoding	Flags
ASR	Arithmetic shift right	16 or 32 bit	N,Z,C
LSR	Logical shift right	16 or 32 bit	N,Z,C
LSL	Logical shift left	16 or 32 bit	N,Z,C
ROR	Rotate right	16 or 32 bit	N,Z,C
RRX	Rotate right with extend	32 bit	N,Z,C

these instructions is given below.

$$\text{LSR}\{S\}\{cond\}\{.size\} \quad Rd, \quad Rm, \quad Rs \text{ or } \#n$$
$$\text{ASR}\{S\}\{cond\}\{.size\} \quad Rd, \quad Rm, \quad Rs \text{ or } \#n$$
$$\text{LSL}\{S\}\{cond\}\{.size\} \quad Rd, \quad Rm, \quad Rs \text{ or } \#n$$
$$\text{ROR}\{S\}\{cond\}\{.size\} \quad Rd, \quad Rm, \quad Rs \text{ or } \#n$$
$$\text{RRX}\{S\}\{cond\} \quad Rd, \quad Rm$$

{}	Represents the optional field.
S	Is an optional suffix. If S is specified, the condition code flags are updated on the result of the operation, see conditional execution below.
cond	This is an optional condition code suffix. See conditional execution below.
.size	The optional suffix to enforce 16-bit or 32-bit instruction size encoding and is formerly called *instruction size qualifier*.
Rd	Specifies the destination register.
Rm	Specifies the register containing the value to be shifted.
Rs	Specifies the register holding the shift length to apply to the value in *Rm*. Only the least significant byte is used.
n	Specifies the shift length using an immediate value. The range of shift length depends on the instruction.

It should be noted that in the above instructions either register *Rs* or immediate value given by #*n* (not both) can be used to specify the number of bit positions by which the contents of *Rm* should be moved. RRX moves the bits in register *Rm* to the right by 1 bit.

Most data processing instructions can optionally update the condition flags in the Application Program Status Register (APSR) according to the result of the operation. Some instructions update all flags, while others only update a subset. If a flag bit is not updated, the original value is preserved. The list of flags that are updated as a consequence of shift and rotate instructions is tabulated in Table 5.1. In addition, Table 5.1 provides details about the instructions supporting a 16-bit or 32-bit or both of these instruction encoding formats.

Instruction Size Qualifier

Many instructions can be represented either with a 16-bit encoding or a 32-bit encoding depending on the operands and destination register specified. We can also force a specific instruction size by using the instruction size qualifier suffix. The .W suffix generates a 32-bit instruction encoding. To force a 16-bit instruction encoding, the .N suffix is employed. These are the two possible instruction size qualifiers. If you specify an instruction width suffix and the assembler cannot generate an instruction encoding of the requested width, it generates an error. If the instruction size qualifier is not specified, the assembler or complier chooses appropriate instruction size. It is also more suitable to leave it to the tools to select appropriate instruction size.

Conditional Execution

An instruction can be executed conditionally, based on the condition flags set by the preceding instruction(s). The detailed discussion regarding conditional execution of instructions is postponed until Chapter 7.

Logical and Arithmetic Shift Right

The logical shift right (LSR) instruction shifts the bits in the register by the specified value and zeros are entered to the register from its left side. Arithmetic shift right (ASR) by n bits moves the left-hand 32-n bits of the register Rm to the right by n places, into the right-hand 32-n bits of the result. It also copies the original bit 31 (denoted by b31 in Figure 5.1) of the register into the left-hand n bits of the result. Figure 5.1 shows an example with two bits shifted to the right. It can be observed that shifting right by 1 bit is equivalent to dividing by 2.

It can be seen from Table 5.1 that carry, zero, and negative flags are updated when LSR and ASR instructions are executed with optional S field specified. When LSRS and ASRS instructions shift the register position by multiple bits, the value of the carry flag C will be the last bit that shifts out of the register from right side. See Figure 5.1 for an illustration. The range of shift length is 1 to 32 for this instruction. Some of the similarities and differences in the execution of LSR and ASR are elaborated in Example 5.1.

Example 5.1. The similarity and the difference between the LSR and ASR is illustrated in this example. Let R3 = 64 and we shift right, the contents of R3, by 3 bits, which effectively implements division by 8.

```
LSR      R4, R3, #3       ; R4 = R3 >> 3
ASR      R5, R3, #3       ; R5 = R3 >> 3
```

Shifting contents of R3 by 3 bits using LSR and ASR leads to the same result, i.e., R4 = R5 = 8. Now let us change the starting value in R3 with R3 = -64 = 0xFFFFFFC0. Performing the LSR and ASR operations by 3 bits leads to R4 = 0x1FFFFFF8 = 536870904, while R5 = 0xFFFFFFF8 = -8. In this case

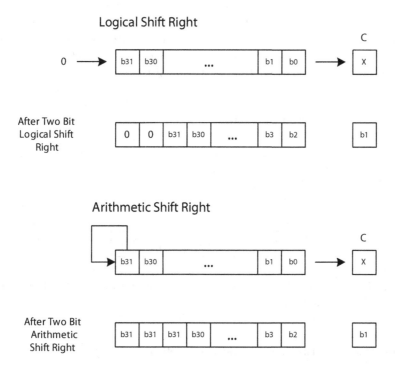

Figure 5.1: Logical and arithmetic shift right operations. The symbol 'C' represents the carry flag in APSR while 'b0' to 'b31' mark the least significant to most significant bits in the register.

we obtain the correct result by using ASR instruction. Now we can conclude that for unsigned integers the LSR and ASR give the same result, however, for signed integers ASR gives the correct result.

Assuming R6 = 3, the above two instructions can also be implemented as

```
LSR   R4, R3, R6    ; R4 = R3 >> R6
ASR   R5, R3, R6    ; R5 = R3 >> R6
```

Finally we illustrate the use of 'S' suffix for the shift right instructions. Assuming R2 = 4 and R3 = 3 the following code shows how the C flag is updated.

```
ASRS  R4, R2, #3    ; R4 = R2 >> 3 and C = 1 after execution
ASRS  R5, R3, #3    ; R5 = R3 >> 3 and C = 0 after execution
```

Logical Shift Left

In logical shift left (LSL) instruction, the bits entering the register from least significant bit (LSB) side are always zero, while the bits exiting the register are discarded. Shift left by one bit position leads to multiplication by 2. It is important to realize that an arithmetic shift left operation is not required because of its functionality being exactly the same as that of logical shift left operation. The shift length can take values between 0 and 31 for this instruction.

In case of LSLS (i.e., when 'S' suffix is appended to update the flags in APSR), the bits leaving the register (from most significant bit [MSB] side) go to the carry flag bit. If the shift operation shifts the register position by multiple bits, the value of the carry flag C will be the last bit that shifts out of the register from left as shown in Figure 5.2.

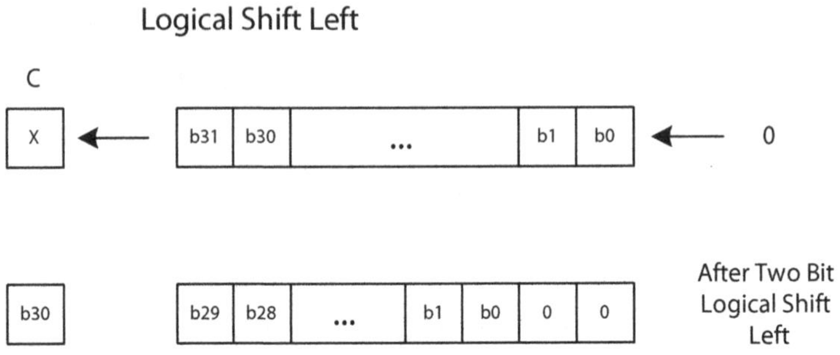

Logical Shift Left

Figure 5.2: Logical shift left operation. The arithmetic shift left operation is the same as the logical shift left operation.

Example 5.2 (Illustration of logical shift left operation.)**.**

```
LSL    R2, R1, #10        ; R2 = R1 << 10
LSLS   R3, R1, R5         ; R3 = R1 << R5
```

In the first instruction the contents of register R1 are shifted left by 10 bit positions. This is equivalent to multiplication by 1024. The flags are not updated after the execution of the first instruction. In the second instruction, register R5 specifies the number of bits by which the contents of register R1 will be shifted. In this instruction, the flags do get updated after the instruction execution.

Rotate Right and Rotate Right Extended

The rotate right (ROR) instruction rotates the contents by a specified value without making carry flag part of the rotation. The variable n is the number of bit positions to rotate and $1 \leq n \leq 31$. A rotate right by two bits is shown in Figure 5.3. The rotate left can be implemented by using rotate right operation with a different rotate value. For example, a rotate left by 4 bits can be replaced with a rotate right by 28 bits, which gives the same result and takes the same amount of time to execute. The range of shift length is 1 to 31 for this instruction. In Example 5.3, we illustrate how the rotate left operation can be implemented using ROR instruction.

Example 5.3. Below we illustrate how ROR can be used for performing rotate left operation. Consider R2 = 8 and we want to rotate left its contents by 3 bits. This can be achieved by performing rotate right by 29 bits.

```
ROR    R2, R2, #29     ; Equivalent to rotate left by 3 bits
```

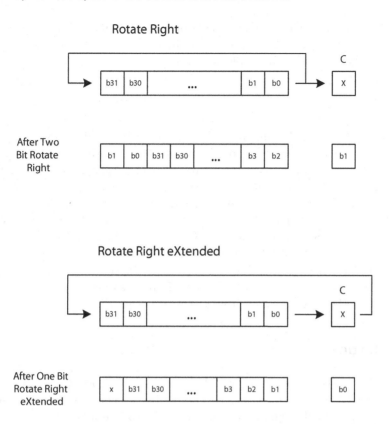

Figure 5.3: Rotate right and rotate right extended operations.

```
                        ; R2 = 64 after execution
```

It is important to realize that due to the barrel shifter the rotate right operation for 29 bit positions takes one processor clock cycle and no issue regarding execution performance inefficiency arise.

The special rotate right extended (RRX) instruction has a slightly different behavior from the usual rotate operations. One key difference is that no count is specified and this instruction always rotates by one bit. Second, 1 bit rotation in RRX involves the C flag as part of the rotation. The LSB of the register moves to C flag and the original value of C flag enters the MSB of the register. Figure 5.3 also illustrates the operation of RRX instruction. The use of RRX instruction is explained in Example 5.4.

Example 5.4. In this example, we illustrate the use of RRX instruction. Consider the scenario where you have a 64 bit number and we are interested in performing logical shift right by 1 bit. Assume that R2 contains high 32 bits and R1 contains low 32 bits of the 64 bit number. Now applying LSRS operation to R2 will shift the LSB of R2 to C flag. Next using RRX with R1 register will move the value of carry flag (which contains the LSB of R2) to the MSB of R1 and shift right the original value of R1 by 1 bit to achieve the desired result. The following code gives an implementation.

```
LSRS   R2, R2, #1      ; R2 = R2 >> 1 and C = LSB of R2
RRX    R1, R1          ; R1 = ((MSB = C) + R1 >> 1)
```

It is important to notice that different shift operations have different shift ranges. The amount by which the register will be shifted is specified either using a 5-bit field in the instruction or it is specified by the bottom byte of a register. However, using 5-bit field in the instruction code has no extra overhead, while using a register requires an extra cycle. When we use 5-bit field for specifying shift range, it allows 32 different values. Since we need to have a 0 value when we don't want to apply any shifts the possible values of LSL range from 0-31. On the other hand, LSR and ASR have a range of 1-32 because LSR #0 or ASR #0 are not required since they are equivalent to LSL #0 and can be replaced by LSL #0 (this aspect is tool dependent and is available only if supported by the tools). The LSR #32 or ASR #32 can be useful if we want to set the register value either to zeroes or ones.

5.1.2   Logical Instructions

The logical instructions support bitwise AND, exclusive OR (EOR), simple OR (ORR), OR negative (ORN), and bit clear (BIC) operations. The general syntax for ORR instruction is given below and is the same for other logical instructions.

$$AND\{S\}\{cond\}\{.size\} \quad \{Rd,\} \quad Rn, \quad Operand2$$
$$BIC\{S\}\{cond\}\{.size\} \quad \{Rd,\} \quad Rn, \quad Operand2$$
$$EOR\{S\}\{cond\}\{.size\} \quad \{Rd,\} \quad Rn, \quad Operand2$$
$$ORN\{S\}\{cond\}\{.size\} \quad \{Rd,\} \quad Rn, \quad Operand2$$
$$ORR\{S\}\{cond\}\{.size\} \quad \{Rd,\} \quad Rn, \quad Operand2$$

The AND, ORR, and EOR instructions perform bitwise AND, OR, and Exclusive OR operations between instruction parameters Rn and $Operand2$. The ORN instruction performs an OR operation of the contents of Rn with the bit wise complement of the value of $Operand2$. The BIC instruction performs an AND operation on the bits in Rn with the bit wise complement of the value in $Operand2$. The bit wise complement of $Operand2$ in ORN and BIC instructions is effectively 1's complement of the value specified by $Operand2$. The result of these operations is stored in optionally specified register Rd, which can be the same as Rn. In addition, when Rd is omitted the result of these operations is stored in the register Rn.

Table 5.2 provides details about the 16/32 bit instruction encoding format as well as the list of flags that are updated as a consequence of an instruction execution. It should be remembered that to update the flags, suffix 'S' should be appended to the instruction. Next we discuss the syntax and the possible values that flexible second operand can admit.

Table 5.2: Instruction encoding and flag(s) affected by logical instructions.

Mnemonic	Brief description	Encoding	Flags
AND	Logical AND operation	16 or 32 bit	N,Z,C
ORR	Logical OR	16 or 32 bit	N,Z,C
EOR	Exclusive OR	16 or 32 bit	N,Z,C
ORN	Logical OR NOT	32 bit	N,Z,C
BIC	Bit clear	16 or 32 bit	N,Z,C

Flexible Second Operand

Many general data processing instructions have a *flexible second operand*. This is shown as *Operand2* or *Op2* in the syntax of an instruction. The flexible second operand can take different values using one of the three possible syntaxes, which are discussed briefly below.

- Constant: An Operand2 constant is specified in the form #constant where constant can be any constant that can be produced by shifting an 8-bit value left by any number of bits within a 32-bit word. For further details on the possible uses of constant Operand2 see ARMv7-M Architecture Reference Manual [ARM(2010)].

- Register: You specify an Operand2 register in the form of *Rm* where *Rm* specifies the register holding the data for the second operand.

- Register with Optional Shift: For this case, Operand2 is specified in the form: *Rm* {, *shift*}, where *Rm* specifies the register holding the data for the second operand and *shift* is an optional shift to be applied to *Rm*. It can be one of the following options.

ASR #n	Arithmetic shift right n bits, $1 \le n \le 32$.
LSL #n	Logical shift left n bits, $0 \le n \le 31$.
LSR #n	Logical shift right n bits, $1 \le n \le 32$.
ROR #n	Rotate right n bits, $1 \le n \le 31$.
RRX	Rotate right one bit, with extend.
	If omitted, no shift occurs, equivalent to LSL#0.

When Operand2 is specified as register with optional shift, the specified value of the shift is applied to the contents of register *Rm* and the resulting 32-bit value is used by the instruction. However, the contents in the register *Rm* remain unchanged.

When *Operand2* is a register with optional shift the shift operation is executed by the *barrel shifter* as can be seen from Figure 5.4. The barrel shifter has the capability of performing a shift operation by any arbitrary number of bit positions (up to 32 bits in the case of Cortex-M processors) in one instruction cycle of the processor. In other words, the shift operation by 3 bits or 30 bits takes the same amount of execution time when using a barrel shifter. Example 5.5 illustrates the use of different bit operation instructions.

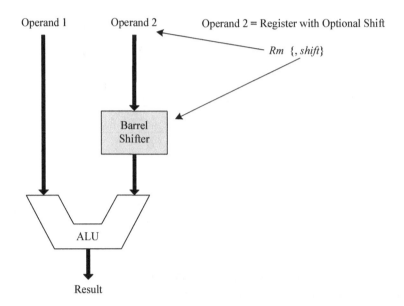

Figure 5.4: Barrel shifter functionality explained when the flexible second operand is a register with optional shift.

Example 5.5 (Illustration of bitwise logical operations).

```
AND    R2, R1, #0xFF00        ; R2 = R1 & 0xFF00
ORR    R3, R1, R5             ; R3 = R1 | R5
BIC    R5, R1                 ; R5 = R5 & (~R1)
EORS   R4, R1, #0x0055        ; R4 = R1 ^ 0x0055
```

The first two instructions simply illustrate the use of AND and OR operations. The comment field shows the C language equivalent representation of the logical operations. In the BIC instruction above we have used only two operands and have omitted the optional destination register. In the last instruction, the flags in APSR are affected based on the result of the operation.

5.1.3   Applications of Bit Operations

In many different applications shift and rotate bit operations can be combined with logical operations to implement many useful functions. Below we illustrate, using examples, some useful functionalities that can be considered fundamental building blocks for many different embedded system applications.

Example 5.6 (Testing n^{th} bit of a variable.). Let R0 contain the value of variable x. We are interested in finding whether the nth bit of variable x is equal to 1 or 0. There is more than one possible way to find the answer to this question. We give two different illustrations.

One possible solution is to shift right R0 by n positions and then perform AND operation with 1. The flags should be updated to reflect the result of AND operation. Below is the assembly code that precisely performs this task for $n = 7$.

```
LSR    R0, #7         ; R0 = R0 >> 7
ANDS   R0, #1         ; R0 = R0 & 1 and flags are updated
```

After the execution of ANDS instruction, we will check the zero flag in the APSR to find out whether the nth bit is 1 or 0.

One drawback of the above implementation is that the original value of the variable in R0 is changed and must be preserved before performing the bit testing. However, the following implementation overcomes this issue.

```
MOV    R1, #1
LSL    R1, #7         ; R1 = 1 << 7
ANDS   R1, R1, R0     ; R1 = R0 & R1 and flags are updated
```

In the second method above, we have used a free register R1 and set its nth bit to 1 to perform the bit test operation.

Bit testing is just one type of operation. Some other useful operations involve setting or clearing specific bit or bits of a variable or register. This task of bit setting or clearing is given the name of *bit masking* . Bit masking is one of the most fundamental operations used in embedded programming. Below we illustrate the bit masking operation using simple examples.

Example 5.7 (Masking the bits). Based on the requirement, different types of bit masking operations can be performed. In some cases, we need to set a bit or bits of a specific register or variable. For that purpose we can either use ORR instruction or can combine it with a shift operation to perform bit setting. The following code illustrates setting the fourth bit (this is bit 3 when LSB is labeled as bit 0) of register R0.

```
MOV    R1, #1
LSL    R1, #4         ; R1 = 1 << 4
ORR    R0, R0, R1     ; R0 = R0 | R1
```

In case of clearing a bit we use BIC instruction along with shift operation. The following code illustrates clearing the fourth bit of register R0.

```
MOV    R1, #1
LSL    R1, #4         ; R1 = 1 << 4
BIC    R0, R0, R1     ; R0 = R0 & ~R1
```

5.2 Basic Arithmetic Instructions

The elementary arithmetic operations include addition, subtraction, multiplication, and division. Different variants of these instructions are implemented by ARM

Table 5.3: Basic arithmetic instructions.

Mnemonic	Brief description	Encoding	Flags
ADC	Add with carry	16 or 32 bit	N,Z,C,V
ADD	Addition	16 or 32 bit	N,Z,C,V
ADDW	Addition (32-bit encoded)	32 bit	N,Z,C,V
RSB	Reverse subtract	16 or 32 bit	N,Z,C,V
SBC	Subtract with carry	16 or 32 bit	N,Z,C,V
SUB	Subtract	16 or 32 bit	N,Z,C,V
SUBW	Subtract (32-bit encoded)	32 bit	N,Z,C,V
MUL	Multiply (32-bit result)	16 or 32 bit	N,Z
MLA	Multiply with accumulate (32-bit result)	32 bit	-
MLS	Multiply with subtract (32-bit result)	32 bit	-
SMLAL	Signed multiply accumulate (64-bit result)	32 bit	-
SMULL	Signed multiply (64-bit result)	32 bit	-
UMLAL	Unsigned multiply accumulate (64-bit result)	32 bit	-
UMULL	Unsigned multiply (64-bit result)	32 bit	-
UDIV	Unsigned divide	32 bit	-
SDIV	Signed divide	32 bit	-

Cortex-M architecture. The differences and similarities of these instruction variants will be highlighted and elaborated using examples.

The list of basic math instructions with their brief description is provided in Table 5.3. The information related to possible instruction encoding formats (16-bit and 32-bit) is provided in the third column of Table 5.3. When the instruction is appended with 'S' field, then it can affect different flag bit(s). The information regarding which flag(s) is updated in the application program status register (APSR), when an instruction is executed, is given in the last column of Table 5.3.

5.2.1 Addition and Subtraction Instructions

The addition operation is implementable using one of the three possible instructions depending on the requirement of operation. The three different instructions available for the addition operation are ADD, ADDW, and ADC. The ADD instruction simply adds two numbers and can be encoded using either 16-bit or 32-bit formats. The ADDW instruction is always 32-bit encoded. The difference between ADD and ADDW is not only in the instruction encoding size but also in the instruction syntax as we will observe below. The ADC instruction implements add with carry and adds the value of carry flag (at the time of instruction execution) to the operands of addition operation.

The subtraction operation is implementable using subtract (SUB, SUBW), subtract with carry (SBC), and reverse subtract (RSB) instructions. The functionality of SUB and SUBW is similar to ADD and ADDW except that the operation performed is subtraction. In SBC the current value of the carry flag is first inverted and then it is subtracted from the result of subtraction. In case of RSB instruction, the order of operands is reversed as will be clarified in the example illustration. The instruction syntax for different addition and subtraction instructions is given below.

$$
\begin{array}{llll}
\text{ADD}\{S\}\{cond\}\{.size\} & \{Rd,\} & Rn, & Operand2 \\
\text{ADC}\{S\}\{cond\}\{.size\} & \{Rd,\} & Rn, & Operand2 \\
\text{SUB}\{S\}\{cond\}\{.size\} & \{Rd,\} & Rn, & Operand2 \\
\text{SBC}\{S\}\{cond\}\{.size\} & \{Rd,\} & Rn, & Operand2 \\
\text{RSB}\{S\}\{cond\}\{.size\} & \{Rd,\} & Rn, & Operand2 \\
\text{ADDW}\{cond\} & \{Rd,\} & Rn, & \#imm12 \\
\text{SUBW}\{cond\} & \{Rd,\} & Rn, & \#imm12 \\
\end{array}
$$

In the above instruction syntax description, most of the fields are the same as described previously for shift and logical instructions. The $\#imm12$ in case of ADDW and SUBW instructions, represents an arbitrary integer value in the range 0-4095.

From the instruction syntax, we can observe that ADDW and SUBW cannot affect the status flags because of the absence of 'S' field. For further details related to some other variants of this syntax refer to ARMv7-M Architecture Reference Manual [ARM(2010)]. Now let us elaborate the use of basic arithmetic instructions using some examples. Example 5.8 explains the syntax for different addition and subtraction instructions.

Example 5.8 (Syntax illustration for basic arithmetic instructions).

```
ADD   R2, R1, R3    ; R2 = R1 + R3
ADC   R2, R1, R3    ; R2 = R1 + R3 + C, here C represent
                    ; value of carry flag
SUBS  R8, R6, #240  ; R8 = R6 - 240, sets the flags on
                    ; the result
RSB   R4, R4, #1280 ; R4 = 1280 - R4,
```

The next example implements the addition of two 64-bit numbers. This is accomplished using two instructions, ADDS performs addition of least significant 32-bit value and updates the flags based on the result. Then ADC instruction is used to add the most significant 32-bits along with the updated value of C flag.

Example 5.9. The following two instructions add a 64-bit integer contained in R2 and R3 to another 64-bit integer contained in R0 and R1. The result of addition is stored in registers R4 and R5.

```
ADDS  R4, R0, R2    ; adding lower 32-bit words
ADC   R5, R1, R3    ; adding higher 32-bit words along with
                    ; carry flag from lower word addition
```

The next example illustrates the use of flexible second operand for an efficient implementation of the linear scaling. For the implementation in Example 5.10, it is required that the scaling coefficient can be represented as 2^m for $1 \leq m \leq 31$.

Example 5.10 (Efficient use of flexible second operand). Consider an expression $y = 32x + c$. Assuming that the variables of the expression are already loaded to the registers with R3 = y, R2 = x, and R1 = c. One possible implementation is given by the following assembly code.

```
LSL    R2, R2, #5      ; Multiplying R2 with 32
ADD    R3, R2, R1      ; Now perform the addition operation
```

An efficient implementation of the above expression is achieved by using the flexible second operand. The single assembly instruction implementation of the above expression is given below.

```
ADD R3, R1, R2, LSL #5  ; Shift left (logical) R2 by five
                        ; bits, add to R1 and finally store
                        ; result in R3. This operation is
                        ; equivalent to R3 = R1 + (R2 << 5)
```

The above-mentioned implementation requires the scaling coefficient to be 2^m. An efficient implementation of linear scaling will be presented using multiplication instruction that allows arbitrary choice of scaling coefficients.

5.2.2   Multiply and Divide Instructions

Multiplication and division operations are also supported at the hardware level in the Cortex-M3 processor. Two different types of multiplication instructions are supported. One set of multiplication instructions uses 32-bit operands and the result of multiplication is also 32-bit. The other set of multiplication instructions produces a 64-bit result as a consequence of 32-bit multiplication. Below we discuss these two different types of instructions separately.

Multiplication Instructions with 32-bit Result

Multiply (MUL), multiply with accumulate (MLA), and multiply with subtract (MLS) instructions use 32-bit operands and produce a 32-bit result. Since both the operands are 32-bit, it is possible that the result of multiplication can be as large as 64 bits. However, the above-mentioned multiplication instructions are limited in the sense that they only write the least significant 32 bits of the result to the destination register. In addition, these instructions do not differentiate whether the operands are signed or unsigned integers. Below is the instruction syntax for these instructions.

$$\begin{aligned}
&\text{MUL}\{S\}\{cond\} \quad \{Rd,\} \quad Rn, \quad Rm \\
&\text{MLA}\{cond\} \qquad\quad\; Rd, \qquad Rn, \quad Rm, \quad Ra \\
&\text{MLS}\{cond\} \qquad\quad\; Rd, \qquad Rn, \quad Rm, \quad Ra
\end{aligned}$$

where

Rn, Rm	These are the two registers containing the values to be multiplied.
Rd	This is the destination register that holds the least significant 32-bits of the result of operation.
Ra	This is the register holding the value to be added to or subtracted from.

The MUL instruction multiplies the values contained by the registers *Rn* and *Rm* and writes the least significant 32 bits of the result in the destination register *Rd*. The MLA instruction multiplies the two operand values from *Rn* and *Rm*, adds the result of multiplication to the contents of *Ra* and finally writes the least significant 32 bits of the result to *Rd*. The MLS instruction first multiplies the contents of *Rn* and *Rm* and then subtracts the least significant 32 bits of the product from the value of register *Ra* and finally writes the least significant 32 bits of the result to register *Rd*. The results of these instructions do not depend on whether the operands are signed or unsigned. Next we illustrate the use of these instructions using examples.

Example 5.11 (Illustration of multiplication instruction syntax with 32-bit result.). Below we exemplify the use of multiply instructions and illustrate the instruction syntax.

```
MUL    R1, R2, R5       ; Multiply, R1 = R2 x R5
MLA    R1, R2, R3, R5   ; Multiply with accumulate,
                        ; R1=(R2 x R3)+R5
MULS   R1, R2, R2       ; Multiply with flag update,
                        ; R1=R2 x R2
MLS    R1, R5, R6, R7   ; Multiply with subtract,
                        ; R1=R7-(R5 x R6).
```

It is important to remember that only the least significant 32 bits of the result are written to a destination register.

Our next example illustrates the use of MLA instruction to implement the arithmetic sum of a series.

Example 5.12 (Finding the sum of the first n integers). The arithmetic sum of the first n integers is given by $S = n(n+1)/2$. For its efficient implementation, let us rewrite the expression as $S = (n^2 + n)/2$. Assuming register R2 $= n$, the last expression is implemented using the following code:

```
MLA   R1, R2, R2, R2   ; Multiply with accumulate,
                       ; R1 = (R2xR2)+R2
LSR   R1, R1, #1       ; R1 = R1/2
```

The equivalent C code for the above assembly program is given below.

```
R1 = (R2*(R2 + 1)) >> 1;
```

Multiplication Instructions with 64-bit Result

The Cortex-M3 processors also support 32-bit multiplication instructions with or without accumulate and produce 64-bit results. These instructions support both signed as well as unsigned operand values and produce correspondingly signed or unsigned results. Specifically, unsigned multiply long (UMULL), unsigned multiply with accumulate long (UMLAL), signed multiply long (SMULL), and signed multiply with accumulate long (SMLAL) instructions are supported by the Cortex-M3 hardware architecture. Below is the general syntax for these instructions.

$$\text{UMULL}\{cond\} \quad RdLow, \quad RdHigh, \quad Rn, \quad Rm$$
$$\text{UMLAL}\{cond\} \quad RdLow, \quad RdHigh, \quad Rn, \quad Rm$$
$$\text{SMULL}\{cond\} \quad RdLow, \quad RdHigh, \quad Rn, \quad Rm$$
$$\text{SMLAL}\{cond\} \quad RdLow, \quad RdHigh, \quad Rn, \quad Rm$$

where
Rn, Rm Are the registers containing the operands to be multiplied.

$RdHigh$, Rd-Low These are the destination registers to hold the 64-bit result of multiplication. For UMLAL and SMLAL instructions these registers also hold the value to be accumulated.

For the UMULL instruction the values from the registers Rn and Rm are treated as unsigned integers. It multiplies the 32-bit integer operands and writes the least significant 32 bits of the result in $RdLow$ while the most significant 32 bits of the result are written to $RdHigh$. The UMLAL instruction uses the values from the registers Rn and Rm as unsigned numbers. It multiplies these integers, adds the 64-bit result to the 64-bit unsigned integer contained in $RdHigh$ and $RdLow$, and finally writes the result back to $RdHigh$ and $RdLow$ registers. The least significant 32-bits of the multiplication result are added to $RdLow$, while the most significant 32-bits of the multiplication result are added to the $RdHigh$ register.

The SMULL instruction treats the values from the registers Rn and Rm as 2's complement signed numbers. It multiplies these operands and places the least significant 32-bits of the result in $RdLow$ while the most significant 32-bits of the result are stored in $RdHigh$. The SMLAL instruction treats the values from Rn and Rm as signed integers in 2's complement form. This instruction first multiplies these operands, adds the 64-bit result of multiplication to the 64-bit signed value contained in $RdHigh$ and $RdLow$, and finally writes the result back to $RdHigh$ and $RdLow$ registers.

The registers $RdHigh$ and $RdLow$ used in these instructions must be different from each other. In addition, these instructions do not affect the condition code flags.

Example 5.13 (Multiplication instructions with 32-bit operands and 64-bit result.).

```
UMULL   R0, R4, R5, R6 ; Unsigned [R4 R0] = R5 x R6
SMLAL   R4, R5, R3, R8 ; Signed [R5 R4] = [R5 R4] + R3 x R8
```

The notation $[x\ y]$ represents that y is the lower 32-bits and x is the upper 32-bits of a 64-bit variable.

Divide Instructions

Unsigned division (UDIV) and signed division (SDIV) are the two integer divide instructions that are supported by the Cortex-M3 architecture. It should be noted that in case of integer division the fractional part (i.e., the remainder) is discarded. The syntax for these two instruction is given below.

$$\text{SDIV}\{cond\} \quad Rd, \quad Rn, \quad Rm$$
$$\text{UDIV}\{cond\} \quad Rd, \quad Rn, \quad Rm$$

The SDIV instruction performs a signed integer division of the number in the register Rn by the value contained in the register Rm. On the other hand, the UDIV instruction is similar to SDIV but performs an unsigned integer division of the number in Rn by the value in Rm. Since these instructions perform integer division, the fractional part (remainder) is discarded and the quotient, as a result of division, is stored in the destination register the Rd. For both of the instructions, if the integer value in Rn is smaller than the value in Rm, the result of division is zero. In addition, these instructions do not affect the condition code flags in the application program status register as can be observed from the absence of the 'S' field in the instruction syntax. The examples below illustrate the use of divide instructions.

Example 5.14 (Illustration of integer division instructions).

```
UDIV   R1, R8, R2    ; Unsigned divide, R1 = R8/R2
SDIV   R1, R2, R4    ; Signed divide, R1 = R2/R4
```

Example 5.15 (Implementation of remainder operation).

In C and other high level languages remainder operation is quite useful for performing certain tasks. In C programming, for variables x and y the remainder is evaluated as $x\%y$. The assembly program equivalent to reminder operation is quite straightforward if the divide instruction produces both the quotient as well as remainder. For instance, in x86 processor architecture the divide instruction produces both quotient and remainder. However, in ARM Cortex-M processor architecture, the divide instruction only produces the quotient. In this case, we need to develop an algorithm to evaluate the remainder. Since the divide instruction in ARM Cortex-M produces the quotient, we can use the expression $x - (x/y)y$ to evaluate the remainder indirectly. Assuming R0 = x and R1 = y the following assembly code evaluates the reminder.

```
SDIV   R2, R0, R1    ; Signed divide, R2 = R0/R1
MLS    R3, R1, R2, R0  ; R3 = R0 - (R2 x R1)
```

Table 5.4: Data movement instructions.

Mnemonic	Brief description	Encoding	Flags
MOV	Move	16 or 32 bit	N, Z, C
MOVW	Move 16-bit constant to register	32 bit	
MOVT	Move 16-bit constant to register top halfword	32 bit	
MVN	Move NOT	16 or 32 bit	N, Z, C

The register R3 will contain the reminder.

5.3 Data Movement Instructions

Data movement inside the processor can be performed using move (MOV and MOVW), move negative (MVN), and move top (MOVT) instructions. Table 5.4 provides details about instruction size(s) and the flags updated by the corresponding instructions.

The instruction syntax for different move operations is provided below. It should be noted that the move instructions with *Operand2* can update the flags, while the others having #*imm*16 as operand cannot update any of the flags. When the optional field {*S*} is specified the instruction can update the 'Z' and 'N' flags but does not affect the 'V' flag. Additionally these instructions do not update the 'C' flag. However, there is a possibility of getting the 'C' flag updated as a consequence of evaluation of *Operand2* and should not be considered as a flag update due to move operation.

$$\text{MOV\{S\}\{cond\}} \quad Rd, \quad Operand2$$
$$\text{MOV\{cond\}} \quad Rd, \quad \#imm16$$
$$\text{MVN\{S\}\{cond\}} \quad Rd, \quad Operand2$$
$$\text{MOVW\{cond\}} \quad Rd, \quad \#imm16$$
$$\text{MOVT\{cond\}} \quad Rd, \quad \#imm16$$

The MOV instruction copies the value from *Operand2* to the register *Rd*. When *Operand2* in a MOV instruction is a register with an optional shift and the shift operation is other than LSL #0, then it preferred that the MOV instruction should be replaced with the corresponding shift instruction. For instance, the instruction

$$\text{MOV\{S\}\{cond\}} \quad Rd, \quad Rm, \quad ASR\#n.$$

should be replaced by the corresponding shift instruction, which is

$$\text{ASR\{S\}\{cond\}} \quad Rd, \quad Rm, \quad \#n$$

and is the preferred syntax for this operation. The MOVW instruction provides the same functionality as that of MOV instruction, but is restricted to using the #*imm*16 as its second operand. The MVN instruction takes the value of *Operand2*, performs a bitwise logical NOT operation on the value, and places the result in the destination register *Rd*. MOVT writes a 16-bit value, #*imm*16, to the top halfword of the destination register, i.e., $Rd[31:16]$. The move operation in MOVT instruction does not affect the lower halfword, i.e., $Rd[15:0]$. Example 5.16 illustrates the use of different move operations described above.

Example 5.16 (Illustration of move instructions).

```
MOVS   R11,  #0x000B  ; Write 0x000B to R11, flags are updated
MOV    R1,   #0xA05   ; Write 0xA05 to R1, do not update flags
MOVS   R10,  R12      ; Move R12 to R10, flags are updated
MOV    R3,   #23      ; Write value of 23 to R3
MOV    R8,   SP       ; Write stack pointer to R8
MVNS   R2,   #0xF     ; Write 0xFFFFFFF0 (bitwise inverse
                      ; of 0xF) to R2 and update flags.
MOVT   R3,   #0xF123  ; Write 0xF123 to upper halfword of R3,
                      ; lower halfword and APSR are unchanged.
```

The execution of MOV instruction is further elaborated in Figure 5.5, which shows the contents of program (code) memory and the state of registers at different steps of instruction execution. Specifically, at time instance labeled '1', the program counter, PC, is holding the address of MOV R0, 0x124 instruction, which is 0x264. At that point in time the contents of register R0 are equal to zero. At the next time instant '2', when this move instruction is executed, the PC will be incremented to 0x268 and the register R0 = 0x124. The 32-bit machine code of the move instruction is also shown in Figure 5.5.

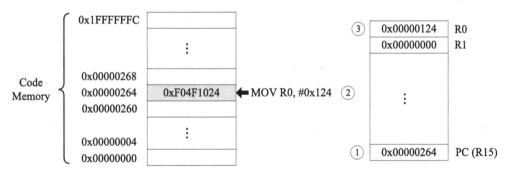

Figure 5.5: Move instruction execution.

In many situations we need 32-bit constants for certain operations. For example, the raw data from a sensor needs to be scaled and an offset is to be added for proper visualization. Now the two coefficients, scaling and offset required for data processing, are actually two constants that will be provided by the application developer. One possible solution for generating 32-bit constants is based on the use of MOVW and MOVT instruction pair. The simultaneous use of these two instructions enables us to generate any 32-bit constant as illustrated by Example 5.17.

Example 5.17 (Illustration of move instructions). In digital systems, the real life analog signals are converted to the equivalent digital signals, which needs to

Table 5.5: Bitfield instructions.

Mnemonic	Brief description	Encoding	Flags
BFC	Bit Field Clear	32 bit	
BFI	Bit Field Insert	32 bit	
CLZ	Count leading zeros	32 bit	
RBIT	Reverse bits	32 bit	
REV	Reverse byte order in a word	16 or 32 bit	
REV16	Reverse byte order in each halfword	16 or 32 bit	
REVSH	Reverse byte order in bottom halfword & sign extend	16 or 32 bit	
SBFX	Signed Bit Field Extract	32 bit	
UBFX	Unsigned Bit Field Extract	32 bit	
SXT	Sign extend	16 or 32 bit	
UXT	Zero extend	16 or 32 bit	

be scaled and may also require an offset to be adjusted for proper visualization. This can be represented as $y = \mathrm{m}x + \mathrm{c}$, where x is the raw data and y represents the processed data using scaling constant 'm' and an offset 'c'. For example, the temperature sensor output is sampled and converted to digital equivalent using analog to digital converter, which can be further mapped to either Celsius (^oC) or Fahrenheit (F) using the above procedure.

Since the use of constants is justified, now we describe a possible solution to generate arbitrary 32-bit constants. The assembly code below illustrates the generation of 0x55334466 constant in register R2 for further use by the application for data processing.

```
MOVW R2, #0x4466    ; move immediate value 0x4466 to lower
                    ; halfword of R2
MOVT R2, #0x5533    ; move immediate value 0x5533 to upper
                    ; halfword of R2
```

5.4   Bitfield Instructions

The bitfield instructions operate on the adjacent group of bits in the operand registers. The provisioning of these operations make Cortex-M architecture attractive for many control and signal processing applications. We have categorized the bitfield instructions further into the following subgroups.

- Bit and byte reversal instructions

- Bitfield clear and insert instructions

- Bitfield extract instructions

- Miscellaneous bitfield instructions

Table 5.5 provides the information related to possible instruction encoding formats (16-bit and 32-bit) for different bitfield instructions. It should be observed that the bitfield instructions do not affect any program status flags. Next we discuss each of the above-mentioned subgroups.

5.4.1 Bit and Byte Reversal Instructions

The bit and byte reversal instructions are used, respectively, to reverse the bit order in a word and the byte order in a word or a halfword data. The general syntax for bit and byte reversal instructions is given below.

$$
\begin{array}{lll}
\text{REV}\{cond\} & Rd, & Rn \\
\text{REV16}\{cond\} & Rd, & Rn \\
\text{REVSH}\{cond\} & Rd, & Rn \\
\text{RBIT}\{cond\} & Rd, & Rn
\end{array}
$$

REV instruction performs byte reversals in a word size data. This instruction can be used to convert a word data from big-endian format to little-endian format or little-endian format to big-endian format. Figure 5.6 illustrates the functionality of REV instruction.

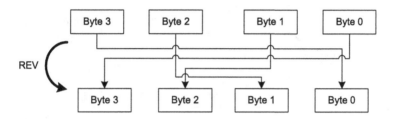

Figure 5.6: Illustration of byte reversal in REV instruction.

REV16 instruction reverses byte order in each halfword independently. It can be used to convert 16-bit big-endian data into little-endian data or little-endian data into big-endian format. REVSH reverses the byte order in the bottom halfword and then performs sign extension to make it 32-bit word data. This instruction can be used to convert 16-bit signed big-endian data into 32-bit signed little-endian data or 16-bit signed little-endian data into 32-bit signed big-endian data. A pictorial illustration of these instructions is given in Figure 5.7.

RBIT instruction reverses the bit order in a 32-bit word. Bit reversal is performed by swapping bit 0 with bit 31, bit 1 with bit 30, and so on. It is different from REV instruction where byte order is reversed by swapping byte 0 with byte 3 and byte 1 with byte 2. Example 5.18 illustrates the use of bit and byte reversal instructions.

5.4.2 Bitfield Clear and Insert Instructions

Bitfield clear (BFC) and bitfield insert (BFI) instructions are used to modify the set of bits in a register. It should be noted that the bits modified by these instructions are always contiguous. The syntax for these instructions is given below.

$$
\begin{array}{lllll}
\text{BFC}\{cond\} & Rd, & \#lsb, & \#width \\
\text{BFI}\{cond\} & Rd, & Rn, & \#lsb, & \#width
\end{array}
$$

where

Rd	Specifies the destination register.
#lsb	Specifies the position of the least significant bit of the bitfield to be modified. *lsb* must be in the range 0 to 31.
#width	Specifies the width of the bitfield and must be in the range 1 to 32-*lsb*.
Rn	Specifies the source register.

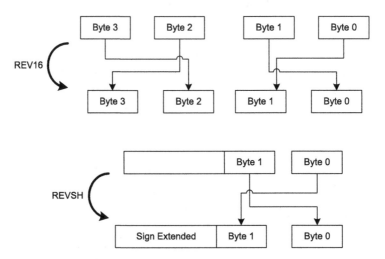

Figure 5.7: Illustration of byte reversal in REV16 and REVSH instructions.

Example 5.18 (Bit and byte reversal instructions illustration). This example illustrates the use of bit and byte reversal instructions.

```
MOVW R1, 0x5678    ; create 32-bit word data
MOVT R1, 0x1234
RBIT R3, R1        ; bit reversal of data
REV  R4, R1        ; byte reversal of data
```

The contents of registers R1 and R3 after the execution of above instructions appear as given below.

```
R1 = 00010010001101000101011001111000 (0x12345678)
R3 = 00011110011010100010110001001000 (0x1E6A2C48)
```

On the other hand the execution of REV instruction updates the contents of register R4 to 0x78563412. Assuming that the register R1 still holds the above mentioned value, the following assembly code describes the use of REV16 and REVSH instructions.

```
MOVW  R2, 0x12E3   ; create 16-bit halfword data
REV16 R5, R1       ; byte reversal in each halfword
REVSH R6, R2       ; sign extended byte reversal
```

The value of register R5 after the execution of REV16 is 0x34127856. With a starting value of 0x12E3 in register R2, the content of register R6 becomes

0xFFFFE312 after the execution of REVSH instruction.

BFC clears the specified bitfield in a register. It clears *width* number of bits, starting at the bit position *lsb*, in the destination register *Rd*, while the other bits of register *Rd* remain unchanged. The BFI instruction performs bitfield copying operation from source register to the destination register. It replaces a set of bits in *Rd* equal to the size specified by the *width* field and starting at the bit position specified by *lsb*. The number specified by the field *lsb* marks the least significant bit position of the bitfield to be replaced in destination register *Rd*. The register *Rn* is the bitfield source with the number of bit specified by the field *width*. However, the starting location of the bitfield is bit[0] for register *Rn*. Similar to the BFC instruction, the other bits in *Rd* remain unchanged in case of BFI as well. Figure 5.8 shows one such illustration where BFI instruction is used to combine bitfields from two different sources. Example 5.19 illustrates the use of bitfield clear and insert instructions.

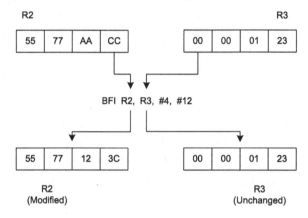

Figure 5.8: Insertion of an arbitrary number of adjacent bits using BFI.

Example 5.19 (Illustration of bitfield clear and bitfield insert instructions.).

```
MOV   R3, #0x123        ; Arbitrary source bitfield data
BFC   R1, #4, #8        ; Clear bit 4 to bit 11 (8 bits) of
                        ; R1 to 0
BFI   R2, R3, #4, #12   ; Replace bit 4 to bit 15 (12 bits)
                        ; of R2 with bit 0 to bit 11 of R3.
```

Let the register R1 = 0x11223344 before the execution of BFC instruction and is updated to 0x11223004 after the execution of BFC instruction. Assuming the register R2 = 0x5577AACC before the execution of BFI instruction in the above code. After the execution of this instruction the contents of register R2 become 0x5577123C. This is also illustrated in Figure 5.8.

Bit manipulations are an essential operation for efficient packet processing in different networking applications. For instance, bitfield modification operations are capable of providing selective access to packed data structures, such as those encountered in network packet headers. The bit-manipulation instructions in the Cortex-M3 processor like BFI and BFC reduce data-bus activity as well as the need to unpack data structures, resulting in faster network processing operations.

5.4.3   Bitfield Extract Instructions

Signed and unsigned bitfield extract (SBFX and UBFX) instructions are used to extract a specified length bitfield from the source register and copy it to the destination register after sign extending it.

$$\text{SBFX}\{cond\} \quad Rd, \quad Rn, \quad \#lsb, \quad \#width$$
$$\text{UBFX}\{cond\} \quad Rd, \quad Rn, \quad \#lsb, \quad \#width$$

In the above syntax for SBFX and UBFX different fields have the similar definitions as described for the bitfield insert instruction in the preceding subsection. However, there is one key difference. The *lsb* field in the case of BFI instruction was the lowest bit address of the bitfield of destination register. However, in the case of SBFX and UBFX instructions, the *lsb* field specifies the lowest bit address of the source register. The lowest bit address is always zero for the destination register in the case of bitfield extract instructions.

SBFX extracts a bitfield from one register, sign extends it to 32 bits, and writes the result to the destination register. UBFX extracts a bitfield from one register, zero extends it to 32 bits, and writes the result to the destination register. Example 5.20 illustrates the use of signed and unsigned bitfield extract instructions. The bitfield extract operations are also illustrated in Figure 5.9.

Automotive and industrial applications are characterized by the need to process large volumes of general-purpose I/O data. Interfaces to sensors or similar external devices are often read in as 8- or 16-bit words, which then have to be manipulated to extract bit-level data. These manipulations are typically done by custom program that takes several cycles to complete. The Thumb2 UBFX (zero extend) and SBFX (sign extend) bit-field instructions, enable this bit level extraction in a single cycle.

Example 5.20 (Signed and unsigned bitfield extract instructions).

```
SBFX R2, R1, #16, #16 ; Extract bit 16 to bit 31 (16 bits)
                      ; from R1 sign extend it to 32 bits
                      ; and finally write the result to R2.
UBFX R3, R1, #6, #10  ; Extract bit 6 to bit 15 (10 bits)
                      ; from R1 and zero extend to 32 bits
                      ; and then write the result to R3.
```

5.4.4   Miscellaneous Bitfield Instructions

In this subsection we discuss some remaining instructions that also perform bitfield operations. Sign extend (SXT) and zero extend (UXT) instructions are used to extend an 8-bit or 16-bit value to, respectively, 32-bit sign or zero extended value. Count leading zeros (CLZ) is another instruction that is used for counting the leading zeros

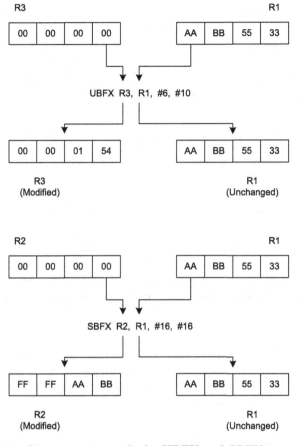

Figure 5.9: Bit extraction with the UBFX and SBFX instructions.

on an integer value. The instruction syntax for sign extend, zero extend, and count leading zeros is given by

SXTB{*cond*}	{*Rd*,}	*Rm*,	{, ROR #*n*}
SXTH{*cond*}	{*Rd*,}	*Rm*,	{, ROR #*n*}
UXTB{*cond*}	{*Rd*,}	*Rm*,	{, ROR #*n*}
UXTH{*cond*}	{*Rd*,}	*Rm*,	{, ROR #*n*}
CLZ{*cond*}	*Rd*,	*Rm*	

where

Rm Specifies the register holding the value of the operand.

ROR #n Is one of the following possible values.

- ROR #8: Value from Rm is rotated right 8 bits.
- ROR #16: Value from Rm is rotated right 16 bits.
- ROR #24: Value from Rm is rotated right 24 bits.
- ROR #n omitted: If ROR #n is omitted, then no rotation is performed.

The SXT and UXT instructions optionally rotate right the value from the register Rm

by 0, 8, 16, or 24 bits and then extract bits from the resulting value. SXTB instruction extracts 8 bits, i.e., bits[7:0] and sign extends it to 32 bits. SXTH instruction extracts 16 bits, bits[15:0] and sign extends to 32 bits. Similarly, the instructions UXTB and UXTH optionally apply the rotate right by specified value, then extract, respectively, bits[7:0] and bits[15:0] and then zero extends to 32 bits. Example 5.21 illustrates the use of sign and zero extend instructions.

Example 5.21 (Illustration of sign and zero extend bitfield instructions.).

```
SXTH R1, R2, ROR #16    ; Rotate R2 right by 16 bits, obtain
                        ; the lower halfword of the result
                        ; and then sign extend to 32 bits.
                        ; Write the result to R1.
UXTB  R2, R1            ; Extract lowest byte from source
                        ; R1 zero extend it and write the
                        ; result to register R2.
```

The SXT and UXT instructions are useful for implementing the type-casting operations, which are frequently used in high level languages. We illustrate this aspect using an example.

Example 5.22 (Variable type casting in C programming using SXT and USX.). Consider a 16-bit variable casted to an 8-bit variable. We use two different types of 8-bit variables, one being signed and the other being unsigned. The below C code listing implements this type casting.

```
unsigned short x = 0x8387;
unsigned char y1 = 0;
signed char y2 = 0;

y1 = (unsigned char) x;
y2 = (signed char) x;
```

When the above code is compiled, the tools have assigned register R2 for variable 'y1' and R3 for variable 'y2'. The register R1 is initialized with variable 'x'. The disassembly of the last two C instructions is listed below.

```
UXTB  R2, R1           ; y1 = (unsigned char) x;
SXTB  R3, R1           ; y2 = (signed char) x;
```

When the above program is executed the value loaded in register R2 is 0x00000087, while register R3 is loaded with 0xFFFFFF87.

The CLZ instruction counts the number of leading zeros in the value provided by the register Rm and returns the result in destination register Rd. The result of CLZ instruction is 32 if all the bits are set to zero in the register Rm while it is zero if bit 31 of source register is equal to 1.

It is possible to use CLZ instruction for data normalization. For example, CLZ instruction followed by a left shift applied to the value in register Rm equal to the value specified by the register Rd normalizes the value of register Rm. It is important to realize that MOVS instruction should be used to indicate the case when Rm is zero, by updating the flag bits in the application program register, specifically the zero flag. Example 5.23 implements the data normalization using CLZ instruction.

Example 5.23 (Data normalization using CLZ instruction.).

```
CLZ   R5, R9
MOVS R9, R9, LSL R5
```

The above code normalizes the data in register R9 to 32-bit value.

5.5 Test and Compare Instructions

The two test instructions, test bits (TST) and test equivalence (TEQ), are used to test the value in a register against *Operand2*. The comparison instructions, compare (CMP) and compare negative (CMN), compare the value in a register with the contents of *Operand2*. All of these instructions update the condition flags on the result, but do not place the result in any register. In other words, these instructions are useful for testing and comparing a variable against another when the actual difference between their values is not important. The list of flags that are updated as a consequence of these instructions is tabulated in Table 5.6. It also contains information regarding the encoding. The general syntax of these instructions is given below.

$$\text{CMN}\{cond\} \quad Rn, \quad Operand2$$
$$\text{CMP}\{cond\} \quad Rn, \quad Operand2$$
$$\text{TEQ}\{cond\} \quad Rn, \quad Operand2$$
$$\text{TST}\{cond\} \quad Rn, \quad Operand2$$

The TST instruction performs a bitwise AND operation on the value in the register Rn and the value of *Operand2*. This is the same as an ANDS instruction, except that the result is discarded. To test whether a bit in register Rn is 0 or 1, use the TST instruction with an *Operand2* constant having the corresponding bit set to 1 while all other bits cleared to 0. The TEQ instruction performs a bitwise Exclusive OR operation on the value in Rn and the value of *Operand2*. This is the same as an EORS instruction, except that the result is discarded. The TEQ instruction is used to test if two values are equal.

The CMP instruction subtracts the value of *Operand2* from the value in Rn. This is the same as a SUBS instruction, except that the result is discarded. The CMN instruction adds the value of *Operand2* to the value in Rn. This is the same as an ADDS instruction, except that the result is discarded. Both CMP and CMN instructions update the N, Z, C, and V flags depending on the result of operation.

The TST and TEQ instructions update Z and N flags depending on the result of operation but do not affect V or C flags. This is in contrast to CMP and CMN instructions which do affect V or C flags as well. However, it should not be confused with the possibility of TEQ instruction affecting the C flag during the calculation of

Table 5.6: Test and compare instructions.

Mnemonic	Brief description	Encoding	Flags
CMN	Compare negative	16 or 32 bit	N,Z,C,V
CMP	Compare	16 or 32 bit	N,Z,C,V
TEQ	Test equivalence	32 bit	N,Z,C
TST	Test	16 or 32 bit	N,Z,C

Operand2. TEQ is also useful for testing the sign of a value. After the comparison, the N flag is the logical Exclusive OR of the sign bits of the two operands.

Example 5.24 (Illustration of test and compare instructions.).

```
TST   R4, #0x6CF
TEQ   R2, R3
TST   R1, R5, ASR #5   ; Arithmetic shift right the contents of R5
   by
                       ; 5 bit positions and then test R5 for being
                       ; equal to R1
```

The above set of instructions illustrates the use of test operation for three different types of *Operand2*. Next we illustrate the use of compare instructions.

```
CMP   R2, R9
CMN   R0, #226
```

5.6 Saturating Instructions

Signals from different sources are first conditioned before they can be processed for information extraction. One of the fundamental operations that is performed in the process of signal conditioning is the *signal amplification*. Signal amplification can be performed either in the hardware (e.g., operation/instrumentation amplifiers are widely used for this purpose) or it can be carried out in the software once the signal is digitized and is available in the form of sampled data.

When the signal amplification factor (also termed amplification gain) is too large, a saturation phenomenon is observed. Signal saturation can be experienced due to hardware as well as software amplification. However, the saturation phenomenon experienced as a result of hardware amplification can be different from the software amplification, as will be explained in detail in this section.

When a signal saturates due to excessive amplification in the hardware, signal clipping is experienced. Figure 5.10 shows the signal before and after amplification. Signal clipping can be clearly observed after amplification in Figure 5.10. This can be regarded as signal saturation due to hardware amplification.

Now let us discuss what happens when signal software amplification is performed. Signal amplification in the software is effectively equivalent to multiplication of the sampled signal data with an integer. Let us assume that the sampled signal is obtained

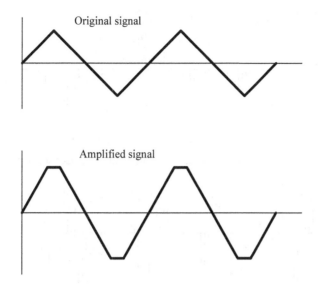

Figure 5.10: Signal saturation due to hardware amplification.

using an analog to digital converter with an 8-bit resolution and is stored in an array with 8-bit element size. Considering the signal is bipolar (i.e., has both positive as well as negative values) the maximum values permitted for the signal range between -128 and +127. During software amplification each sample of the signal (which corresponds to an entry in the data array) will be multiplied with a number and will be stored in another array. If the result of multiplication of an arbitrary sample value exceeds either the lower or upper limit, saturation will happen. To see this, let us consider one data sample with a value of 87 and is amplified by a gain value of 2. The result of amplification is 174, which exceeds the upper limit of 127 and effectively has become a negative number due to the sign bit set to 1 and is equal to -82. This effect is shown in Figure 5.11, which is the saturation due to software amplification. This effectively is equivalent to the overflow condition and highly distorts the data. The saturation instructions discussed here effectively avoid this overflow condition by saturating the data similar to the saturation due to hardware amplification. This is also illustrated in Figure 5.11.

Now we discuss the saturation instructions provided by the Cortex-M3 architecture. Specifically, we discuss signed saturate (SSAT) and unsigned saturate (USAT) instructions. The instruction syntax for the signed and unsigned saturate instructions has the following format.

$$\text{SSAT}\{cond\} \quad Rd, \quad \#n \quad Rm\{, \text{ shift } \#s\}$$
$$\text{USAT}\{cond\} \quad Rd, \quad \#n \quad Rm\{, \text{ shift } \#s\}$$

where

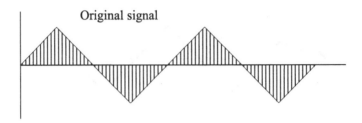

Original signal

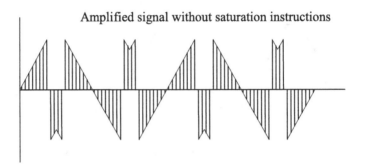

Amplified signal without saturation instructions

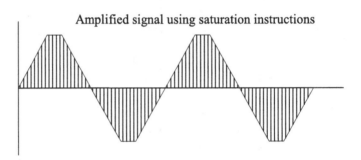

Amplified signal using saturation instructions

Figure 5.11: Signal saturation due to software amplification.

$\#n$	Specifies the most significant bit position to which the saturation action is performed to. The value of n ranges from 1 to 32 for SSAT and it ranges from 0 to 31 for USAT.
Rm	Specifies the register that contains the value to be saturated.
{shift $\#s$}	This is an optional shift applied to register Rm before saturating. The shift operation must be one of the following:

- ASR $\#s$ with a range of s from 1 to 31.
- LSL $\#s$ with a range of s from 0 to 31.

Table 5.7 gives information regarding the instruction encoding and the condition code flag dependency of the saturation instruction.

The SSAT and USAT instructions are used to saturate a given value, respectively, to a signed or unsigned n-bit final result. The SSAT instruction applies the specified shift, then saturates to the signed range $-2^{n-1} \leq x \leq 2^{n-1} - 1$. The USAT instruction applies the specified shift and then saturates to the unsigned range $0 \leq x \leq 2^n - 1$.

Table 5.7: Saturation instructions.

Mnemonic	Brief description	Encoding	Flags
USAT	Unsigned saturation	32-bit	Q
SSAT	Signed saturation	32-bit	Q

For signed n-bit saturation using SSAT, the following possibilities exist.

- If the value to be saturated is less than -2^{n-1}, the returned result is -2^{n-1}.

- Similarly, if the value to be saturated is greater than $2^{n-1} - 1$, the returned result is $2^{n-1} - 1$.

- If none of the above two conditions exist, the result returned is the same as the value provided for saturation.

Likewise for unsigned n-bit saturation using USAT, the following possibilities exist.

- If the value to be saturated is less than 0, the result returned is 0.

- Similarly, if the value to be saturated is greater than $2^n - 1$, the result returned is $2^n - 1$.

- If none of the above two conditions is satisfied, the result returned is the same as the value provided for saturation.

If the result returned by the instruction is different from the value provided for saturation, we term this as saturation has occurred. If saturation occurs, the instruction sets the 'Q' flag to 1 in the APSR. Otherwise, it leaves the Q flag unchanged. These instructions do not affect the condition code flags. To clear the Q flag to 0, you must use the MSR instruction. To read the state of the Q flag, the MRS instruction can be used. The use of MSR and MRS instructions is explained in Chapter 6.

Example 5.25 (Signed and unsigned saturation instructions.).

```
SSAT R1, #16, R0          ; Saturating a 32-bit signed
                          ; value to 16-bit signed value
USAT R3, #16, R2          ; Saturating a 32-bit unsigned
                          ; value to 16-bit unsigned value
SSAT R5, #16, R5, LSL #8  ; Logical shift left value in R5
                          ; by 8, bits then saturate it as
                          ; signed 16-bit value and write
                          ; it back to R5
USAT R4, #8, R5, ASR #8   ; Saturate value in R5 as an
                          ; unsigned 8 bit value and write
                          ; it to R4.
```

The first two instructions in Example 5.25 can be used to saturate a signed and unsigned 32-bit value, respectively, to a 16-bit signed and unsigned value. Tables 5.8 and 5.9 show the result of execution of signed saturate and unsigned saturate instructions for different given values.

Table 5.8: Signed saturation instruction execution with 16-bit saturated result.

Input value	Output value	Q Flag update
0x00012000	0x00007FFF	Flag set
0xFF3F0000	0xFFFF8000	Flag set
0x00008000	0x00007FFF	Flag set
0xFFFF8000	0xFFFF8000	Unaffected
0x00007FFF	0x00007FFF	Unaffected
0xFFFF7FFF	0xFFFF8000	Flag set

Table 5.9: Unsigned saturation instruction execution with 12-bit saturated value.

Input value	Output value	Q Flag update
0x00012000	0x00000FFF	Flag set
0xFF3F0000	0x00000000	Flag set
0x00000824	0x00000824	Unaffected
0x80000180	0x00000000	Flag set
0x70000180	0x00000FFF	Flag set

Saturation instructions can also be used for data type conversions. For example, they can be used to convert a 32-bit integer value to a 16-bit integer value. However, C compilers might not be able to directly use these instructions, so intrinsic function or assembler functions (or embedded/in-line assembler code) for the data conversion could be required.

5.7   Summary of Key Concepts

Data processing instructions have been covered in this chapter in detail with an emphasis on the instruction syntax and their effective use. In particular, the following key components, from the perspective of Cortex-M assembly programming, have been covered.

- Data processing instructions are used for performing different operations on the data including bit operations, basic math operations, operations performed on selective data bit-fields, data movement, data comparison and testing, data saturation, etc.

- The assembly instruction syntax broadly explains the mandatory and optional fields of an instruction.

- The optional suffix {S}, when specified, results in updating of condition code flags based on the result of execution of an instruction.

- Some of the instructions have an optional condition code suffix, which can be used to execute the instructions conditionally.

- Another optional suffix is the instruction size qualifier, which is specified to enforce either 16-bit or 32-bit size for instruction encoding.

- Since logical shift left (LSL) and arithmetic shift left (ASL) instructions effectively perform the same operation, only LSL instruction is supported.

- There is no rotate left instruction, as its operation can be implemented by using rotate right (ROR) without any extra execution overhead due to the use of barrel shifter.

- The use of barrel shifter allows to optionally shift or rotate the operand 2 by an arbitrary number of bit positions (up to 31/32 bits) in one instruction cycle.

- Basic math instructions supported include, addition, subtraction, multiplication, and division.

- Two different types of multiplication instructions are supported, one produces 32-bit result while the other produces 64-bit result.

- The divide instruction discards the remainder and only provides the quotient as the result of operation.

- MOVW and MOVT instruction pair can be used to construct 32-bit constants.

- Byte reversal instructions are useful for changing the data endianness.

- Bitfield clear and insert instructions can be used for selective data processing in a register.

- Test and compare instructions are used to only update the flags based on the result of operation. These instructions discard the result of operation.

- Saturating instructions are useful to reduce the distortion due to saturation in sampled data acquisition.

- Unipolar samples result in unsigned data, while bipolar samples produce signed data; as a result, both signed and unsigned saturating instructions are supported.

Review Questions

Question 5.1. Under what conditions will the LSL instruction set the Z (zero) flag? Give an example.

Question 5.2. Consider the execution of an addition instruction. Before the execution of this instruction both C and V flags are cleared. If the instruction uses the S suffix, then what values of its operands can result in setting of both C and V flags?

Question 5.3. Why does Thumb2 instruction set architecture not support ASL (arithmetic shift left) and ROL (rotate left) instructions?

Question 5.4. Consider the case where an instruction can be either encoded as 16-bit or 32-bit. In such a case, how can you control the size of instruction encoding?

Question 5.5. How can an instruction update the condition code flags?

Question 5.6. What are different possible formats for flexible second operand of an assembly instruction?

Question 5.7. How does the barrel shifter improve the execution speed?

Question 5.8. How can bit operation instructions be used to clear bits 2 to 6 of a register? In addition, how can bits 3 to 5 be set?

Question 5.9. Explain the use of MOV instruction to construct a 32-bit constant.

Question 5.10. Which assembly instruction is best suited to implement the following expressions?

$$r = y + 64x$$

$$r = z - xy.$$

Question 5.11. How does signed bit field extract (SBFX) instruction differ from the UBFX counterpart?

Question 5.12. Is it possible for TST and TEQ instructions to affect the C flag?

Question 5.13. Illustrate the use of byte reversal instruction to transform from big-endian to little-endian format.

Question 5.14. How does saturating instructions reduce signal distortion?

Exercises

Exercise 5.1. Write an efficient assembly code that can implement the following expression.
$x = z/8 - y.$

Exercise 5.2. Write an assembly program that multiplies two 8-bit unsigned numbers using only shift and addition operations.

Exercise 5.3. Write an assembly code that can perform 64-bit subtraction. Assume that the registers R1, R0 contain first operand while registers R4, R3 hold the second operand that is to be subtracted from the first operand.

Exercise 5.4. You are given a number in say register R0. We are interested in finding out whether the number can be represented as n^{th} power of 2. If the given number is indeed n^{th} power of 2, then find the value of n also.

Exercise 5.5. Consider the addition instruction with flexible second operand Operand2, given below.

```
ADD   Rd , Rn , Rm , LSL #n
```

It is desired to modify this instruction to the following new syntax.

```
ADD   Rd , Rm , LSL #n , Rn
```

This new syntax can be called addition instruction with flexible first operand *Operand1*. What minimum change in the hardware architecture is required for the implementation of this instruction?

Exercise 5.6. Write the equivalent assembly instructions for each of the following C instructions.

```
y = y + x*32;
z = y + x*x;
```

Exercise 5.7. Write an assembly code to evaluate

- The expression $-x^2 + 16x + 9$. Assume register R0 $= x$ and the variable x is a 16-bit variable. What is the least number of instructions required for efficient implementation?

- The expression $(x-y)(x+y)$ and store the result in register R2. Assume register R0 $= x$ and R1 $= y$. The variables x, y are 16-bit each. Maximum credit will be given to the efficient implementation using the least number of instructions.

Exercise 5.8. Write an assembly program to generate the sequence of integers $\{1, 3, 9, 27, 81, 243\}$, iteratively. The sequence can be generated using the relation, $s_t = 3s_{t-1}$, with initial value $s_0 = 1$ and executing the relation for five iterations. The result at each iteration should be stored in the destination register, R3. After five iterations, when the program terminates, the contents of the destination register should look like as shown in Figure 5.12.

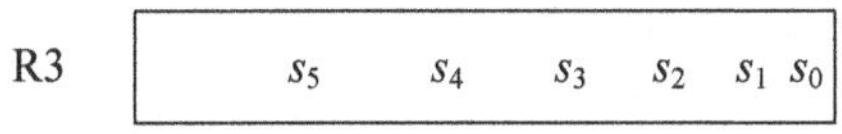

R3 $\boxed{\quad s_5 \quad s_4 \quad s_3 \quad s_2 \quad s_1 \; s_0 \;}$

Figure 5.12: Packed data representation for destination register R3.

Exercise 5.9. Write a sequence of minimum assembly instructions to extract bits 17:11 of register R0 and insert them to register R1 leaving the remaining bits of R1 unchanged.

Exercise 5.10. Give two different assembly procedures to clear register R5 to zero. You may not use any register other than R5.

Exercise 5.11. Write an assembly program that multiplies register R4 by 18 without using MUL instruction.

Chapter 6

Memory Access Instructions

Overview

Memory holds the data besides the code of a program. While carrying out operations given in a program, a processor may need to exchange data with the memory as well as peripherals. This chapter describes load and store instructions as well as their different variants, which are utilized by the Cortex-M processor to exchange data with the memory. Specifically, load and store instructions will be discussed in the context of different memory addressing modes.

6.1 When to Interact with Memory

In embedded systems, data transfers among the memory processor as well as peripherals is an essential task that is performed quite frequently. For instance, existing data can be loaded to the memory during the application initialization, which can then be processed by the user program during the associated task execution. In another scenario, the raw data can be obtained from a peripheral device by the processor and then can be stored to the memory after processing. In Cortex-M architecture, the peripherals are memory mapped. In case of memory mapped peripherals, data read and write operations are performed similar to the memory read/write operations. The only distinguishing aspect is that a different address space, based on memory address map, is used to interact with the peripherals.

Besides processing, sometimes data needs to be transferred from the processor to the memory so that it can be stored temporarily and be retrieved later on when required. In the above-mentioned examples, data transfers are performed between the processor registers and memory (either data or code memory). These data transfers are carried out using memory access instructions, which will be discussed in detail in this chapter.

Table 6.1: Basic memory access instructions.

Mnemonic	Brief description	Encoding	Flags
LDR	Load register using immediate offset	16 or 32 bit	No change
STR	Store register using immediate offset	16 or 32 bit	No change
LDM	Load multiple registers	16 or 32 bit	No change
STM	Store multiple registers	16 or 32 bit	No change
ADR	Generate address relative to PC	16 or 32 bit	No change
POP	Pop registers from stack	16 or 32 bit	No change
PUSH	Push registers onto stack	16 or 32 bit	No change

Unlike CISC architecture, any processor based on RISC architecture allows only a few instructions for accessing the memory. In Cortex-M architecture, memory access instructions include load-store and push-pop. Memory access instructions are crucial as transfer of data occurs frequently between memory and processor in digital systems. Therefore, we can state that as the processor is at the heart of a computer system, so are the memory access instructions in an instruction set.

In addition to data transfers between processor and memory, transfer of data also occurs within the processor. Examples include register to register transfers and the transfer of immediate values to registers. These kinds of transfers are accomplished using MOV instructions, which have already been discussed in Chapter 5.

Load and store instructions were introduced briefly in Chapter 4. Recall that a load register (LDR) instruction transfers data from memory to processor registers, while a store register (STR) instruction transfers data from registers to memory. Variants of these instructions allow data transfers of size byte, halfword, word and double-word. For temporary storage of data (to the memory) that we need to retrieve later, we use a PUSH instruction that stores the data to a specific memory region called a stack. To retrieve data to a processor register, which has already been stored onto the stack, a POP instruction is used. These memory access instructions with their brief descriptions are given in Table 6.1. The LDM and STM instructions are used to load or store multiple data. The ADR instruction is used to generate a relative address with respect to program counter. This chapter explores these memory access instructions, their corresponding addressing modes, as well as any variants of load and store instructions in detail, so as to employ them in our real-life application development.

Conventionally, memory accesses can be performed using different *addressing modes*. In Cortex-M architecture, two modes are associated with the memory access instructions, namely *immediate offset addressing* mode and *register offset addressing* mode. Next we discuss, in detail, these two memory addressing modes for simple load and store operations.

6.2   Load and Store Instructions

By accessing a specific memory location, the microprocessor can read or write information to or from the memory. If the processor is required to access a large memory segment sequentially, then one possible implementation is to maintain a pointer to the starting address of that memory segment. This pointer, which is the starting memory address of the memory segment, is made available in one of the registers of the processor. This address is termed the *base address*. An *offset* from this address can be used to generate a relative address and this mode of memory access is called *offset addressing*. Conventionally, the base address in relative addressing is maintained using a general purpose register. When the based address is derived as the current value of program counter register, then this relative addressing is specifically termed *PC-relative addressing* and will be discussed later in this chapter. In Cortex-M processors, this offset based addressing can be implemented in the following two possible ways.

- Immediate offset addressing

- Register offset addressing

6.2.1 Immediate Offset Addressing

When the offset is an immediate value from the base address stored in a register, the resulting addressing mode is called immediate offset addressing. The general syntax for load and store instructions in immediate addressing mode is given below.

$$\text{LDR}\{type\}\{cond\} \quad Rt, \quad [Rn \ \{, \ \#offset\}]$$
$$\text{STR}\{type\}\{cond\} \quad Rt, \quad [Rn \ \{, \ \#offset\}]$$

The load register instruction is used to fetch 32-bits of information from the memory and put the value in the destination register Rt. The working of store register instruction is opposite that of LDR. In case of STR instruction, Rt is the source operand while a memory location addressed by Rn along with immediate offset value determines the destination address in the memory. In LDR and STR instructions, the register Rn holds the base address. To explicitly identify this special usage of the register, it is surrounded by square brackets. The optional *#offset* field is used to specify an offset from the base address register Rn. Its value is either added to or subtracted from the value of Rn to evaluate the memory address. The *#offset* field can also be zero in which case the register Rn value is used as the memory address. Furthermore, when LDR and STR instructions using immediate offset addressing mode are executed, the contents of the register Rn remain unchanged. The immediate offset can take any value between -255 and +255 for 16-bit instruction encoding, while in the case of 32-bit instruction encoding, the offset can take values between -4095 and 4095.

By default the LDR and STR instructions perform 32-bit data transfers. However, it is possible that an application might require to perform data transfers of sizes other than 32 bits. In Cortex-M processor architecture, different possible data variable sizes are 8, 16, 32 bits or multiples of 32 bits. As a result, an additional attribute of LDR and STR instructions is to handle data transfers of different sizes, between the processor and memory. This is accomplished using the optional *type* specifier with the LDR and STR instructions. Table 6.2 lists different optional *type* specifiers that can be appended to the LDR and STR instructions. Using the optional type specifier

B and SB allows to perform 8-bit unsigned and signed data transfers, respectively, between the register and memory. Similarly, the type specifier H and SH is used to perform unsigned and signed halfword data transfers.

Table 6.2: Optional modifier to specify data type when accessing memory.

{*type*}	Memory Access Type
	When this field is omitted then either signed or unsigned 32-bit word access is performed.
B	Unsigned 8-bit byte. Zero extended to 32 bits for load instructions.
SB	Signed 8-bit byte. Sign extended to 32 bits for load instructions.
H	Unsigned 16-bit halfword. Zero extended to 32 bits for load instructions.
SH	Signed 16-bit halfword. Sign extended to 32 bits for load instructions.

When 8-bit or 16-bit data is transferred from memory to register (which is always 32-bit), we need to pay attention toward extending the data size to 32 bits. The information can be signed or unsigned. Extending 8- and 16-bit data to 32-bit equivalent requires to take the sign into account; otherwise, data will be corrupted. When an 8- or 16-bit unsigned value is loaded to a register, the most significant bits of the register are filled with 0s. This is termed zero extending the data. On the other hand, when we load a register with 8- or 16-bit signed value, the sign bit is extended to fill the most significant bits. This is called sign extension. Transferring 32 bits or its multiples between memory and processor registers, however, does not raise any issues regarding data size extension.

For instance, when an 8-bit number, say -6 (or 0xFA) is loaded to a 32-bit register, the number needs to be sign extended. The sign extension will yield 0xFFFFFFFA in the register, which is still -6 in decimal. If we load an unsigned number, say 6 (or 0x06) to a 32-bit register, then zero extended 32-bit value will result in 0x00000006. The same principle is followed for 16-bit signed and unsigned numbers, when extended to 32-bit equivalent. On the other hand, no extension is required when an 8-bit or 16-bit value is stored from the register to the memory.

When optional *type* field is not specified, 32 bits are exchanged between memory and processor registers. The execution of these instructions can be made conditional with the help of optional *cond* field, which will be discussed in Chapter 7.

Example 6.1 (Examples of immediate offset addressing.). Here, we present examples of load and store instructions in the addressing mode with immediate offset.

```
LDR    R0, [R1]        ; Loads R0 from address in R1
STR    R2, [R9,#0x7]   ; Stores R2 to a memory location
                       ; with address R9+7
```

To gain insight into the working of LDR instruction, let's refer to the illustration given in Figure 6.1. Execution of the instruction LDR R0, [R1] will result in the transfer of the 32-bit value pointed to by R1 in the destination register R0. As a first step, the LDR instruction is fetched from the code memory region addressed by the program counter, PC. This instruction is then decoded by the processor. After decoding the instruction, the execution phase starts. The first sub-step in the execution phase is to use the contents of register R1 as an address, which is sent on the data bus address lines to read the contents from the data memory. Note that here we have assumed that R1 contains a valid address. For the LDR instruction, a read signal is sent on the control bus to the data memory and as a result 32 bits are copied by the memory (from the corresponding address) on the data lines of the data bus. Finally, the data from the data bus is copied to the register R0 and this phase completes the execution of the instruction. During the execution of this instruction, the contents of register R1 are not changed.

It should be noted that since the data memory, code memory as well as the peripherals are the part of the memory map for Cortex-M, hence the address register (e.g., R1 in Figure 6.1) in LDR instruction can be used for pointing toward any object in a valid memory region of the memory map. In Thumb2 instruction set architecture, the following two different variants of this addressing mode are defined, which will be discussed next.

- Pre-indexed immediate offset addressing

- Post-indexed immediate offset addressing

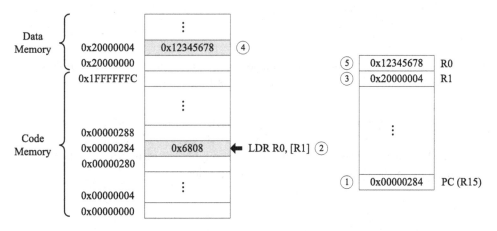

Figure 6.1: Illustration of the steps involved in the execution of LDR instruction. Specifically, LDR, R0, [R1] instruction is executed to load one data word from the memory address provided by register R1 to register R0. Encircled labels 1 to 5 mark the sequence of steps performed during the execution of instruction.

Pre-Indexed Offset Addressing Mode

In pre-indexed offset addressing mode, the register containing the base address is updated first. This updated address is used by the LDR or STR instruction during execution. This updating of the base address register is unlike the previous addressing mode, where the register holding the base address is not changed during instruction execution.

$$\text{LDR}\{type\}\{cond\} \quad Rt, \quad [Rn, \#offset]!$$
$$\text{STR}\{type\}\{cond\} \quad Rt, \quad [Rn, \#offset]!$$

The presence of '!' after the closing or right square bracket ']' indicates that the contents of the register containing the base address are permanently updated and distinguishes the pre-indexed addressing mode from the immediate addressing mode. It should be noted that the #offset is an optional field in case of an immediate offset addressing mode, while it is a mandatory field for the pre-indexed offset addressing mode. The use of pre-indexed offset addressing mode is illustrated in Example 6.2. In Example 6.2, register R10 in the first instruction is updated first by #4 and then, from the updated memory address, the 32-bit data is fetched and copied to R0. The second instruction updates the contents of register R9 as R9 = R9+#0xA and then stores 32-bits of R2 to the memory location pointed to by register R9.

Example 6.2 (Examples for pre-indexed addressing mode.). The example illustrates the syntax for pre-indexed addressing mode for memory access operations.

```
    LDR R8, [R10, #4]!   ; R10 = R10+4 is performed first and
                         ; then load operation is performed
    STR R2, [R9,#0xA]!   ; First, R9 = R9+0xA and then store
                         ; R2 to memory with address R9
```

The use of pre-indexed offset addressing is further illustrated using Example 6.3 to reverse the ordering of elements in a character string.

Example 6.3 (Pre-indexed offset addressing.). Assume src and dst are defined in memory sections with READONLY and READWRITE attributes, respectively. The output of this program results in copying the string from src location in memory to the dst by reversing the string order.

```
; An assembly program that reverses a string
Stack        EQU 0x400

    AREA STACK, NOINIT, READWRITE, ALIGN=3
MyStackMem   SPACE Stack

    AREA    RESET, READONLY
    EXPORT  __Vectors

__Vectors
    DCD  MyStackMem + Stack ; stack pointer for empty stack
    DCD  Reset_Handler      ; reset vector

    AREA    MyData, DATA, READWRITE
dst SPACE   0x10

    AREA    MYCODE, CODE, READONLY
src DCB "HELLO WORLD", 0

    ALIGN
```

```
      ENTRY
      EXPORT Reset_Handler

Reset_Handler
  LDR   R0, =src
  LDR   R2, =dst
  LDRB  R1 , [R0]

loop
  CMP   R1, #0              ; First enumerate the number of
  BEQ   copy                ; elements in the source array
  ADD   R4, #1
  LDRB  R1 , [R0, #1]!
  B     loop

copy
  CMP   R4, #0
  BEQ   End_Prog

copy1
  LDRB  R1, [R0, #-1]!      ; Start with the last element of
  STRB  R1, [R2]            ; source array and store element
  ADD   R2, #1              ; by element
  SUBS  R4, #1
  BNE   copy1

End_Prog
  B     End_Prog
  END
```

Post-Indexed Offset Addressing Mode

If the base address contained in the address register is updated after the load or store operation has been performed, then this addressing mode is called the post-index offset addressing mode. The general instruction syntax for post-indexed addressing mode is given by,

$$\text{LDR}\{type\}\{cond\} \quad Rt, \quad [Rn], \quad \#offset$$
$$\text{STR}\{type\}\{cond\} \quad Rt, \quad [Rn], \quad \#offset$$

Example 6.4 (Examples for post-index addressing mode.). Given below are examples of post-index addressing. In the first instruction, data from R1 is copied to a memory location whose address is specified by R3. Then, R3 is updated as R3=R3+5.

```
STR  R1, [R3], #5   ; STR is performed with R3 and then
                    ; R3 is updated as R3 = R3+5.
LDR  R0, [R6], #4   ; First perform LDR with R6 and then
                    ; update R6 as R6 = R6+4.
```

The next example further illustrates the post-indexed memory addressing using data transfer from code (read only) memory region to data (read write) memory region. The executable for Example 6.5 should be built using a project environment, which includes both the 'startup.s' file (provided in Appendix A.1) and the example assembly source code in a separate source file. The label __main is entry point and is called from within startup.s file.

Example 6.5 (Copying single data at a time with post-indexing.). Assume Src to be an array of numbers in the memory section with READONLY attribute while Dst is an array in the READWRITE memory section. We want to copy the elements of Src to Dst using post-indexing.

```
; Post indexing based data copying from flash to RAM memory.
    THUMB
    AREA DataBlock1, DATA, READONLY
Src   DCD 101,121,232,144,25,116,7,98,110,52

    AREA DataBlock2, DATA, READWRITE
Dst   DCD 0,0,0,0,0,0,0,0,0,0

    AREA CodeBlock, CODE, READONLY
Count EQU  0x0A     ; number of words to copy

    EXPORT __main   ; the first instruction to call
    ENTRY
__main
  LDR R0, =Src      ; R0 = pointer to first source element
  LDR R1, =Dst      ; R1 = pointer to first destination element
  MOV R2, #Count    ; R2 = number of words to copy

Loop1
  LDR R3, [R0], #4  ; load a word from the source
  STR R3, [R1], #4  ; store it to the destination
  SUBS R2, R2, #1   ; decrement the counter by one
  BNE Loop1         ; if there is more to copy

  ALIGN
  END
```

6.2.2 Register Offset Addressing

In this addressing mode, the base address is contained in a register while the offset value is also in a register. This is unlike immediate offset addressing mode, where the offset is an immediate value. In the following, we present the general syntax used for the load and store instructions with register offset addressing mode.

$$\text{LDR}\{type\}\{cond\} \quad Rt, \quad [Rn, Rm \{, LSL \#n\}]$$
$$\text{STR}\{type\}\{cond\} \quad Rt, \quad [Rn, Rm \{, LSL \#n\}]$$

In the above instruction syntax, Rn contains the base address of the memory, to or from which an offset contained in Rm is added or subtracted. The offset can be shifted by up to 3 bits by left shift logical operation. The {, LSL #n} field is optional where the parameter n can take values in the range 0 to 3. Example 6.6 details the use of LDR and STR instructions with register offset addressing mode. The registers Rn and Rm preserve their original values after the instruction execution.

Example 6.6 (Examples of register offset addressing mode.). Load and store operations using register offset addressing are illustrated in this example. The first instruction stores R7 to memory location with address obtained as R4 + 4*R2. The second instruction performs a memory read operation, where register R3 is loaded with the memory contents from address location given by R2 + 2*R0.

```
STR   R7, [R4, R2, LSL #2]    ; R7 is stored to memory
                              ; location [R4 + R2*4]
LDR   R3, [R2, R0, LSL #1]    ; R3 is loaded from [R2 + R0*2]
```

6.2.3   Aligned and Unaligned Memory Accesses

For Cortex-M processor, memory accesses can be of 8, 16, 32 or multiples of 32 bits. The starting memory address of a data element is crucial, which can result in aligned or unaligned access. A 32-bit word object will result in an aligned word access if the first two bits of its starting address are zero or if it is divisible by 4. Otherwise, the 32-bit data access is unaligned. Similarly, if the starting address of a 16-bit halfword object is divisible by 2 or its LSB is zero, we say the 16-bit object is half-word aligned. Otherwise, its access will be unaligned. Memory accesses of size 1 byte are, however always aligned.

Usually aligned memory accesses are faster than the unaligned accesses. But on the other hand, memory is better utilized with unaligned access. As a result, there is a tradeoff between memory size required for certain data and the speed at which this data can be accessed by the processor. In Cortex-M processor, not all of the read and write memory instructions can perform unaligned access operations. Those instructions, which support unaligned access, are given in Table 6.3.

Table 6.3: Instructions which support unaligned memory access.

Instruction	Description
LDR	Load a 32-bit word
LDRH	Load a 16-bit unsigned half-word
LDRSH	Load a 16-bit half-word with sign extension
STR	Store a 32-bit word
STRH	Store a 16-bit half-word

From Table 6.3, we observe that no sign extension is required when storing a 16-bit signed halfword to the memory. Therefore, unlike LDRH and LDRSH, only STRH is available. Table 6.4 tabulates the set of those assembly instructions that are used for data transfers of size 1 byte.

Table 6.4: 8-bit byte instructions.

Instruction	Description
LDRB	Load 8-bit unsigned byte
LDRSB	Load 8-bit signed byte (sign extend bit 7 over bit 8 to 31 positions)
STRB	Store 8-bit byte

Example 6.7 (Examples of aligned and unaligned LDR and STR.). Below the use of load and store instructions for aligned and unaligned memory access is illustrated.

```
LDRB R8, [R5, #12]  ; no problem at all with byte transfers
LDRH R0, [R3, #3]   ; LDRH can be unaligned access
STRB R0, [R1, #1]   ; no problem at all with byte transfers
STRH R0, [R4, #3]   ; STRH can be unaligned access
```

If the address values in registers R3 and R4 (for second and fourth instructions above) are odd numbers, then these instructions result in an unaligned access.

6.2.4   Unprivileged Load and Store Instructions

Recall from Chapter 2 that the Cortex-M processor can access the memory either in privileged or unprivileged mode. Load and store instructions with unprivileged access can be used to mandate the applications running on top of the operating system to have only unprivileged access. Unprivileged access only permits a restricted memory access to the application program.

$$\text{LDR}\{type\}\text{T}\{cond\} \quad Rt, \quad [Rn \ \{, \#offset\}]$$
$$\text{STR}\{type\}\text{T}\{cond\} \quad Rt, \quad [Rn \ \{, \#offset\}]$$

The optional *type* field is the same as defined in Table 6.2. As far as the working of these load and store instructions is concerned, they perform the same function as the memory access instructions with immediate offset. However, the only difference is that these instructions have unprivileged access even when they are part of a software with privileged access rights. When used in unprivileged software, behavior of these instructions is the same as normal memory access instructions with immediate offset. The use of unprivileged instructions is illustrated in Example 6.8.

Example 6.8 (Unprivileged load and store instructions.). Usage of unprivileged LDR and STR instructions is given below.

```
STRBT R4, [R7]       ; Store least significant byte in R4
                     ; to an address in R7, with
                     ; unprivileged access.
LDRHT R2, [R2, #8]   ; Load halfword value from an address
                     ; equal to sum of R2 and 8 into R2,
```

```
                    ; with unprivileged access.
```

6.3 LDR with PC-Relative Addressing Mode

The LDR instruction can also be used to perform memory read and write operations using PC-relative or PC-offset memory addressing mode. In PC-relative addressing mode, the memory address is generated using an offset from the current value of PC. PC-relative memory addresses can be specified using labels, which are also known as PC-relative expressions. A label is used for representing an instruction address or *literal data* in code memory region. Use of literal data will be explained later in this section. General syntax of LDR with PC-relative addressing mode is given below.

$$\text{LDR}\{type\}\{cond\} \quad Rt, \quad label$$

The LDR instruction, using the above-mentioned syntax, is called pseudo assembly instruction. This is due to the fact that it is converted to an equivalent PC-relative addressing based assembly instruction, during the compilation process. When assembled, the assembler calculates the *label* address offset with respect to this instruction location (i.e., current value of PC). It is required that the *label* must be within an address offset of ±4095 from the current instruction address. Otherwise, the assembler generates an error. The assembler may also permit us to directly write the label as [PC, #number]. The expression [PC, #number] generates the label address by adding or subtracting an offset equal to #number to the current value of PC.

Example 6.9 (Illustration of LDR with PC-relative addressing.). Let's assume that var1 is a 32-bit data object declared in an area with READWRITE attribute while var2 with two elements, each of 32 bits, is declared in an area with READONLY attribute, then the following instructions are valid.

```
LDR R0, =var1     ; loads the address of var1 in R0
LDR R7, var2      ; loads R7 with first element of var2
LDR R4, =var2     ; loads R4 with the address of var2
LDR R3, var2 + 4  ; loads R3 with second element of var2
```

In the above listing, the second and fourth instructions are valid if the var2 is within an offset of ±4095 from the address of corresponding instructions. In addition, the first and third instructions are always valid, independent of the memory addresses of var1 and var2 as explained below in this section.

6.3.1 How Does LDR with PC-Relative Addressing Work?

As illustrated in the above example, reading data from memory with READONLY attribute (or ROM) space can be done directly with an instruction such as LDR R7, var2, provided the data (var2 in this case) is within an offset of ±4095 from the address of the corresponding instruction. However, loading registers with the data from memory with READWRITE attribute or I/O will in general violate the offset value restriction and requires two instructions for data loading. This working of LDR with PC-relative addressing is illustrated next using an example.

Example 6.10 (LDR with PC-relative addressing works in two steps.). Here, we show how LDR with PC-relative works when data is accessed from a memory section with READWRITE attribute.

```
LDR R1, =var1    ; R1 is initialized with address of
                 ; var1 using LDR R1, [PC, #offset]
LDR R0, [R1]     ; R0 holds the value pointed to by R1
```

The first instruction uses PC-relative addressing and loads the register R1 with the address of the variable named var1. The 32-bit address, loaded to R1, is physically stored in the code memory region at an address obtained by adding the offset to the current value of PC. This address in R1 is used in the second instruction to read the current value of var1 from the RAM memory. This is depicted in Figure 6.2. Let's assume that var1 is 32 bit, with current value 5 and is located in the RAM region at address 0x20000000. First, we initialize R1 with the address of var1 using LDR R1, =var1. The pseudo instruction LDR R1, =var1 is replaced by the corresponding assembly instruction LDR R1, [PC, #offset]. During this process, the address of var1, (i.e., 0x20000000) will be stored in the form of literal data at an address obtained as PC+offset in the code memory region.

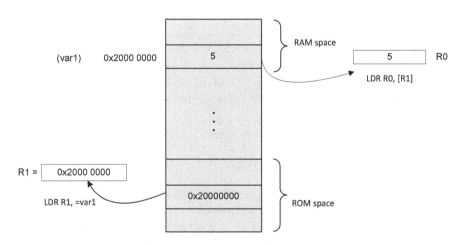

Figure 6.2: The working of LDR with PC-relative addressing.

6.3.2   Generating 32-Bit Constants with LDR Instruction

The LDR Rd,=const is a pseudo assembly instruction that can be used to construct any arbitrary 32-bit constant value in single instruction. This pseudo-instruction is useful for generating 32-bit constants, which are out of range of the MOV and MVN instructions. The LDR pseudo-instruction is the most efficient single instruction to generate an arbitrary 32-bit immediate value. The LDR pseudo instruction can be converted to two different assembly instructions depending on the value of the immediate constant as discussed next.

1. If the immediate value can be constructed using a single MOV or MVN instruction, the assembler is responsible for generating the appropriate instruction.

2. If the immediate value cannot be constructed using single MOV or MVN instruction, the assembler:

   - Places the value in a literal pool (which is a special region in the code memory and is responsible to hold constant values).

   - Generates an LDR instruction with a PC-relative address that reads the constant from the literal pool.

Example 6.11 (LDR pseudo-instruction translation to equivalent appropriate instruction.). The example below illustrates how the LDR pseudo-instruction can be used to construct any arbitrary 32-bit constant. Consider the two possible scenarios. In one case we generate a 16-bit constant while in the other case we generate a 32-bit constant. The corresponding LDR pseudo-instructions are given below.

```
LDR   R8, =0x1234      ; generate 16-bit constant
LDR   R9, =0x12345678  ; generate 32-bit constant
```

The above two instructions are converted to the corresponding appropriate assembly instructions by the complier and are listed below.

```
MOVW R8, #0x1234       ; For 16-bit constant LDR pseudo
                       ; instruction is converted to
                       ; MOVW instruction
LDR   R9, [PC, #offset] ; For 32-bit constant LDR pseudo
                        ; instruction is converted to
                        ; PC relative addressing
                        ; instruction
```

The offset in the second instruction can be any 12-bit signed number. The 32-bit immediate constant is stored in the literal pool at an address PC + offset.

It must be ensured that there is a literal pool within range of the LDR instruction generated by the assembler. It will be explained below how the LTORG assembler directive can be used to create a literal pool within range when the code section becomes too large.

6.3.3   Literal Pools

The assembler uses literal pools to hold certain constant values that are loaded to the registers. The assembler can place a literal pool at the end of each section. The end of a section is defined either by the END directive at the end of the assembly program or it is marked by the AREA directive at the start of a subsequent section. The END directive at the end of an included file does not result in the end of the section automatically.

LTORG Assembler Directive

LTORG is an assembler directive that is used to instruct the assembler for creating a literal pool immediately at the end of the current section. The assembler creates a literal pool at the end of each code section, if required. The end of a code section can be determined due to the use of AREA directive at the start of the following code section, or it can also be determined using the end of the assembly program. Sometimes these default literal pools are out of the range of some of the LDR pseudo-instructions. Using LTORG assembler directive ensures that a literal pool will be created within specific range.

Multiple literal pools might be required for large programs. Placing LTORG assembler directive after the unconditional branches or after the return instructions automatically stops the processor from attempting to execute these constants by misinterpreting them as instructions. The data alignment by the assembler in literal pools follows word-aligns.

Creating a Literal Pool

In large code sections, placement of a default literal pool might be out of range of one or multiple LDR instructions. The offset from the PC to the constant must be:

- In forward direction and should be less than 1 KB, when using 16-bit Thumb LDR instruction,

- Less than 4 KB in either direction when using 32-bit Thumb2 code.

When an LDR Rd,=const pseudo-instruction requires an immediate value to be placed in a literal pool, the assembler can perform any of the following actions based on the situation encountered.

- First of all, the assembler checks the availability as well as addressability of that value in any of the previous literal pools. If this is the case, it addresses the existing constant.

- If the value is not already available, the assembler attempts to place the value in the next literal pool.

The assembler generates an error message, in case the next literal pool is out of range. In such a situation, one must use the LTORG directive to place an additional literal pool in the code. Place the LTORG directive at the appropriate location following the failed LDR pseudo-instruction, and it should be within -4KB to +4KB (32-bit Thumb2) or in the range 0 to +1KB (16-bit Thumb). As mentioned earlier, literal pools should be placed where the processor does not attempt to execute them as instructions. For instance, placing them after unconditional branch instructions or after the return instruction at the end of a subroutine are some of the possible choices. Example 6.12 illustrates how this can be implemented. The implementation in Example 6.12 also requires the use of startup.s file to build the executable.

> **Example 6.12** (Placement of literal pools.). This example shows how literal pools can be placed in a program. Two functions are implemented as part of the program and a large table of size 4200 is created at the end of the area segment. In function 1 (labeled as func1) the 32-bit constant, 0x77556655, is placed in the literal pool and an offset from program counter to its address in the literal pool is created by the tools during compilation. However, the tools can place a literal pool at the end of this code segment, which will be at an offset larger than 4095. This scenario is constructed by commenting the LTORG directive before LargeTable. To resolve the the compiler generated error, you should uncomment the LTORG. Using LTORG directs the tools to create a literal pool at the end of func2 before the LargeTable.

```
    AREA       MYCODE, CODE, READONLY
    EXPORT __main
    ENTRY                   ; Marks first user instruction
__main                      ; to execute
  BL    func1               ; Branch to first subroutine
  BL    func2               ; Branch to second subroutine
stop
  B     stop

func1
  LDR   R0, =42             ; performs MOV R0, #42
  LDR   R1, =0x77556655     ; LDR R1, [PC, #offset], 32-bit
                            ; constant stored in literal pool
                            ; at an address equal to #offset
  LDR   R2, =0x33456788     ; same here
  BX    LR

func2
  LDR   R3, =0x77556655     ; Use PC-relative addressing here
  LDR   R4, =-0x2345        ; Use PC-relative addressing here

  BX    LR
;   LTORG                   ; Uncommenting this line will
                            ; resolve compilation error related
                            ; to literal pool

LargeTable
  SPACE   4200              ; Starting at the current
                            ; location, clears a 4200 byte
                            ; area of memory to zero
  END
```

One can use LDR instruction with those labels that are outside the current section. The assembler uses a relocation directive at the time of assembling the source file. The linker is instructed by the relocation directive to resolve the address at the linking step. The address remains valid wherever the linker can place the section containing the LDR instruction and the associated literal pool within ±4095 offset. In Example

6.12, the instructions listed as part of the comments are the equivalent instructions to the user code and are generated by the assembler. These are the instructions that are executed eventually at the time of program execution.

6.4 The ADR Instruction

The ADR instruction can be used to generate PC-relative addresses. Its general syntax is given below.

$$\text{ADR}\{cond\} \quad Rd, \quad label$$

Here, Rd specifies the destination register while *label* has been defined in the previous section. The ADR adds an immediate value to the PC and writes the result to the destination register. A PC-relative address helps in generating position-independent code as the addresses are relative to PC. The permissible values for the label should be within the range of -4095 to 4095 from the current value of PC. Example 6.13 illustrates the use of ADR instruction.

Example 6.13 (Example of ADR instruction.)**.** The example below shows the usage of ADR.

```
ADR   R2, TxtMsg  ; Write address of a memory location
                  ; named TxtMsg to R2
```

6.4.1 Comparison of LDR and ADR Pseudo-Instructions

From the above discussion, we see that both LDR and ADR instructions can be used to load addresses to registers. The syntax and working of these instructions, however, are different. In case, we are loading a register with a program address value using LDR, the assembler will automatically set the LSB to 1. Let's us first illustrate this concept using Example 6.14.

Example 6.14 (Working of LDR pseudo-instruction for code labels.)**.** Let's assume progAdd is a label in a program and its address is 0x2000. The example below shows a segment of the program to explain the working of LDR.

```
LDR R0, =progAdd  ; R0 is initialized to 0x2001
ADD R4, R5

progAdd                ; assume address here is 0x2000
  MOV R3, R2
```

In this example, we can see the LDR instruction loads 0x2001 to register R0, keeping in view that 0x2000 is the actual address of label progAdd. The assembler will convert the LSB to 1 to indicate that it is Thumb code. When the address is a data address, LSB will not be changed. To further explain this concept, let us look at another illustration in Example 6.15 .

Example 6.15 (Working of LDR pseudo-instruction for data labels.). Let's assume dataAdd is a label in a data segment with address 0x3000. The example below shows a segment of the program to explain the working of LDR.

```
; dataAdd is the label of a 32-bit data object whose address
; is 0x3000
  LDR R4, =dataAdd ; R4 set to 0x3000
  SUB R4, R5
dataAdd                ; address here is assumed to be 0x3000
  DCD 0x0              ; dataAdd contains data
```

With ADR, the address value of a program code is loaded into a register without setting the LSB. This is demonstrated using Example 6.16.

Example 6.16 (Working of ADR pseudo-instruction.). Let's assume progAdd is a label in a program and its address is 0x2000. The example below shows a segment of the program to explain the working of ADR.

```
; progAdd is a label in the program code whose address
; is 0x2000
  ADR R0, progAdd   ; R0 is initialized to 0x2000
  ADD R4, R5
progAdd               ; address here is 0x2000
  MOV R3, R2          ; progAdd is a label in the program code
```

The similarities and key differences in the execution of LDR and ADR instructions are illustrated in Example 6.17.

Example 6.17 (LDR and ADR pseudo-instruction comparison.). Both LDR and ADR instructions can be used to obtain the address. The address obtained can be of a label in the code memory (used to define a constant) or data memory (used to define a global variable). Both LDR and ADR instructions are translated to equivalent PC-relative immediate offset addressing mode. The maximum value of the immediate offset can only be within -4KB to +4KB. If the label, for which the address is being obtained, is in the code memory then it is quite possible to have it within -4KB to +4KB offset from the current instruction. For this scenario using ADR instruction to obtain the address is more memory efficient compared to using LDR instruction. This is explained using the following code listing.

```
      AREA    MYCODE, CODE, READONLY
data6  DCD    0x2233, 0x5566, 0x8899

  LDR  r8, =data6
  ADR  r9, data6
```

The instruction LDR r8, =data6 uses 8 bytes for its implementation. Four bytes are required for the encoded instruction in the code memory and another

4 bytes are required for storing the PC-relative offset to the literal pool where the address is stored. On the other hand, the instruction ADR r9, data6 requires 4 bytes only because it creates a PC-relative offset to the label data6 directly (assuming it is with in -4KB to +4KB offset).

When the offset is large, ADR instruction will result in an error; however, using LDR with =label syntax will always work using the literal pool and be able to generate the address of the label. The label can be anywhere in the code or data memory.

From the examples discussed above, we can conclude that the key differences in the working of LDR and ADR are

- Setting of LSB when loading the label address of an instruction from code region using LDR instruction. LSB is unaffected when ADR is used to generate the address of an instruction using its label.

- As far as the syntax is concerned, we should note that there is no equal sign (=) in the ADR statement.

- The assembler puts the address in program code space from where LDR obtains that immediate data using a PC relative and loads it into a register. On the other hand, ADR tries to generate an address value by adding or subtracting an offset directly to the PC. The impact of this strategy is that the target address label should be in a close range. Using ADR can however generate smaller code sizes (by saving 4 bytes per instruction) compared to LDR.

6.5 Double and Multiple Word Memory Accesses

In the previous sections, we have looked at the memory read and write accesses involving byte, halfword, or word size of data. In case memory data access becomes larger than one word, two solutions can arise. One of them is to use the existing word size memory access instruction repeatedly. But a more efficient solution, which is also supported by the Cortex-M3 instruction set, allows either double or multiple load and store operations to be performed in single memory access instruction. We discuss double and multiple memory access instructions below.

6.5.1 Double Word Load and Store Instructions

These instructions allow memory accesses for two data words. All three addressing modes, immediate offset, pre-index offset, and post-index offset, are permitted for double word memory access instructions. The general syntax for load double word instructions with immediate offset, pre-index offset, post-index offset, and pc-relative offset, respectively, is given below.

$$\begin{array}{llll} \text{LDRD}\{cond\} & Rt1, & Rt2, & [Rn\ \{,\ \#offset\}] \\ \text{LDRD}\{cond\} & Rt1, & Rt2, & [Rn,\ \#offset]! \\ \text{LDRD}\{cond\} & Rt1, & Rt2, & [Rn],\ \#offset \\ \text{LDRD}\{cond\} & Rt1, & Rt2, & label \end{array}$$

The above syntax is equally applicable for store double word instructions by replacing LDRD with STRD. In the above syntax *Rt1* and *Rt2* are two destination registers for load instruction and these registers will become the source registers for store instruction. The use of LDRD and STRD is illustrated in Example 6.18.

Example 6.18 (Examples of double load and store instructions.). The following example shows the working of LDRD and STRD with different addressing modes.

```
LDRD  R1, R0, [R2, #0x20]    ; immediate offset addressing
                             ; mode
STRD  R3, R4, [R7], #-16     ; post indexed offset
                             ; addressing mode
```

The first instruction in Example 6.18 loads register R1 with a word at an address obtained by adding an offset of 32 bytes to the value in register R2, and loads a word to R0 from an offset of 36 bytes. The second instruction stores R3 to an address in R7, and the register R4 to an address equal to R7+4 and then decrements R7 by 16.

6.5.2 Multiple Word Load and Store Instructions

Multiple registers can be loaded or stored using assembly instructions, load multiple registers (LDM) and store multiple registers (STM), respectively. The general syntax for these instructions is given below.

$$\begin{array}{ll} \text{LDM}\{addrmode\}\{cond\} & Rd\{!\}, \quad reglist \\ \text{STM}\{addrmode\}\{cond\} & Rd\{!\}, \quad reglist \end{array}$$

The optional suffix {addrmode} can be either increment address (IA) after each access or decrement address before (DB) each access. The former is the default. The {cond} is an optional condition code, which will be discussed in the next chapter. The register Rd holds the memory base address, from where multiple registers are either loaded from or stored to. The *reglist* is the non-empty list of one or more registers to be loaded or stored. Register ranges can also be described for this operand. In case there are more than one register or register range, the registers should be separated by commas. Table 6.5 lists different combinations that can be obtained from the general syntax introduced above.

In the above instructions, exclamation mark {!} field is optional and specifies whether the address register Rd should be updated after the instruction has been executed. This difference is illustrated in Example 6.19.

Table 6.5: Different combinations of LDM and STM.

{Instruction}	Description
LDMIA Rd!, reglist	Read multiple words from memory specified by Rd
STMIA Rd!, reglist	Store multiple words to memory location specified by Rd
LDMIA.W Rd!, reglist	Read multiple words from memory location specified by Rd
LDMDB.W Rd!, reglist	Read multiple words from memory location specified by Rd
STMIA.W Rd!, reglist	Write multiple words to memory location specified by Rd
STMDB.W Rd!, reglist	Write multiple words to memory location specified by Rd

IA means address increment after while DB will decrement address before (DB) each transfer.

Example 6.19 (Significance of {!} for LDM and STM instructions.). The following example shows the working of STMIA with and without {!}. Let's assume that the register R8 equals 0x7000 as the base address.

```
STMIA.W R8!, {R0-R3} ; R8 changed to 0x7010 after store
                     ; (increment by 4 words)
STMIA.W R8 , {R0-R3} ; R8 unchanged after store
```

In the above examples, R0 is the first register to be copied to the memory, then R1 and in the end R3 is stored in the memory. The LDM instructions work in the opposite manner as compared to their STM counterparts. We may also encounter instructions such as LDMFD and STMFD. The LDM and LDMFD instructions are synonyms for LDMIA while STMFD is a synonym for STMDB.

For LDM, LDMIA, LDMFD, STM, and STMIA, the accesses occur in order of increasing register numbers, with the lowest numbered register using the lowest memory address and the highest numbered register using the highest memory address. For LDMDB, LDMEA, STMDB, and STMFD, the accesses occur in order of decreasing register numbers, with the highest numbered register using the highest memory address and the lowest number register using the lowest memory address. Therefore, the order in which we write the registers does not matter. This fact is illustrated in Example 6.20.

Example 6.20 (Does the order of registers in reglist matter?). The following two instructions produce the same result, showing that the order of registers in reglist does not matter.

```
STMIA R8, {R3,R6,R7} ; Three registers stored to memory
STMIA R8, {R7,R3,R6} ; The order of storing registers does
                     ; not change by shuffling the register
                     ; list
```

Memory to memory data transfers can be performed efficiently using LDM and STM instructions. Example 6.21 illustrates how data from one memory segment to another memory segment can be transferred using LDM and STM with post-indexed addressing.

Example 6.21 (Copying multiple data using LDM and STM with post-indexing.). Assume that src is an array with 10 elements in a memory with READ-ONLY attribute while dst is a label in a READWRITE memory section. We wish to transfer the elements of src to dst using LDM and STM instructions.

```
        AREA DataBlock1, DATA, READONLY
src    DCD 1,2,3,4,5,6,7,8,9,1,2,3,4,5,6,7,8,9

    AREA DataBlock2, DATA, READWRITE
dst    DCD 0,0,0,0,0,0,0,0,0,0,0,0,0,0,0,0

    AREA CodeBlock, CODE, READONLY
counter EQU 18   ; counter is equal to number of words in dst

    EXPORT Start
    ENTRY                    ; Marks first user instruction to execute
Start
  LDR   R0, =src           ; R0 = pointer to source block
  LDR   R1, =dst           ; R1 = pointer to destination block
  MOV   R2, #counter       ; R2 = number of words to copy
  MOVS  R3,R2, LSR #3       ; R3 = number of eight word multiples to
      copy to dst
  BEQ   stop               ; Stop if less than eight words
  STMFD SP!, {R4-R11}      ; Save current values of these registers

copy
  LDMIA R0!, {R4-R11}      ; Load 8 words from the src
  STMIA R1!, {R4-R11}      ; Store them at the destination
  SUBS  R3, R3, #1         ; Decrement the counter
  BNE   copy               ; more to copy?
  LDMFD SP!, {R4-R11}      ; Restore the original values

stop
  B stop
  END
```

6.6 Stack Memory Access with PUSH and POP

The stack is part of main memory and is used to store data objects in last-in-first-out buffering format. In ARM Cortex-M processor, the stack always operates on 32-bit data. The stack pointer contains an address that points to a 32-bit data at the top of the stack. As we push data objects onto the stack, the addresses are decremented. The most recent item known as the "top of the stack" is actually the data object stored at the lowest address. Hence, the stack grows in the downward direction and we say that it employs a full-descending stack management. The assembly instructions to access the stack memory region are PUSH and POP. A PUSH instruction copies one or more data objects from a register or a list of registers onto the stack, whose starting address is determined by SP. The POP instruction works in the opposite manner. The general syntax for these two instructions is given below.

$$\text{PUSH}\{cond\}\quad reglist$$
$$\text{POP}\{cond\}\quad reglist$$

To push a data object on the stack, the stack pointer is first decremented by 4, and then the 32-bit information is stored at the address specified by SP. To pop a data object from the stack, the 32-bit information pointed to by SP is first retrieved, and then the stack pointer is incremented by 4. The SP register points to the last item pushed, which is also the next item to be popped. It is possible to use SP instead of R13, when writing an assembly program. It is important to note that with PUSH and POP instructions SP gets adjusted automatically.

In the case of PUSH, the register or set of registers is source operand while the memory region reserved for the stack in the RAM is the destination. On the other hand, for POP instruction a register or set of registers is the destination and memory is the source. Please note that the contents of the source operands are not changed, while those of the destination operands get updated. Example 6.22 illustrates the use of stack with simple single register PUSH and POP operations.

> **Example 6.22** (Illustration of working of stack PUSH and POP.). To have a better understanding about the working of the stack operations, let's consider that the following sequence of instructions is executed by the processor.
>
> ```
> PUSH {R0}     ; push the 32-bit value of R0 onto the stack
> PUSH {R1}     ; push the 32-bit value of R1 onto the stack
> PUSH {R2}     ; push the 32-bit value of R2 onto the stack
> POP  {R3}     ; retrieve a 32-bit value from the stack and
>               ; store in R3
> POP  {R4}     ; retrieve a 32-bit value from the stack and
>               ; store in R4
> POP  {R5}     ; retrieve a 32-bit value from the stack and
>               ; store in R5
> ```

Here, let us assume that R0 initially contains the value 10, R1 contains 20, and R2 contains 30. The left part of Figure 6.3 shows the initial state of the stack. The stack status is also illustrated for each of these instructions in Figure 6.3. Each stack entry in this figure represents a 32-bit storage space in the RAM. After the execution of three PUSH instructions in Example 6.22, when the first POP instruction is executed the register R3 is loaded with 30. Similarly, the next two POP instructions load registers R4 and R5 with values 20 and 10, respectively.

The instruction PUSH {R0} first decrements SP by 4, and then stores the contents of R0 to a memory location pointed to by the updated value of SP. The right-most part of the figure shows the stack after the push occurs three times. The stack contains the numbers 10, 20, and 30 with 30 on the top of the stack. The instruction POP {R3} first loads the value from the stack memory pointed to by SP into R3, and then increments SP by 4. After the pop occurs three times, the stack returns to its original state and registers R3, R4, and R5 contain 30, 20, and 10, respectively.

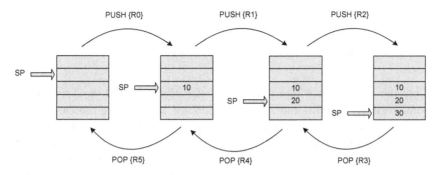

Figure 6.3: Illustration of stack working.

Now consider the case when multiple registers are pushed onto or popped from the stack. In this case, the PUSH instruction will store the lowest numbered register on the lowest memory address and the highest numbered register will be copied to the highest memory address. In other words, when multiple registers are pushed onto the stack using single PUSH instruction, the registers are pushed in descending order of register number. For instance, when registers R1 and R2 are pushed together, R2 will be pushed first followed by R1. In case of POP instruction with multiple registers, the registers are loaded in ascending order of register numbers. For the above-mentioned scenario, register R1 will be popped first followed by register R2. The register ordering, when multiple registers are used with PUSH and POP instructions, is illustrated using Example 6.23.

Example 6.23 (Examples of PUSH and POP instructions.). The following instructions illustrate the use of PUSH and POP operations using multiple registers.

```
PUSH {R7, R8}              ; push R7 and R8 onto stack and
                          ; SP = SP-8
PUSH {R0-R7, R12, R14}    ; push R0-R7, R12, R14 onto stack
                          ; and SP = SP-40
POP  {R0-R7, R12, R14}    ; pop R0-R7, R12, R14 from stack
                          ; and SP = SP+40
POP  {R7, R8}             ; pop R7 and R8 from the stack
                          ; and SP = SP+8
```

Two registers, R7 and R8, are pushed by the first instruction in Example 6.23. When this instruction is executed, register R8 is pushed first followed by register R7. The register ordering in the instruction does not affect the ordering in which the registers are pushed onto the stack, i.e., writing PUSH {R8, R7} does not affect the order in which the registers are pushed onto the stack. The same is true for the POP operation performed with multiple registers. After executing the two PUSH instructions followed by two POP instructions in Example 6.23, the register contents of all the registers used by the four instructions remain intact.

6.6.1   Why Do We Need a Stack?

There are multiple uses of the stack. In the following, we discuss some of the common stack uses.

- Consider a scenario where during the execution of a task it is required to use some registers, say R0 and R1, and we do not want to lose the current contents of these registers. What we normally do is to push the current contents of R0 and R1 onto the stack, use them for processing the task and after completing the task, we can pop them from the stack. The PUSH and POP instructions are used to save register contents to stack memory at the start of a subroutine and then restore the registers from stack at the end of the subroutine, respectively. This scenario is depicted in Figure 6.4. For proper stack usage, each PUSH instruction must have a corresponding POP instruction.

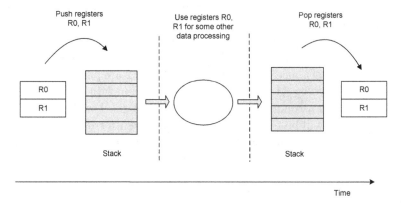

Figure 6.4: Stack is used to store register values temporarily.

- Stack can also be used for parameter passing purposes. Conventionally, few general purpose registers are used for passing the parameters while making a function call. If the available registers for parameter passing are not enough, then stack is used to pass parameters during function calls. The parameters are pushed onto the stack before making a function call and are popped inside the function.

- Stack memory is also used for declaring local variables. When a function is being executed, stack memory is allocated for its local variables, which is deallocated while returning from the function.

- In addition to parameter passing, stack can also be used for storing the return address. In Cortex-M processor, the LR (R14) register is meant to hold the return address. Once a function is called, it is good practice to store the LR register to the stack before starting the function execution. Let us illustrate the reason for doing so. Assume that during the execution of the main function, another function named 'func1' is called. The control will break from this point and starts executing the instructions of 'func1'. But once we are done with 'func1', we have to return to the main function and execute any of the remaining instructions. Now consider the possibility that while executing 'func1', another function named 'func2' needs to be called. At this point, the address in LR register, to return from 'func1', will be overwritten by the new return address corresponding to 'func2'. If the LR register is saved to the stack before calling 'func2' then returning from 'func1' will not be a problem. This scenario is termed nested function calls.

It is important to realize that implementing nested functions has a direct impact on the size of the stack. If the number of nested levels is x and we are only pushing LR register in each nested function call, then the stack space required will equal $4x$. However, if a certain subset of registers is pushed onto the stack, say y number of registers are pushed onto the stack each time a nested function call is made, then stack size requirement for implementing the nested function calls only is given by $4xy$. Based on this fact it is important that the stack size is selected carefully for the worst possible scenario to avoid any stack overflow condition, which is discussed below in this section.

- Another usage of the stack is in case of interrupts. A number of registers are pushed automatically on the stack when an interrupt occurs. Similarly, the pushed registers will be popped automatically when exiting an interrupt handler and the SP is adjusted accordingly. The nested function call scenario also applies to the nested interrupts in the context of stack memory usage. Nested interrupts can occur, due to the possibility of assigning different priority levels to the interrupts from different sources.

6.6.2 Rules for Stack Usage

To simplify the access to stack memory, special instructions, PUSH and POP, are supported by the instruction set architecture. When we use these instructions, the SP is incremented or decremented automatically. However, the user is still required to keep track of a few things for proper stack usage. In that context, the following basic rules should be followed when using a stack.

1. The push and pop operations should always be performed inside the memory region specified for stack.

2. The number of push operations should always have a matching number of pop operations following a specified order.

The violation of the first rule is possible due to *stack overflow* or *stack underflow*, when performing a stack read or write operation. The stack overflow occurs when a push operation is performed, while the stack is already completely filled and no more room is available. On the other hand, the stack underflow occurs when the pop operation is performed while the stack is empty already. An overflow can occur due to a mismatch in the number of push and pop operations leading to the violation of the second rule mentioned above. In addition, an overflow can also be the result of a large number of push operations (performed without any pop operation in between) requiring memory space larger than the size of the stack. This can happen due to nested function calls as well as recursion. On the other hand, a stack underflow usually occurs due to corruption of the stack pointer. The stack overflow as well as underflow may result in data corruption or a bus fault when a read from or write to an invalid address is performed.

6.7 Summary of Key Concepts

In this chapter, we have discussed in detail different addressing modes to perform memory read and write operations. Specifically, the following key concepts were introduced and discussed.

- Cortex-M's LDR instruction allows it to read 32-bit memory data and load it into a register. On the other hand, the STR instruction writes a 32-bit data from a register to memory. Optional modifiers such as B, SB, H and SH are employed to override the default 32-bit value for the transfers between processor and memory. B is to interact with unsigned 8-bit byte and performs zero extension to 32 bits. SB is for signed 8 bit byte and performs sign extension. H is for unsigned 16 bit halfword access while SH stands for signed 16 bit halfword.

- Different addressing schemes to access data are supported by Cortex-M. In immediate addressing, the offset that is added to the register contents to calculate the exact memory location is an immediate value. If the offset is placed in a register, we call it register addressing.

- Cortex-M's aligned and unaligned memory access support. The starting address of the data from where memory data is read can be aligned or unaligned. Alignment depends on the data size that is being transferred. Aligned memory accesses are typically faster while better memory utilization can be achieved with unaligned accesses.

- LDR with PC-relative Addressing Mode. In this mode, the data is accessed from a memory address, which is generated by adding an offset from the current value of PC.

- Literal Pools. The assembler uses literal pools to hold certain constant values. The LTORG assembler is an assembler directive that is used to instruct the assembler for creating a literal pool immediately at the end of current section. We can create multiple literal pools as per the need of a program.

- Double and Multiple Word Memory Accesses supported by Cortex-M. These instructions allow more than 32-bit data to be transferred between the memory and the processor registers. LDRD, STRD, and LDM,STM, are the mnemonics designed for the transfer of double and multiple words respectively.

- Towards end of the chapter, stack memory access is discussed. Stack is a place in the RAM to hold return addresses when functions are invoked, local variables defined in a function and parameters passed to a function. PUSH and POP are the Cortex-M instructions with which we can transfer data between memory and processor.

Review Questions

Question 6.1. How can LDR/STR instructions be used to perform byte and halfword size memory read/write operations?

Question 6.2. What different memory addressing modes are supported by Cortex-M based processor?

Question 6.3. In immediate offset addressing, what is the maximum possible value of immediate offset that can be used in case of (a) 16-bit encoding, (b) 32-bit encoding?

Question 6.4. How does the working of LDRH and LDRHT differ? Under what conditions will the operation performed by these two instructions not differ at all?

Question 6.5. Which memory, code, or data is used to create literal pools?

Question 6.6. What could be done if the literal pool required by a pseudo LDR instruction is out of range?

Question 6.7. We are interested in generating a 32-bit constant. In addition, we want to avoid the use of a literal pool. How can this dual objective be achieved?

Question 6.8. What is the maximum number of read cycles that may be required to access an unaligned word?

Question 6.9. How does LDR and ADR pseudo-instructions compare in terms of their code memory usage?

Question 6.10. When multiple registers are read, using LDM instruction, does the order of registers provided in the register list matter?

Question 6.11. Do we need to initialize the stack memory region with zeros before using it?

Question 6.12. Is it possible to access the stack memory region without using PUSH/POP instructions? If yes, give an example, while in case of no justify the reasons.

Question 6.13. How is memory allocated for local and global variables?

Exercises

Exercise 6.1. Consider a scenario where a large data array of ASCII values is to be accessed. We are interested in reading every 16th element of the array (i.e., element 0, 16, 32, ...). Which memory addressing mode(s) will perform this operation efficiently?

Exercise 6.2. Write an assembly program that has an array of four elements, i.e., {1, 2, 3, 4} in the code area, where each element is halfword size. The program should read this array element by element, take the square of each element, and then store the result in another array of four elements in the data area.

Exercise 6.3. Determine the operation performed by the program given below. What will be the contents of registers R0, R1, and R2 when the program enters infinite loop at the end? Draw a memory map clearly showing start and end addresses of the stack region allocated by this program.

```
        PRESERVE8
        THUMB
Stack   EQU        0x00000100
        AREA       STACK, NOINIT, READWRITE, ALIGN=3
StackMem           SAPCE
        AREA       RESET, DATA, READONLY
        EXPORT    __Vectors
__Vectors
    DCD StackMem + Stack              ; stack pointer value
    DCD Reset_Handler                 ; reset vector
    ALIGN

    AREA MYCODE, CODE, READONLY
    ENTRY
    EXPORT Reset_Handler
Reset_Handler
        MOV    R0 ,#0xF0F0              ; User code starts here
        MOVT   R0 ,#0x5555
        MOV    R2, #31

Count   ANDS   R5,R0,#1
        CBNZ   R5, SKIP
        ADD    R1,#1
SKIP    LSR    R0,#1
        SUBS   R2,#1
        CBZ    R2, STOP
        B      Count
STOP
        B      STOP
        END                           ; End of the program
```

Exercise 6.4. Assume that src1 DCD 0xFF00FF00 and src2 DCD 0x00AA00AA are located consecutively in the READONLY area of memory. Also, assume that src1 starts at address 0x00000008 and SP= 0x20000400. For each of the following instructions, calculate the values of registers involved. Draw the status of the STACK area of memory for those instructions where we use it.

```
    LDR     R0 , =src1
    ADD     R0 , #1
    LDR     R1 , [R0]
    PUSH    {R1}
    ADD     R0 , #4
    LDRSH   R2 , [R0]
    PUSH    {R2}
    MOV     R3 , SP
    ADD     R3 , #3
    LDR     R4 , [R3]
    SUB     R4 , #0x004C004C
    STR     R4 , [R3]
    POP     {R2}
    POP     {R1}
```

Exercise 6.5. Consider the assembly program given in the following listing. What will be the contents of registers R1 and R2 after the execution of the program?

```
; Memory segment for main function
 AREA MyMain, CODE, READONLY
data  DCD    0x87654321
 EXPORT   __main                ; Entry point of user program
                                 ; and is called from startup.s
__main                           ; file
 LDR    R0, =data
 LDRSH R1, [R0, #1]!
 LDRSB R2, [R0, #2]

Stop
 B   Stop
 END
```

Exercise 6.6. Calculate the memory addresses used by the following instructions for R3 = 0x20000000 and R4 = 0x20.

```
 STRH     R9, [R3, R4]
 LDRB     R8, [R3, R4, LSL #3]
 STRB     R6, [R3], R4, ASR #2
```

Exercise 6.7. A matrix A is given below along with its memory storage illustration. This matrix is stored row wise with memory addresses marked. For instance, the first element, 0x11, of the matrix is stored at address 0x20001000.

$$A = \begin{bmatrix} 11 & 12 & 13 & 14 & 15 & 16 \\ 17 & 18 & 19 & 20 & 21 & 22 \\ 23 & 24 & 25 & 26 & 27 & 28 \\ 29 & 30 & 31 & 32 & 33 & 34 \\ 35 & 36 & 37 & 38 & 39 & 40 \\ 41 & 42 & 43 & 44 & 45 & 46 \end{bmatrix}$$

0x20001020	46	45	44	43
⋮		⋮		
0x20001014	34	33	32	31
⋮		⋮		
0x20001008	22	21	20	19
0x20001004	18	17	16	15
0x20001000	14	13	12	11

Write an assembly code to implement the reading of the second column of the matrix and storing it to a destination array after scaling each element by 2.

Exercise 6.8. What will be the contents of the destination registers after the execution of each of the following instructions? You can assume the label 'StackMem' is initialized with a value of 0x20000000.

```
; Allocate space for the stack.
      AREA    STACK, NOINIT, READWRITE, ALIGN=3
StackMem   SPACE    0x00000100

; Initialize the two entries of vector table.
      AREA    RESET, DATA, READONLY
      EXPORT  Vectors
Vectors
      DCD   StackMem + 0x00000100   ; stack pointer value
```

```
        DCD    Reset_Handler                  ; reset vector

; Main user program
    AREA      |.text|, CODE, READONLY, ALIGN=2
    ENTRY                      ; ENTRY marks the starting point
    EXPORT Reset_Handler
Reset_Handler
  MOV    R4, #3
  MVN    R5, #2
  MOVT   R6, #1
  UMULL R0 , R4 , R5 , R6
  PUSH {R4,R0,R6,R5}
  MOV    R1, SP
  ADD    R1, #2
  LDR    R3, [R1]
  ADD    R1, #2
  LDR    R8, [R1]
  ADD    R8 , R8 , R3 , LSL #5
  STR    R8, [R1]
  STR    R3, [R1]
  END                             ;End of the program
```

Exercise 6.9. Write an assembly program, which can perform reversal for each word of a sentence. For the implementation purpose, let us consider the following source string stored in a 'scr' array.

"Welcome to Microprocessor Systems Course."

The output string, after the program execution, when stored to destination array 'dst', should look like as given below.

"emocleW ot rossecorporciM smetsyS esruoC."

The program should terminate after encountering a full-stop.

Chapter 7

Branch and Control Instructions

Overview

The software development process effectively translates the desired task to a corresponding explicit set of instructions that can be executed by the processor. The development of software program would have been straightforward, provided the desired task has been a contiguous sequence of instructions. However, it is highly unlikely to have such software programs appear in real-life applications. Rather a user program of moderate complexity will involve decision making, conditional execution, calling subtasks or functions as well as executing multiple tasks sequentially using time multiplexing. To handle these complexities associated with the real-life software programs, the processor architecture supports branch and control instructions. These instructions allow the software developer to control the flow of program execution at will.

The Cortex-M processor supports different types of branch and control instructions with varying complexity. In this chapter, we will discuss different types of branch and control instructions along with their use to implement higher level conditional constructs. In particular, this chapter will provide detailed discussion related to the following branch and control instructions.

1. Branch instructions (conditional and unconditional)

2. Function calls (conditional and unconditional)

3. Combined compare and conditional branch

4. Conditional execution of instructions (IF-THEN instruction)

5. Table branch

The key difference between a simple branch instruction and a function call is that each function call requires a return address before calling the function. The combined compare and conditional branch instruction effectively implements two operations in one instruction to improve the branch operation efficiency. The IF-THEN (IT) block and table branch instructions allow to map if-else and switch-case types of higher level conditional constructs directly to these assembly instructions. In addition, some special instructions are also discussed in this chapter.

7.1   Introduction to Conditional Execution

Decision making as well as conditional execution are the key attributes of the branch and control instructions. Every 'if' statement, 'switch' statement, or a conditional loop found in high-level languages uses branch and control instructions supported by the processor. Following are some possible scenarios, where branch and control instructions can be used to implement required program flow control.

- The conditional instructions in a high level language make use of branch and control instructions.

- There can be an unconditional branch causing the user program to execute a specific code segment.

- A conditional branch will call a function or execute a specific set of instructions only when a certain condition is fulfilled.

- A function call requiring the processor to execute a specific code implemented by the function.

- A return from a function call will return to the next instruction following the instruction that called the function.

- An interrupt suspends the normal program execution and starts executing an interrupt service routine to make use of branch and control instructions.

- A return from an interrupt service routine will return to the instruction where the program execution was before the interrupt has occurred.

- A hardware or software error or bug will usually cause a bus fault and stops the normal execution.

- The system exceptions as well as software interrupts use branch and control instructions.

In ARM Cortex-M architecture, it is not only the branch instructions that can be executed conditionally but many other data processing as well as data transfer instructions can also have conditional execution. This is in contrast to conventional processor architectures and is one of the distinguishing features of the ARM processor architecture. This special feature is implemented using the IT (IF-THEN) block instruction, which is quite useful to deal with simple conditional execution scenarios. Conventionally, when a branch is performed, the pipeline is first flushed and then filled again to start execution, which degrades the processor execution performance (compared to sequential execution) and is commonly referred to as *branch penalty*.

The IT block instruction avoids the penalties associated with the branch instruction by avoiding a change in the program flow.

In this chapter, we will use the conditional branch instructions to implement if-then, while-loop, as well as for-loop control structures. We will present assembly language implementations for these conditional constructs. In addition, the advantages of using IT block compared to conventional branch instructions will also be elaborated using illustrations.

We have selected C programming as the high level language and many of the conditional executions will be presented simultaneously in assembly as well as C. The introduction to C language provided in this chapter will be utilized to write more complex interfacing programs in subsequent chapters. For a quick review of C language, we refer the reader to Appendix B.

7.2   Branch Instructions

The most basic branch instructions, supported by Cortex-M processor, are branch (B) and branch indirect (BX). The instruction B is also called branch immediate and BX is also termed branch register due to the type of the operand used by these instructions. Another associated pair of instructions include branch with link (BL) and branch indirect with link (BLX). This pair of instructions is also differentiable based on the type of the operand being either immediate (in case of BL instruction) or register (in case of BLX instruction). For BL and BLX instructions, the address of the next instruction, following the branch instruction, is saved to register LR (link register or R14) at the time of execution of the branch instruction, which is used subsequently at the time of returning. The general syntax for these instructions is given below.

$$B\{cond\} \qquad label$$
$$BX\{cond\} \qquad Rm$$
$$BL\{cond\} \qquad label$$
$$BLX\{cond\} \qquad Rm$$

In the above instruction syntax, {*cond*} is one of the condition codes tabulated in Table 7.1. When the optional condition code is omitted, these branch instructions become unconditional branches. On the other hand, use of any of the conditional codes makes these branch instructions become conditional branches. The *label* specifies the address of the instruction to which the branch has to be performed. The compilation tools (assembler or compiler) are responsible to evaluate the required value of the offset from this label to the current value of the program counter (R15 or PC) (i.e., the address of the current B instruction being executed).

For the instructions BX and BLX, operand *Rm* is a register containing the address to which the branch has to be performed. Bit 0 of the address contained in register *Rm* should be set to 1, which indicates that the code execution is in Thumb mode as explained later in this section. However, the address to which the branch is performed to is obtained by resetting the bit 0 to 0. These instructions do not affect any of

Table 7.1: Condition code suffix and their dependency on flags.

Code Suffix	Flags	Description
EQ	Z = 1	Equal (==)
NE	Z = 0	Not equal (! =)
CS or HS	C = 1	Unsigned higher or same ($\geq$)
CC or LO	C = 0	Unsigned lower (<)
MI	N = 1	Negative
PL	N = 0	Positive or zero
VS	V = 1	Overflow occurred
VC	V = 0	Overflow did not occur
HI	C = 1 and Z = 0	Unsigned higher (>)
LS	C = 0 or Z = 1	Unsigned lower or same ($\leq$)
GE	N = V	Signed greater than or equal ($\leq$)
LT	N != V	Signed less than
GT	Z = 0 and N = V	Signed greater than
LE	Z = 1 and N != V	Signed less than or equal
AL	any value	Always. This is the default setting

Table 7.2: Branch instructions.

Mnemonic	Brief description	Encoding	Flags
B	Branch	16 or 32 bit	-
BX	Branch indirect	16 bit	-
BL	Branch with link	32 bit	-
BLX	Branch indirect with link	16 bit	-

the flags in the program status register. Table 7.2 provides the details regarding instruction encoding and flag dependency of these branch instructions.

From Table 7.2 we can observe that BX and BLX instructions are encoded using 16-bit encoding. Conventionally we have observed that 16-bit encoded instructions can only use registers from low register group. However, BX and BLX are an exception to that rule and can use both low as well as high registers as operand despite encoded as 16-bit. For example, BX LR instruction, used frequently for returning from function calls or subroutines, is 16-bit encoded and uses LR (R14) register as an operand. The branch instructions with label operand evaluate an offset with respect to the current PC value. This offset is stored as part of the encoded instruction and as a result defines the limit for the offset range. For instance, B instruction can be encoded as either 16-bit or 32-bit (see Table 7.2) and as result has different offset ranges. The offset ranges for different branch instructions are tabulated in Table 7.3.

In case of BX and BLX instructions, the LSB of the register Rm (containing the address to branch to) determines the next operating state of the processor. Since for the Cortex-M processors, the processor state is always Thumb state, this bit should be set to 1. If it is zero, the program will cause a usage fault exception because it is equivalent to switching the processor state to ARM state. On the other hand, an ARM processor supporting both Thumb and ARM states will have this bit set to either 1 or 0 depending on the current mode of execution.

Table 7.3: Branch ranges.

Instruction	Branch instruction offset range
B	-256 to 254 bytes for 16-bit encoding (outside IT block) -2 KB to +2 KB bytes for 16-bit encoding (inside IT block) -1 MB to +1 MB bytes for 32-bit encoding (outside IT block) -16 MB to +16 MB bytes for 32-bit encoding (inside IT block)
BL	-16 MB to +16 MB bytes
BX	Any arbitrary value in the register
BLX	Any arbitrary value in the register

7.2.1 Implementing Function Calls

When BL and BLX instructions are used for performing a branch operation, the address of the next instruction following the branch instruction is saved to LR register (the link register or R14). This address is used at the time of returning from the called function. The return from the function can be implemented using instruction BX LR, which causes program control to return to the calling process. It is important to realize that we do not use BL or BLX instructions when returning from a function or subroutine. Example 7.1 illustrates the implementation of a function call.

Example 7.1 (Implementing function calls using BL or BLX instructions.). This example illustrates the use of BL and BLX instructions to implement the function call. A function 'addition' is implemented to perform addition of the parameters passed using registers R0 and R1, while the result is returned using R2 register. The function is first called using BL instruction, while it is called a second time using BLX. To return from function 'addition' the instruction BX LR is used.

```
    AREA      MYCODE, CODE, READONLY

    EXPORT __main   ; the first instruction to call
    ENTRY
__main
  MOV     R0, #0x18
  MOV     R1, R0
  BL      addition        ; Branch to function 'addition'
  LDR     R3, =addition
  BLX     R3
  SUB     R2, R1, R2
  B       stop

addition
  ADD     R2, R1, R0      ; R2 = R1 + R0
  BX      LR
stop
  B       stop
  END
```

It is also possible to implement nested function calls using BL and BLX instructions. Here we concentrate only on the use of BL instruction for nested function call implementation and it is equally applicable for the BLX instruction as well. Consider the scenario where a function, say 'func1', is called using BL instruction and we are about to execute the first instruction in that function. At this point, the LR register contains the return address from this function. Now if we call another function, say 'func2', from within 'func1' using BL or BLX instruction, then the LR register will be loaded with the new return address to return from 'func2'. If this is permitted to happen, then the previously stored returned address to return from 'func1' will be overwritten and will be destroyed. As a result, it will not be possible to return from function 'func1'. To avoid this problem, we first need to store the current value of LR register before updating it again by calling 'func2'. A good common practice would be to push the LR register to stack at the entry point of 'func1' and pop it just before returning from function 'func1'. If we pop the return address to LR, then we will require another instruction, BX LR, to return from this function. These two operations (pop to LR and branch using LR) can be implemented using single instruction POP PC. The above-mentioned approach should be practiced irrespective of the current level of nesting. Example 7.2 illustrates the implementation of nested functions.

Example 7.2 (Implementing nested functions using either BL or BLX instructions.). Here we implement nested function calls by using BL instruction. The need for using the stack in implementing nested function calls is also illustrated in this example.

```
    AREA     MYCODE, CODE, READONLY

    EXPORT __main  ; the first instruction to call
    ENTRY

__main
  MOV    R0, #0x18
  MOV    R1, R0
  BL     multiply_add   ; Branch to function 'multiply_add'
  SUB    R2, R1, R2
  B      stop

multiply_add
  PUSH   {LR}            ; First preserve the return address
  BL     addition        ; Call another function
  MUL    R2, R2, R0
  POP    {LR}            ; Retrieve return address from stack
  BX     LR

addition
  PUSH   {LR}
  ADD    R2, R1, R0      ; R2 = R1 + R0
  POP    {PC}
```

```
stop
  B       stop
  END
```

7.2.2 Implementing Branch Operations Indirectly

In Example 7.2 we have illustrated the use of instruction POP PC to implement the return operation from a function call, which conventionally would have been implemented using a BX LR instruction. The use of POP PC shows how some other assembly programming instructions can be used to implement the branch operation indirectly. We can also carry out a branch operation using MOV instructions and LDR instructions. The next example illustrates indirect branch operations using these instructions.

Example 7.3 (Indirect Branch.). This example illustrates the use of MOV, LDR, and POP instructions to perform branch operations indirectly.

```
MOV    R15, R0       ; Branch to an address in R0
LDR    R15, [R0]     ; Branch to an address in memory
                     ; specified by R0
POP    {R15}         ; Do a stack pop operation to change
                     ; the program counter.
```

Some other assembly programming instructions, e.g., ADD and SUB, can also update the PC register. However, it is recommended that these instructions should not be used to create a branch operation.

7.3 Conditional Branch Execution

The Cortex-M processor, like other processors, supports instructions to make decisions. Conditional branch instructions are used to determine whether a branch should be carried out and are implemented with the help of Cortex-M processor application program status register (APSR). There are four bits, namely, overflow (V), carry (C), negative (N), and the zero (Z) flags, which indicate the status of the current operation performed by the microprocessor. Conditional branch instructions can be broadly categorized in three groups: (1) the single-flag branch instructions, (2) the signed branch instructions that are used when operands are treated as signed numbers, and (3) the unsigned branch instructions, which are used when the operands are interpreted as unsigned numbers. As an example, in making the decision whether two numbers such as 0x7FFF FFFF and 0x8000 0000 are equal, the microprocessor needs to take only one flag (the Z flag in this case) into account. As another example, let's take the case of testing whether 0x7FFF FFFF is greater than 0x8000 0000. This requires to decide whether these are signed or unsigned numbers. If signed, then 0x7FFF FFFF is greater than 0x8000 0000 but when these numbers are interpreted as unsigned numbers, 0x8000 0000 is greater than 0x7FFF FFFF. In making such decisions we may need to consider multiple flags.

Irrespective of their category, branch instructions work depending on the current values of the flags but do not affect the flags. We next look at how we can implement various decision making structures in order to control the flow of software programs.

7.3.1   Single-Flag Branch Instructions

From Table 7.1, we can find that BEQ, BNE, BCS, BCC, BMI, BPL, BVS, and BVC are examples of single-flag branch instructions. These single-flag instructions can be used in different scenarios for making decisions. We present couple of examples in this section.

The Z-Flag Usage: Implementing Equality and Inequality

The Thumb2 instruction set uses condition codes EQ (equal) and NE (not equal) with branch instructions to check equality (==) and inequality (!=) of data objects. The EQ code results in true condition when Z flag is 1 while NE code gives true condition for $Z = 0$. Example 7.4 illustrates the use of EQ and NE condition codes.

Example 7.4 (Flow control with EQ and NE condition codes.). Consider a scenario where a variable is initialized with nonzero value and is then decremented by 2 after each iteration of conditional execution. This conditional execution is illustrated by the following C code.

```
x = 10;              // Variable initial value

loop:
  x = x - 2;
  if(x != 0)         // Test the condition
      goto loop;
```

The equivalent assembly program is given below. We assume that R3 register holds the value of x.

```
MOV    R3, #10

loop
  SUBS   R3, R3, #2
  BNE    loop
```

The loop runs until the value in R3 becomes zero, at which point the condition code NE (not equal to zero) controlling the branch becomes false. The variable initial value should be even for proper loop termination.

Let's look at another example where EQ and NE codes are used to test the equality and inequality conditions.

Example 7.5 (Polling a device.). Here, we assume that bit2 of a register associated with a device indicates the status of that device. Let's say a printer indicates its status, as idle or busy using that bit. Based on current status, we decide the assignment of another task to the printer. The following code segment implements sending of more data to the device when bit2 is 1, indicating that the device is idle. This is a commonly used technique for the communications between microprocessor and I/O devices and is called polling. We will discuss polling in detail in subsequent chapters.

```
      LDR    R0, =IO_STATUS_REG

Test_bit
    LDR    R1, [R0]
    TEQ    R1, 0x0004
    BEQ    Send_more_data
    B      Test_bit

Send_more_data
    ...
    B      Test_bit
```

The C Flag Usage: Counting the Parity of a Number

Using the BCC or BCS single-flag branch instructions, we can count the number of 1s in a given number. The next example illustrates this idea.

Example 7.6 (Counting the parity of a number.). Let's assume we have a variable named var whose parity is required to be determined. We can use BCC or BCS to count the number of 1s in var and store the count in a variable named parity. The following code fragment gives an implementation for this usage.

```
    LDR    R0, =var
    LDR    R1, [R0]
    MOV    R5, #32        ; Initialize the counter to 32
loop                      ; assuming a word size variable
    RORS   R1, #1
    BCC    skip
    ADD    R4, #1         ; Increment if bit is 1
    SUBS   R5, #1
    BNE    loop
    B      store
skip
    SUBS   R5, #1
    BNE    loop
```

```
store
  LDR   R2, =parity
  STRB  R4, [R2]
stop
  B     stop
```

7.3.2   Unsigned Conditional Branch Instructions

When we treat numbers as unsigned in a program, we use unsigned conditional branch instructions to control the flow of the program. In order to implement the unsigned conditional branch, the carry and zero flags are used. For some comparisons, either the zero or carry flag is used, while for others their combination is used. The use of carry and zero flag as well as their combinations for different unsigned comparisons is summarized in Table 7.1.

One example illustrating the use of unsigned conditional branch is when we need to differentiate characters from the standard and extended ASCII character sets, which is given next. The unsigned conditional branch instructions are given in Table 7.4.

Example 7.7 (Standard ASCII and extended ASCII character representations.). When characters are represented with standard ASCII code, the most significant bit of the byte is always zero. Using signed and unsigned branch instructions would not make any difference. However, the extended ASCII character codes range from 0x80 to 0xFF and thus should be paired up with unsigned branch instructions as the last bit is turned on.

```
; Determining how many characters in a string named str are
; from standard ASCII and the extended ASCII character sets.
      AREA    MyData, DATA, READWRITE
e_ascii DCB  0
ascii   DCB  0

      AREA    MYCODE, DATA, READONLY
src DCB "HELLO",'Îš',0

    AREA    |.text|, CODE, READONLY
    EXPORT __main  ; the first instruction to call
    ENTRY

__main
  LDR   R0, =src
  LDRSB R1, [R0]     ; loading R1 with LDRB will not be an
      issue
  MOV   R5, # 8

loop
  CMP   R1, #0x80
```

```
  BLO     not_e_ascii    ; BLT will give wrong results
  ADD     R2, #1
  LDRSB   R1, [R0, #1]!
  SUBS    R5, #1
  BEQ     store
  B       loop
not_e_ascii
  ADD     R3, #1
  LDRSB   R1, [R0, #1]!
  SUBS    R5, #1
  BEQ     store
  B       loop
store
  LDR     R4, =e_ascii
  STRB    R2, [R4]
  LDR     R4, =ascii
  STRB    R3, [R4]
stop
  B         stop
  END
```

Table 7.4: Unsigned conditional branch instructions.

{*Instruction*}	Description
BLO label	Branch to label if unsigned less than (same as BCC)
BHS label	Branch to label if unsigned greater than or equal to (same as BCS)
BLS label	Branch to label if unsigned less than or equal to
BHI label	Branch to label if unsigned greater than

7.3.3 Signed Conditional Branch

When in a program we are using signed numbers, we need signed conditional branch instructions to control the flow of the program. In order to implement the signed conditional branch, the negative, overflow, and zero flags are used. The signed conditional branch instructions are given in Table 7.5. An example illustrating the use of signed conditional branch instructions to count the negative numbers in an array of numbers is shown in Example 7.8.

Table 7.5: Signed conditional branch instructions.

{*Instruction*}	Description
BLT label	Branch to label if signed less than
BGE label	Branch to label if signed greater than or equal to
BGT label	Branch to label if signed greater than
BLE label	Branch to label if signed less than or equal to

Example 7.8 (Counting negative numbers in an array.). Here, array is the name of an array, which consists of 5 elements. We are interested in determining the count of negative numbers in this array, whose count will be stored in the count variable.

```
    AREA   MyData, DATA, READWRITE
count  DCB  0

    AREA    MYCODE, DATA, READONLY
array  DCB -65, 127, -128, 13, -1

    AREA    |.text|, CODE, READONLY

    EXPORT __main  ; the first instruction to call
    ENTRY

__main
  LDR    R0, =array
  LDRSB  R1, [R0]
  MOV    R5, #5              ; counter for number of iterations

loop
  CMP    R1, #0x0
  BLT    negative
  LDRSB  R1, [R0, #1]!
  SUBS   R5, #1
  BEQ    store
  B      loop

negative
  ADD    R3, #1             ; increment the counter for
                           ; negative number
  LDRSB  R1, [R0, #1]!      ; Loading data using pre-indexed
                           ; addressing
  SUBS   R5, #1
  BEQ    store
  B      loop

store
  LDR    R4, =count
  STRB   R3, [R4]           ; store the count of -ve numbers

stop
  B      stop
  END
```

Using the conditional and unconditional branch instructions we can implement various high-level decision making structures. The next sections discuss in detail different branching structures.

7.4 Implementing Branching Structures

The execution flow of a program can be altered using branching structures. We discuss the following three basic branching structures.

- If-else branching

- Loop based branching

- Switch-case based branching

In the following we discuss these three basic branching structures along with their different variants. The use of these branching structures is illustrated with examples.

7.4.1 Implementing if

The *if* branching structure tests a condition and based on the result of condition testing, a set of instructions is executed. If the condition is true, the instructions inside the *if* block are executed. The implementation of *if* structure is illustrated using Example 7.9.

Example 7.9 (Converting a lower case letter to its respective upper case.). Let us assume that a variable contains a character from 'a' to 'z'. We are interested in converting this lower case letter to its corresponding upper case form. The C code to perform this task using *if* block is given below.

```
// verify that the variable is in range
 if ('a' <= x) && (x <= 'z'){
    x = x - 32;
 }
```

Next, we list the equivalent assembly program, which performs the same task.

```
 CMP    R0, #'a'       ; Less than 'a' condition is tested
 BLT    stop
 CMP    R0, #'z'       ; Greater than 'z' condition is tested
 BGT    stop
 SUB    R0, #0x20

stop
 B      stop
```

7.4.2 Implementing if-else

In case of *if-else* structure, there is a condition to test, while there are two different instruction blocks. When the condition is true, the set of instructions corresponding to *if* block is executed, otherwise the instructions inside the *else* block are executed.

Omitting the else block reduces the *if-else* structure to *if* block. Example 7.10 illustrates the use of *if-else* block.

Example 7.10 (Testing whether a number is even or odd.). Let's assume a variable that contains an arbitrary integer value. Determine whether the number is an even number or an odd number. In case it is an even number, then write 'E' to another variable, else write 'O' to that variable. The C program that implements this task is provided below.

```
if((x%2)==0)    // if remainder is zero, then even
    y = 'E';
else
    y = 'O';    // otherwise it is odd
```

The assembly equivalent of this C code is given next. It is assumed that variables 'x' and 'y' are in registers R0 and R1, respectively.

```
    TST   R0, #0x1      ; Test if LSB of R0 is 1
    BEQ   even
    MOV   R1, #'O'
    B     stop

even
    MOV   R1, #'E'

stop
    B     stop
```

7.5 Implementing Loops

Branching structures enable a program to take different paths in a program. On the other hand, looping structures allow a program to execute a set of instructions for a number of times. Here we discuss two different looping structures.

7.5.1 The *for* Loop

A *for* loop is used whenever we know in advance the number of times a set of instructions should be executed. Example 7.11 illustrates the implementation of *for* loop.

Example 7.11 (Working of the for loop.). Let's copy an asterisk to each element of an array named Array with 50 elements. The C code to implement this functionality is given below.

```
for(i = 0; i <50; i++)
    Array[i] = '*';
```

The equivalent assembly program is given below.

```
    MOV   R0, #'*'
    LDR   R2, =Array
    MOV   R1, #0

loop
    CMP   R1, #50           ; Check for loop termination
    BGE   stop              ; Terminate the loop if count is 50
    STRB  R0, [R2], #1      ; Store '*' to the current address
    ADD   R1, #1
    B     loop
stop
    B     stop
```

7.5.2 The *while* loop

Consider a scenario where the number of loop iterations is not known in advance. Rather the loop is iterated until a certain condition is fulfilled. For such situations, we use *while* loop. Example 7.12 illustrates the use of *while* loop.

Example 7.12 (Working of the while loop.). Let's count the characters in an array that is terminated by a new line character. The C code implementation is given below.

```
count = 0;
i = 0;

    while(ARRAY[i] != '\n')   // Test the condition for loop
        count++;
```

The equivalent assembly code is given below.

```
    LDR   R2, =ARRAY
    MOV   R1, #0            ; initialize the count to 0
    LDRB  R0, [R2], #1

loop
    CMP   R0, #0x0A         ; check loop termination condition
    BEQ   stop
    ADD   R1, #1
    LDRB  R0, [R2], #1
    B     loop
stop
    B     stop
```

Let's look at another example illustrating the working of *while* loop.

Example 7.13 (Read a string named src one character at a time until null terminated. Store it to another array named dst with its capital 'L's replaced by 'l's.).

```
    AREA   MyData, DATA, READWRITE
dst SPACE  0x10

    AREA   MyCode, DATA, READONLY
src DCB "HELLO WORLD", 0

    AREA   |.text|, CODE, READONLY

    EXPORT __main          ; the first instruction to call
    ENTRY
__main
  LDR  R0, =src
  LDR  R5, =dst
  LDRB R1, [R0]            ; Retrieve first element of source
  MOV  R2, #'l'            ; string

while
  CMP  R1, #0
  BEQ  stop               ; Terminate the loop
  CMP  R1, #'L'           ; Search for 'L'
  BNE  continue
  STRB R2, [R5], #1       ; Store 'l' when 'L' is found
  LDRB R1, [R0, #1]!
  B    while

continue
  STRB R1, [R5], #1       ; Store elements other than 'l'
  LDRB R1, [R0, #1]!
  B while

stop
  B stop
  END
```

7.6 Implementing Switch-Case

The condition testing can be performed for multiple possible values, resulting in multi-way branching. For instance, a variable can be tested for three different possible values and correspondingly one of the three different code segments is executed. This situation can be implemented using *switch-case* conditional structure. Example 7.14 illustrates the use of switch-case conditional construct.

Example 7.14 (Testing whether a number is positive, negative, or zero.). Assume variable x is initialized with some arbitrary value. If x is less than 0, write -1 to another variable, say y. In case its value is 0, write 0 to y. If x is greater than 0, then write 1 in y. The C program implementing this function is given below.

```
switch (x){           // Switch construct with three cases
  case '<0' :
       y = -1;
  case '==0' :
       y = 0;
  case '>0' :          // Note there is no default case
       y = 1;
}
```

One possible equivalent assembly program is given below. Assume variable x is loaded to register R0. In this program, three different conditional branches are used after testing the value of variable x.

```
CMP   R0 , #0
BLT   negative
BEQ   zero
BGT   positive

negative
  MOV  R1 , -1
  B    stop

zero
  MOV  R1 , 0
  B    stop

positive
  MOV  R1 , 1

stop
  B    stop
```

When the switch-case construct has multiple cases, it requires multiple test or compare operations followed by conditional branch instructions. However, this type of implementation will execute slowly if the number of cases is large.

7.7   Combined Compare and Conditional Branch

In Cortex-M3/M4, new assembly instructions to perform comparison with zero and branch conditionally are introduced. Specifically, CBZ (compare and branch if zero) and CBNZ (compare and branch if nonzero) are the two assembly instructions introduced. The general syntax for these two instructions is give by

$$\text{CBZ} \quad Rn, \quad label$$
$$\text{CBNZ} \quad Rn, \quad label$$

The CBZ instruction does not affect the condition flags. Otherwise it is equivalent to two consecutive instructions, CMP (compare with zero) followed by conditional branch BEQ. Similarly, CBNZ does not change condition flags but is otherwise equivalent to CMP with zero followed by BNE. In other words, APSR value is not affected by the CBZ and CBNZ instructions. The compare and branch instructions only support forward branches.

Example 7.15 (Usage of CBZ.). Let's assume that a variable i is initialized to 6. We keep on calling a function named funcA until i becomes zero.

```
// Iterative function calling
i=6;
while (i != 0 ){
funcA(); // Call a function
i--;
}
```

This can be compiled to get the following equivalent assembly program.

```
  MOV R0, #6          ; Set loop counter
loop1
  CBZ R0, loop1exit   ; If loop counter = 0 then exit loop
  BL  funcA           ; Call a function
  SUB R0, #1          ; Loop counter decrement
  B   loop1           ; Next loop iteration
loop1exit
```

The usage of CBNZ is similar to CBZ, apart from the fact that the branch is taken if the Z flag is not set (result is not zero).

Example 7.16 (Usage of CBNZ.). Let's assume that an email id is stored in an array named emailid. We are interested in determining whether it contains the special character '@'.

```
    status = strchr(emailid, '@');

    //if emailid does not contain @, then status is 0.
    if (status == 0){
        generate_error_message();
    }
```

This can be compiled into the following.

```
    ...
    CBNZ R0, email_id_ok    ; R0 contains a character from
```

```
                                    ; the emailid string.
      BNE generate_error_message
email_id_ok
   ...
```

Table 7.6 provides the details regarding instruction encoding and flag dependency of these instructions.

Table 7.6: Combined compare and branch instructions.

Mnemonic	Brief description	Encoding	Flags
CBZ	Compare and branch if zero	16 bit	-
CBNZ	Compare and branch if not zero	16 bit	-

7.8 Recursive Functions

A recursive function is a function that calls itself. Recursive functions are an integral part of high-level languages such as C, C++, but the use of stack (by the tools) to implement recursive functions is hidden from the programmer. In case of assembly programming, the programmer has to actually implement those stack operations itself, which provides an opportunity to understand why recursion uses more stack compared to calling a function multiple times iteratively.

Example 7.17 (Write a program that will compute factorial of a positive integer N.). This example illustrates the difference between recursion implemented in C language and assembly language. First of all, we present the C code.

```
unsigned long Fact(unsigned long n) {
  if(n<=1) return 1;
  return n*fact(n-1);
}
```

For the recursive implementation above, an equivalent assembly program is given below.

```
  AREA Mydata, DATA, READWRITE
Result  DCD    0

  ; Separate memory segment for function call
  AREA MyFunction, CODE, READONLY

Func1
  PUSH {LR}                       ; Assume sufficient stack
       size
  CMP R0, #1                      ; is available
  BEQ Done
  PUSH {R0}
```

```
    SUB RO, RO, #1
    BL Func1
    POP {R1}
    MUL RO, RO, R1
Done
  POP {LR}
  BX LR


  ; New memory segment for main function
  AREA MyMain, CODE, READONLY
  EXPORT  __main               ; Entry point of user
     program
                               ; and is called from startup
__main                         ; file
  MOV RO, #4
  BL Func1
  LDR R1,=Result
  STR RO, [R1]
Stop
  B  Stop
  END
```

7.9   Passing Parameters to Functions

Unlike high-level languages, assembly language functions do not have associated parameter lists due to which it is up to the programmer to devise strategies for passing parameters to functions. In programming languages, there are two well-known mechanisms for parameter passing to functions, namely (a) call by value and (b) call by reference. In the first strategy, the actual data values are passed to the called function, while in the latter method the address of the data is passed to the function. The latter method is particularly useful when dealing with data arrays. The next two examples illustrate the working of these methods.

Example 7.18 (Illustration of passing parameters by using call by value method.). For R4 = M + N - R3, where M and N are 32-bit data. We implement addition as a function and pass the operands through the call by value method.

```
    AREA   DATA1, READONLY
M   DCD    5
N   DCD    3

    AREA      MYCODE, CODE, READONLY
    EXPORT __main  ; the first instruction to call
    ENTRY

__main
  LDR    R6, =M
  LDR    R7, =N
```

```
     LDR    R0, [R6]
     LDR    R1, [R7]
     BL     addition          ; Branch to function 'addition'
         and
                              ; pass R0 and R1 as parameters
     MOV    R3, #0x3
     SUB    R4, R1, R3
     B      stop
addition
     ADD    R1, R1, R0        ; R1 = R1 + R0
     BX     LR
stop
     B      stop
     END
```

Next, we illustrate call by reference implementation for the above example.

Example 7.19 (Illustration of parameter passing using the call by reference method.). For R4 = M + N - R3, where M and N are 32-bit data. We implement addition as a function and pass it the operands through the call by reference method.

```
     AREA    DATA1, READONLY
M    DCD     5
N    DCD     3

     AREA     MYCODE, CODE, READONLY
     EXPORT __main   ; the first instruction to call
     ENTRY

__main
     LDR    R6, =M
     LDR    R7, =N
     BL     addition          ; Branch to function 'addition'
     MOV    R3, #0x3
     SUB    R4, R1, R3
     B      stop
addition
     LDR    R0, [R6]
     LDR    R1, [R7]
     ADD    R1, R1, R0        ; R1 = R1 + R0
     BX     LR
stop
     B      stop
     END
```

7.10   If-Then Conditional Instruction Block

One of the fundamental features of Cortex-M architecture is the possibility of executing instructions conditionally. Conditional branch or jump instructions are commonly found in other processor architectures, but ARM Cortex-M architecture extends the use of condition codes with many other instructions as well. For instance, ADD and SUB instructions can be executed conditionally by using the optional condition code. Conditional execution of an instruction is implemented by Thumb2 ISA using If-Then (IT) block. The IT block instruction is very useful for handling small conditional codes. It avoids branch penalties because there is no change to program flow. The IT instructions allow up to four succeeding instructions to be conditionally executed and they collectively form an IT block. Conditional instructions, except for conditional branches, must be inside an IT instruction block.

$$\text{IT}\{x\{y\{z\}\}\} \quad cond$$

In the above instruction syntax

- x, y, and z are the optional conditional execution switches for second, third, and fourth instructions in the IT block.

- *cond* is the base condition for the IT instruction block. The first instruction following IT instruction is executed if the *cond* is true.

Each of the optional condition switches x, y, and z can be either T (THEN) or E (ELSE). When condition switch T is used, then *cond* is applied to the corresponding instruction, whereas use of condition switch E applies the inverse *cond* to the corresponding instruction. The *cond* operand in IT instruction uses the same condition codes as conditional branch. It is possible to use AL (the always condition) for *cond* in an IT instruction. If this is done, all of the instructions in the IT block must be unconditional, and each of x, y, and z must be T or omitted but not E.

In IT instruction blocks, the first instruction must be the IT instruction itself, detailing the choice of condition switches along with the condition it checks. The first conditional instruction after the IT instruction must be TRUE. The condition codes for second through fourth instructions following the IT instruction can be either TRUE or FALSE. The use of condition code with an instruction is specified with a two-letter suffix, such as EQ or CC, appended to the instruction mnemonic. The conditional instructions following the IT instruction in the IT block must specify either the condition code *cond* or its logical inverse as part of their instruction syntax. Table 7.7 provides details regarding instruction encoding and flag dependency of IT instruction.

Table 7.7: IT conditional instruction.

Mnemonic	Brief description	Encoding	Flags
IT	If-then conditional instruction	16 bit	-

The list of the suffixes that can be added to instructions to make them conditional instructions is given in Table 7.1. The condition code suffix appended to an instruction inside an IT block requires the processor to test the condition code based on the current status of the flags. If the condition test of a conditional instruction in IT block fails, the instruction

- Does not execute

- Does not write any value to its destination register

- Does not affect any of the flags

- Does not generate any exception

This feature often eliminates the need to branch, avoiding pipeline stalls and results in an improved execution performance. It can also increase code density. By default the data processing instructions do not affect the condition code flags but can be made to by suffixing S. On the other hand, the comparison instructions, e.g., CMP, TST, do affect the flags implicitly.

The execution of an instruction in IT block is conditionally dependent on the status of the condition flags, which are updated by a priorly executed instruction. The prior instruction that has updated the flags can either be the immediately preceding instruction that updated the flags, or there may be an arbitrary number of intermediate instructions that did not update the flags. Below is an example illustrating the use of IT block instruction.

Example 7.20 (Illustration of IT block instruction.). Different IT block constructions are illustrated in this example. To construct an IF block only without and ELSE, the optional execution conditions, $x, y,$ and z must be T as illustrated in the listing below.

```
CMP       R0, R1        ; Compare parameters loaded to R0
                        ; with R1
ITTT      LT            ; If R0 < R1
ADDLT     R2, R0, R1    ; add the two operands
MULLT     R3, R2, R0    ; multiply the sum with first
                        ; operand in R0
ASRLT     R3, R3, #1    ; divide the result by 2
```

The next illustration details the construction of both IF and ELSE blocks using IT block instructions. The conditional instructions that belong to IF should have the T execution condition (i.e., same as the one used by the IT instruction itself), while the ELSE should use the E execution condition (i.e., condition code which is inverse of the condition code used by the IT instruction).

```
CMP       R0, R1        ; Compare parameters loaded to R0
                        ; with R1
ITTEE     LT            ; If R0 < R1
ADDLT     R2, R0, R1    ; part of IF
MULLT     R3, R2, R0    ; part of IF
SUBGE     R2, R0, R1    ; part of ELSE
ASRGE     R3, R3, #1    ; part of ELSE
```

From Example 7.20, it can be observed that the IT block can have fewer than four conditionally executed instructions. Minimum is one. It should be ensured that the number of T and E occurrences in the IT instruction matches the number of conditionally executed instructions after the IT instruction.

It is important to note that data processing instructions, encoded using 16-bit encoding, do not update APSR when they are used inside an IT instruction block. If the suffix S is added to the conditionally executed instructions inside the IT block, then 32-bit encoding of the instruction would be used by the assembler. The next example illustrates further the use of IT block for translating the if-else construct from C language to the equivalent assembly program.

Example 7.21 (Implementing if-else using IT block.). This example explains the use of IT instruction for implementing the if-else conditional block. For a given signed integer we are interested in finding its magnitude. The following C program implements it using if-else conditional construct.

```
if (x < 0) {
    y = -x; }
else {
    y = x; }
```

This can be implemented using IT block efficiently as given in the listing below. Assuming register R0 hold the variable x and the absolute value in variable y is returned using register R3.

```
CMP    R0, #0       ; Compare R0 with 0
ITE    LT           ; Condition to check if R0 is negative,
RSBLT  R3, R0, #0   ; take absolute value of negative number
MOVGE  R3, R0       ; do nothing and return original value
```

In the above illustration we have initialized the register R0 with a negative value and its absolute value is obtained in register R3.

In case an exception or interrupt has happened, while the processor was executing an IT instruction block, the execution status for that IT block is stored to the stacked xPSR register (specifically in the IT/Interrupt-Continuable Instruction [ICI] bit field). Once the execution of the interrupt service routine has been completed, the IT block execution is resumed and the remaining instructions of the block can continue the execution. In case of multi-cycle instructions (for example, multiple load and store) inside an IT block, if an exception occurs during its execution, the whole instruction is abandoned and restarted after the interrupt process is completed. The following restrictions apply to the IT block instruction execution.

- A branch or any instruction that modifies the PC must either be outside an IT block or must be the last instruction inside the IT block.

- Do not branch to any instruction inside an IT block, except when returning from an exception handler.

- All conditional instructions except conditional branches must be inside an IT block. Conditional branch instructions can be either outside or inside an IT block. The conditional branch instructions have a larger branch range when they are used inside an IT block.

- Each instruction inside the IT block must specify a condition code suffix that is either the same or logical inverse of the condition code used by the IT instruction.

- The first conditional instruction inside the IT block should always use the condition code the same as the one used by the IT instruction itself.

Finally, we illustrate the use of IT block for implementing nested if-else construct. For this purpose, the conditional instructions inside IT block use the suffix S to update the flags inside the IT block and let the following conditional instruction execute based on the updated values of the status flags. Example 7.22 illustrates the nested if-else implementation using IT block.

Example 7.22 (Implementing if-else using IT block.).

This example explains the use of IT instruction for implementing the nested if-else conditional block. First we give the C program that implements the nested if-else construct.

```
if(R0 > R1) {
    if(R1 > R2) {
        if(R2 > R3) {
            R4 = 0x123;
        }
    }
}
```

This can be implemented using IT block efficiently as given in the listing below.

```
CMP      R0, R1       ; Compare R0 with R1
ITTT     GT           ; Condition to check if R0 > R1
SUBSGT   R5, R1, R2   ; check if R1 > R2
SUBSGT   R5, R2, R3   ; check if R2 > R3
MOVGT    R4, #0x123
```

7.10.1   Illustrating the Advantages of IT Instruction

In this subsection the advantages of using IT block will be highlighted. For that purpose, an example based approach will be followed. The example discussed here will illustrate two assembly programing implementations of the Greatest Common Divisor (GCD) algorithm (based on Euclid) [ARM(2014)]. The example will illustrate that how the usage of conditional execution based on IT block can provide both the code density as well as execution speed improvement. We first see the working of the algorithm using a C programming based implementation given below.

```
// GCD implementation based on Euclid algorithm
int gcd(int a, int b)
{
    while (a != b)
    {
        if (a > b)
            a = a - b;
        else
            b = b - a;
    }
    return a;
}
```

Listing 7.1: The GCD algorithm implementation in C

Next in Example 7.23 two different implementations of the GCD algorithm in assembly language programming are discussed.

Example 7.23 (Implementing if-else using IT block.). This example implements the GCD algorithm. The implementation below is based on branch instructions and does not use any conditional instructions.

```
gcd
    CMP    R0, R1
    BEQ    stop
    BLT    less
    SUBS   R0, R0, R1   ; could be SUB R0, R0, R1 for ARM
    B      gcd
less
    SUBS   R1, R1, R0   ; could be SUB R1, R1, R0 for ARM
    B      gcd
stop
```

This can be implemented using IT block efficiently as given in the following listing.

```
gcd
    CMP     R0, R1
    ITE     GT
    SUBGT   R0, R0, R1
    SUBLE   R1, R1, R0
    BNE     gcd
```

Now we analyze the code and execution efficiency of these two implementations. As can be observed from Example 7.23, seven instructions are required for implementing the GCD algorithm using conditional branch instructions without using IT block. The memory space usage for this implementation requires 14 bytes. On the other hand, we only require five instructions to implement the GCD algorithm when IT block is used and the corresponding memory usage is 10 bytes.

It is important to note that branch instructions for Cortex-M3/M4 based processors can take from 2 to 4 cycles for their execution, depending on the alignment and width of the target instruction. For the micro-controller selected in this text, it takes 3 cycles for branch execution. On the other hand, many of the data processing instructions require one cycle for their execution. The extra cycles required for the branch operation are attributed to the fact that branch instructions require refilling of the pipeline. In case of conditional branch, if the condition is false then branch does not occur and in this case the instruction only takes one cycle for its execution.

The execution performance for the GCD algorithm is superior for the IT block based implementation compared to the conditional branch based implementation. The results for execution performance are tabulated in Table 7.8.

Table 7.8: GCD algorithm execution performance for two different implementations.

	Conditional Branch	IT block
R0 = 1, R1 = 2	13 cycles	12 cycles
R0 = 10, R1 = 24	45 cycles	40 cycles

7.11 Table Branch Instructions

The table branch instruction allows to implement the switch-case conditional construct using a branch table. Instruction syntax for the table branch byte (TBB) as well as table branch halfword (TBH) is illustrated below.

$$TBB \quad [Rn, Rm]$$
$$TBH \quad [Rn, Rm, LSL\#1]$$

Rn   This parameter specifies the register that contains the address of the table of branch lengths. If Rn is specified as PC, then the address of the table is the address of the byte immediately following the TBB or TBH instruction.

Rm   This parameter specifies the index register, containing an index to the table. For halfword tables, LSL #1 doubles the value in Rm to form the right offset, with respect to the table base address.

For TBB, the branch offset is twice the unsigned value of the byte returned from the table. For TBH, the branch offset is twice the unsigned value of the halfword returned from the table. These instructions cause a PC-relative forward branch using a table of single byte offsets for TBB, or halfword offsets for TBH. The branch occurs to the address at that offset from the address of the byte immediately after the TBB or TBH instruction.

These instructions do not change the flags. When any of these instructions is used inside an IT block, it must be the last instruction of the IT block. Table 7.9 provides details regarding instruction encoding and flag dependency of table branch instructions.

Table 7.9: Table branch instructions.

Mnemonic	Brief description	Encoding	Flags
TBB	Table branch byte instruction	32 bit	-
TBH	Table branch half-word instruction	32 bit	-

Example 7.24 (Illustration of table branch instructions.).

```
ADR.W  R0,  Table_Byte
TBB    [R0, R1]                ; R1 is index, R0 is the base
       address

Case1                          ; an instruction sequence follows
Case2                          ; an instruction sequence follows
Case3                          ; an instruction sequence follows

Table_Byte
  DCB    0                     ; Case1 offset calculation
  DCB    ((Case2-Case1)/2)     ; Case2 offset calculation
  DCB    ((Case3-Case1)/2)     ; Case3 offset calculation
```

Next table branch halfword instruction is illustrated.

Example 7.25 (Illustration of table branch instructions.).

```
TBH    [PC, R1, LSL #1]        ; R1 is the index, PC is used
                               ; as base of branch table
Table_Halfword
  DCI  ((CaseA - Table_Halfword)/2)   ; CaseA offset
       calculation
  DCI  ((CaseB - Table_Halfword)/2)   ; CaseB offset
       calculation
  DCI  ((CaseC - Table_Halfword)/2)   ; CaseC offset
       calculation

CaseA                          ; an instruction sequence follows

CaseB                          ; an instruction sequence follows

CaseC                          ; an instruction sequence follows
```

7.12   Special Instructions

There are some special assembly instructions that are used for accessing special registers.

7.12.1 MSR and MRS Instructions

The access to special resisters is not permitted using either MOV or LDR/STR instructions. Rather, special assembly instructions are provided for this purpose. The first instruction, MRS, is responsible for moving the contents from a special register to one of the processor general purpose registers. The second instruction, MSR, performs a move from the general purpose register to a special register. These two instructions provide access to the special registers in the Cortex-M3/M4. Below is the syntax for these instructions.

$$\text{MRS}\{cond\} \quad Rd \qquad spec\_reg$$
$$\text{MSR}\{cond\} \quad spec\_reg \quad Rn$$

where *spec_reg* could be one of the special registers listed in Table 7.10. The MRS and MSR instructions can be used in privileged mode only with the exception of APSR special register which can be read in unprivileged mode as well. Otherwise, the operation will be ignored, and the returned read data (if MRS is used) will be zero.

Table 7.10: List of special registers accessible by MSR and MRS instructions.

Register	Description
APSR	Application program status register. The status registers are read using MRS instruction.
IPSR	Interrupt status register.
MSP	Main stack pointer. However, stack pointer(s) is(are) also accessible using other assembly instructions as well.
PSP	Process stack pointer.
PRIMASK	Priority mask register.
BASEPRI	Base priority register.
FAULTMASK	Fault mask register.
CONTROL	Control register.

7.12.2 CPS Instruction

The change processor state (CPS) instruction is used to set or reset the interrupt masking special purpose registers. The instruction syntax is

$$\text{CPS}action \quad flag$$

where *action* is one of the following

IE   Clears the special purpose register to enable the interrupts.

ID   Sets the special purpose register to disable the interrupts.

and *flag* is one or more of the following flags

I   PRIMASK special register is set or clear depending on the value of *action* field.

F   FAULTMASK special register is set or clear depending on the choice of *action* field.

The CPS instruction can only be used in privileged mode and cannot be executed conditionally and hence cannot be inside an IT block. The use of this instruction is illustrated in Example 7.26.

Example 7.26 (Use of CPS instruction for setting and clearing of interrupt masking registers.).

```
CPSID I    ; Set PRIMASK register to disable interrupts as
           ; well as configurable fault handlers
CPSIE F    ; Clear FAULTMASK register to enable interrupts
           ; and fault handlers
```

Table 7.11 provides details regarding instruction encoding and flag dependency of table branch instructions.

Table 7.11: Special instructions.

Mnemonic	Brief description	Encoding	Flags
MRS	Move to register from special register	32 bit	-
MSR	Move to special register from register	32 bit	-
CPS	Change processor state	32 bit	-

7.13   Summary of Key Concepts

The branch and control instructions play the most important role in constructing the program flow control. The focus of this chapter has been to first introduce the branch and control instructions followed by the development of conditional constructs. Specifically, the following key concepts have been introduced.

- Branch immediate (direct) instructions (B and BL) use relative address and evaluate the label address using relative offset with respect to the current value of program counter.

- Branch indirect instructions (BX and BLX), on the other hand, use absolute address provided using register operand.

- Using any of the optional condition codes, the execution of all four branch instructions can become conditional.

- Each condition code is based on either single or multiple status flags.

- Function calls can be implemented using BL or BLX instructions.

- For nested function calls it is necessary to preserve the LR before calling a function.

- It is also possible to implement branch operations with MOV, LDR, and POP instructions by using PC (R15) as the destination register with these instructions.

- Using conditional branch instructions, high level conditional constructs including *if-else*, *for*, and *while* loops as well as switch-case are constructed.

- Combined compare and conditional branch instruction improves the execution efficiency by combining two instructions, namely compare and conditional branch.

- The IT (if-then) block instruction provides an assembly version for *if-then-else* conditional construct.

- The execution of assembly instructions, other than branch instructions, can be made conditional using IT block instruction.

- The IT block instruction improves execution performance by avoiding branch penalty.

- The table branch instruction provides an assembly equivalent for switch-case conditional construct.

Review Questions

Question 7.1. What is the difference between branch (B) and branch with link (BL) instructions?

Question 7.2. How can an instruction be executed conditionally?

Question 7.3. Which instructions can be executed conditionally without using IT block?

Question 7.4. What is the significance of register LR when BL and BLX instructions are executed?

Question 7.5. Under what circumstances should the LR register be pushed onto the stack?

Question 7.6. Let us assume that an indirect branch occurs using BX or BLX instruction. Which fault occurs if the LSB of register Rm, holding the address to branch to, is cleared to 0?

Question 7.7. Both BX and BLX instructions have only 16-bit encoding. Is it possible to use high register as operand with these instructions?

Question 7.8. What is the key limitation of B and BL instructions compared to BX and BLX instructions?

Question 7.9. List at least three different ways to perform a branch operation without using any of the four branch instructions (i.e., B, BL, BX, BLX).

Question 7.10. Which branch instructions can be used to implement a function call?

Question 7.11. Illustrate the implementation of *if-elseif-else* construct using compare and conditional branch instructions.

Question 7.12. What will happen if a new value is loaded to register R15 and the bit 0 of the new value is 0?

Question 7.13. A C function to generate a sequence of numbers can be developed using either iterative or recursive implementation. Which of the two implementations will require larger stack memory space?

Question 7.14. Why does an IT block support only four instructions per block? What would be the disadvantage if the maximum permissible number of IT block instructions is increased to seven?

Exercises

Exercise 7.1. Is it possible to implement a nested function without using stack memory? If the answer is yes, then provide an example; otherwise, justify the reason for not being able to do so.

Exercise 7.2. Illustrate the implementation of *do-while* loop using conditional branch instructions.

Exercise 7.3. Is it possible to use IT block instruction to implement *if-elseif-else* construct? If yes, give an example; otherwise, justify your answer.

Exercise 7.4. Correct the errors if any in the following assembly code implementing an IT block.

```
CMP     R0,  #9
IT      GT
ADDGT   R1,  R0,  #55
ADDLE   R1,  R0,  #48
ADDCS   R2,  R3,  #47
```

Exercise 7.5. Consider src DCB 0xA, 0x5, 0xF, 0xB, 0x3 and dst DCB 0x0, 0x0, 0x0, 0x0, 0x0. The scr is defined in READONLY segment and dst is defined in READWRITE segment. Write an ASSEMBLY code to convert each element of src to its corresponding ASCII value and store it to the dst. To achieve this goal, you are required to develop an algorithm that uses IT instruction. The ASCII codes of 0xA, 0x5, 0xF, 0xB, 0x3 are 0x41, 0x35, 0x46, 0x42, 0x33.

Exercise 7.6. You are given a number in, say, register R0. We are interested in finding whether the number can be represented as n^{th} power of 2. Write an assembly program to find this.

Exercise 7.7. In computer programs, a string is an array of ASCII characters, which is terminated by zero (also called null-terminated string). Table 7.12 provides a list of ASCII characters along with ASCII values.

- Write an assembly function 'numP' (not an entire assembly program) to count the number of paragraphs in a string. R0 holds the address of the string. Return the answer in R1.

Table 7.12: List of useful ASCII characters and their codes.

Description	ASCII code	ASCII character
Small alphabets	0x61 – 0x7A	'a' to 'z'
Capital alphabets	0x41 – 0x5A	'A' to 'Z'
Numbers	0x30 – 0x39	'0' to '9'
Space	0x20	' '
Full stop	0x2E	'.'
New paragraph	0x0A	New line
End of string	0x00	

- Write an assembly procedure/function 'numW' to count the number of English words (ending with spaces or full-stops) in a string. R0 holds the address of string. Return the answer in R1.

Exercise 7.8. Consider the following IT (IF-THEN) block based assembly program.

(a) Write the equivalent C program. The C program variable names can be the same as the register labels.

```
CMP       R0 , R1
ITTEE     GE
SUBSGE    R3, R0, R2
MOVGE     R0, #25
SUBSLT    R4, R2, R0
MOVLT     R0, #15
```

Listing 7.2: Assembly program illustrating Exercise 6.

(b) For the initial values of R0 = 20, R1 = 10 and R2 = 5, what will be the final value of R0 after the execution of the above program?

(c) Repeat part (b) for R0 = 10, R1 = 20, and R2 = 5.

Exercise 7.9. Why will the loop in the code listing below execute infinitely many times? What change will terminate the loop after finite iterations?

```
    MOV      R0, #0x10
loop
    SUB      R0, #1
    BNE      loop
```

Listing 7.3: Assembly program illustrating Exercise 7.

Exercise 7.10. Describe the similarities and differences between a nested function call and an ordinary function call. In this context, discuss the use of stack memory for recursive and iterative function calls.

Exercise 7.11. Write an assembly code which will implement the functionality of the following C code.

```
unsigned int x;
int y, z;
```

```
if(x == 3 || y  > -1)
{
    z++;
}
```

Exercise 7.12. Consider an IT block with four conditional instructions. We are interested in comparing the execution of two different instruction sequences inside IT block. Let us assume the four instructions are organized as ITTEE and ITETE. Compare the execution for these two constructs for (a) when S flag is omitted with all the instructions and (b) when S flag is used.

Part III

Interfacing

Chapter 8

Fundamentals of Input-Output Interfacing

Overview

Cortex-M based microcontrollers integrate the ARM processor core with many peripheral modules, providing a highly capable single chip solution for different applications, including motion control, medical instrumentation, test and measurement equipment, security and surveillance systems, factory automation, automotive and transportation, gaming, renewable energy, intelligent lighting control, to name a few. One of the fundamental operations in these applications is to perform digital actuation or control. Another important aspect is the availability of an interface allowing the user to interact with the system by configuring parameters and visualizing the system status. Most of these requirements are fulfilled by using general purpose input-output (GPIO) interfacing.

This chapter starts with a basic introduction of GPIO interfacing covering the details of GPIO pin related attributes. We will then introduce the ARM Cortex-M based microcontroller, TM4C123 from Texas Instruments, that we have chosen for this text. Multiple factors are considered, while choosing a microcontroller and some of the important ones are listed below.

- Integration of essential interfaces on the microcontroller is one of the key factors when selecting a microcontroller. In addition, the execution speed and memory sizes are also important.

- The availability of low cost and user friendly evaluation platform is another important aspect.

- Availability of economical programming and debugging tools is also highly critical, when selecting a microcontroller. Many of Cortex-M microcontroller manufacturers have started integrating programming and debugging interfaces on the evaluation platforms (for instance, Texas Instrument's, ST Microelectronics, etc.), which is a highly attractive feature.

- In addition, the availability of an easy to use accompanying firmware as well as documentation for programming and debugging purposes can be highly beneficial. In this context, many of the microcontroller chip manufacturers do provide device driver libraries.

The TM4C123 microcontroller is one of the most suitable choices, which provides the best compromise when considering the above-mentioned factors. Once a microcontroller is chosen, the next natural step is to understand its basic wiring for minimum functionality. For that purpose we discuss wiring details related to power, clock, and reset as well as debugging and programming interfaces. This discussion is followed by the introduction of different interfaces that are available on TM4C123. Then we detail the basic steps to configure a microcontroller pin as GPIO pin including enabling of clock, setting pin direction. The use of GPIO pins is illustrated for both output and input functionalities. Specifically, the use of GPIO pins as output and input is illustrated by controlling an LED and scanning the status of a switch. We later extend the use of GPIO pins as parallel ports to interface seven-segment as well as LCD displays for output port interfacing and matrix type keypad for both input and output port interfacing.

8.1   Basic Microcontroller GPIO Interfacing

Different types of interfaces exist on a microcontroller for connecting physical devices. Some of the physical devices are source of information (e.g., sensors and transducers) and require an input type of interface from the microcontroller perspective, while some other devices are consumers of information (e.g., displays) and require microcontroller output interface for connectivity. The microcontroller physical pins are highly flexible and can be configured either as an input or an output. Many of these input-output pins can also be configured for alternate functionalities, as will be discussed in subsequent chapters. A general purpose input output port is a physical pin in a microcontroller that can be configured by software to become either digital input or digital output. These GPIOs are also called bidirectional GPIOs.

A GPIO pin when configured as a digital output allows to translate logical levels within the program to corresponding voltage levels on the associated pin by performing a write operation to the GPIO address. The voltage levels on the pin configured as an output allow the microcontroller to exert an ON/OFF control to the device connected to that pin. When GPIO is configured as an input pin, the software can read external digital signals into the program to get the current state of the connected device. A read cycle access from the corresponding GPIO address returns the current state of the pin, which is configured as an input at that time. When configuring a GPIO pin as input, it is very important to ensure that the voltage levels of the external signal do not exceed the voltage tolerance levels of the GPIO pin. If the external device operating voltage is different from the GPIO voltage levels, a level shifter is required.

8.1.1 GPIO Features

Traditionally GPIO pins in microcontrollers had few configurable parameters and were limited in the features supported. For example, GPIO pin direction was one such configurable feature that was supported by almost all the microcontrollers. With the advancement of fabrication process and cost reduction, more flexibility and versatility has been introduced by integrating many other GPIO features. The advantages provided by this flexibility have been at the expense of more complex configuration or initialization. It is also very important to understand that not all the parameters are of importance for a given application. In general, the associated set of features for GPIOs vary quite widely. On one extreme, they are a mere group of pins that can be switched either individually or as a group to function as input or output. While in more complex GPIO modules, each pin is flexibly configurable to source different logic levels along with controllable current drive strengths as well as multiple resistor values for internal pull-ups.

Figure 8.1 shows the block diagram explaining the digital input-output operation and direction control. The collection of multiple digital GPIO pins is called a parallel port. Below we list some of the important features that have been integrated by many microcontrollers.

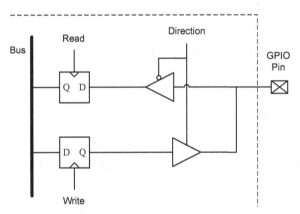

Figure 8.1: Simple block diagram illustrating the digital input-output implementation.

- GPIO with Pull-up/Pull-down Resistors: There are multiple uses of the pull-up resistors. One key purpose of pull-up resistors is to ensure that the system inputs settle at the required logic level when the external devices interfaced with the microcontroller are in high-impedance mode, appearing as disconnected. When no external device connected to a GPIO is active, then the pull-up resistor pulls the GPIO pin up to the logic high level. As soon as a connected device becomes active, the high logic level set by the pull-up resistor on the GPIO pin is overdriven by the corresponding logic level of the device output. Another possible use of pull-up resistors is to interface a device with a microcontroller (or another device), when the operating supply voltages for them are different. Figure 8.2 shows the presence of an external pull-up resistor. It should also be noted that the GPIO pin shown in Figure 8.2 is of type open-drain.

- GPIO Current Sourcing and Sinking Capability: Each GPIO has limited current sourcing (in case of output) and sinking (in case of input) capability. Some of the complex GPIOs have multiple configurable levels of current sourcing capability.

- GPIO as Input with Higher Voltage Tolerance: In some microcontrollers, GPIOs, when configured as inputs, can tolerate voltage levels higher than the operating supply voltage of the microcontroller. For instance, the microcontroller used in this book, TM4C123, has GPIOs which can tolerate 5 V input signals despite its operating voltage of 3.3 V.

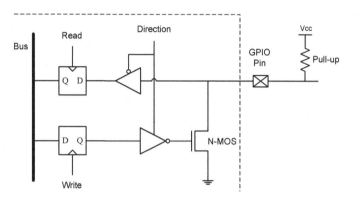

Figure 8.2: The GPIO pin diagram illustrating the use of external pull-up.

Some other important features are slew rate control for output configuration and Schmitt triggered type for inputs. Integration of these features in the GPIO is shown in Figure 8.3.

8.1.2   Multiplexing Functionalities on GPIO Pin

Most of the GPIO pins in a microcontroller can be configured for an alternate hardware function. It should be noted that only one of the alternate functions can be configured at a given time. The user program can switch among different alternate functionalities during execution.

8.2   Cortex-M-Based TM4C123 Microcontroller

The Cortex-M microcontroller selected for hardware-based programming and interfacing illustrations is TM4C123 from Texas Instruments. This microcontroller belongs to the high-performance ARM Cortex-M4F based architecture and has a broad set of peripherals integrated. The key features of the TM4C123 microcontroller are listed below.

- Clock frequency: Processor clock frequency up to 80 MHz with floating point unit (FPU).

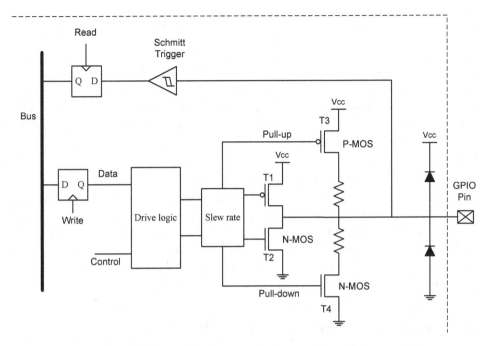

Figure 8.3: The GPIO pin illustrating the integration of advanced features.

- System timer SysTick: SysTick is 24-bit, clear-on-write, decrementing timer. Its flexible control allows its use for the purpose of system time base generation.

- Nested vectored interrupt controller: The microcontroller TM4C123 also includes a nested vectored interrupt controller (NVIC). The NVIC along with Cortex-M processor can prioritize and handle the interrupts in handler mode. On the occurrence of an interrupt, the state of the processor is automatically stored onto the system stack and is restored from the stack when exiting from the interrupt service routine. The fetching of the interrupt vector and saving of the state (i.e., storing the registers onto stack) are performed simultaneously, which provides efficient interrupt entry and as a result the interrupt latency is reduced. The Cortex-M4F processor in TM4C123 also supports the tail-chaining functionality that further reduces interrupt latency. A collection of 7 system exceptions and 65 peripheral interrupts are supported by TM4C123. The microcontroller supports 8 priority levels, which can be configured for these exceptions and interrupts.

- Debugging interface using JTAG/SWD: The TM4C123 microcontroller provides a JTAG and SWD (serial wire debug) based debugging interface for programming and debugging purpose.

8.2.1   TM4C123 Microcontroller Block Diagram

A high-level block diagram of the TM4C123 microcontroller is shown in Figure 8.4. Note that there are two on-chip buses, AHB and APB, which connect the processor core to the peripherals. These buses are constructed from the system bus using a bus matrix. It should be recalled that this bus matrix is different from the one used inside the processor block. The bus matrix inside the processor block is responsible for connecting the instruction and data buses from the processor core to I-Code, D-Code, and System buses.

The advanced peripheral bus (APB) is the low speed bus. The more complex advanced high-performance bus (AHB) gives improved performance than the APB bus and should be used for interfacing those peripherals, which require faster data transfer speeds. The block diagram also shows different peripheral modules connected to these two buses. The details regarding different peripheral modules will be provided in the next section. It should be noted that only the GPIO and direct memory access (DMA) modules have connectivity available to both the buses.

8.2.2   TM4C123 Microcontroller GPIOs

Specifically the microcontroller used in the hardware platform is TM4C123GH6PM, which has a 64 pin package. Out of these 64 pins, 43 pins are available for the purpose of GPIO pins. These GPIO pins are grouped into six ports labeled PortA to PortF. PortA to PortD are eight-pin ports, while PortE is six pin and PortF is five pin. Each of the pins on these ports can be configured as GPIO. Some of the port pins also have special peripheral functionalities multiplexed along with GPIO functionality and can be configured for that purpose as well. When configured as GPIO these port pins have the following capabilities.

- Internal weak pull-up or pull-down resistors.

- Configurable current sourcing capability for levels of 2 mA, 4 mA, and 8 mA, when the pin is configured for one of the digital communication peripheral interfaces. Up to four GPIO pins, when configured as inputs, can sink 18 mA for high current applications.

- Slew rate control capability is provided for 8 mA output drive.

- Each port pin can be configured as open drain.

- Some of the port pins are capable of tolerating 5V when configured as inputs.

8.2.3   Minimum Connectivity for TM4C123

As discussed above, the TM4C123 microcontroller used in the hardware platform has a 64 pin package, of which 43 pins are available as GPIOs. The remaining 21 pins on the microcontroller package are used for some dedicated purposes. Out of these 21 pins 13 are used for power supply connections and are labeled as VDD and GND. Four other pins are reserved for connecting external clock source connectivity. The

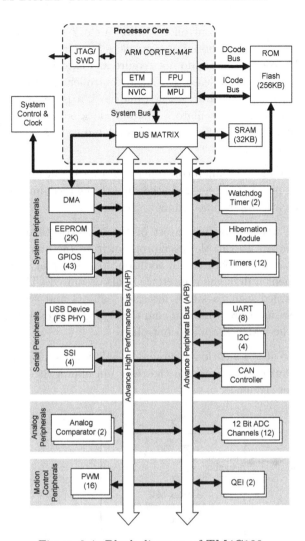

Figure 8.4: Block diagram of TM4C123.

remaining four pins are dedicated for reset, wakeup, hibernation, and an external battery connection. The programming and debugging interface, based on JTAG/SWD, of this microcontroller is multiplexed with GPIO pins. These GPIO pins are by default configured for the debugging functionality. Based on this fact, as a precautionary measure it is highly recommended that these GPIO pins should not be configured to be used as GPIOs.

A minimum wiring of the microcontroller is required for its proper functionality. In general, the following four hardware connections are required for proper functioning of the microcontroller.

- Power supply: A DC power supply of appropriate voltage level and current capacity needs to be connected to the Vcc and GND pins. It is a good practice to use bypass capacitors to filter any ripple or noise in the DC power supply.

- Clock source: Each microcontroller requires a clock source for its proper working. Nowadays, some of the microcontrollers have built in multiple oscillators and some of those oscillators do not require any external components for their clock generation. In that case, the microcontroller can function without any clock related wiring. However, for a stable clock it is better to use an external crystal with the built in oscillator. A stable clock is required for asynchronous communication interfaces. Further details related to this aspect will be covered in Chapters 10 and 11.

- Reset connection: Each microcontroller has a reset pin for hardware reset capability. This pin should be wired according to the recommendation of the chip manufacturer for proper functioning.

- Programming and debugging: Most of the microcontrollers have an on-chip program memory. To run the user program from the on-chip memory, a programming interface is required. In addition, for debugging the user program in real environment a debugging interface is also required. These two operations can be performed using a single interface. JTAG and serial wire debug are the two widely used programming and debugging interfaces and are also made available on the TM4C123 microcontroller.

Figure 8.5 shows the above-mentioned minimum connections for TM4C123 microcontroller. The push button connected to the reset pin allows the user to activate the microcontroller reset sequence.

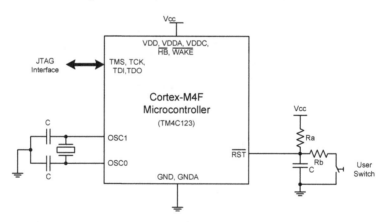

Figure 8.5: Minimum connections required for TM4C123 microcontroller.

8.2.4   The Hardware Development Board

The microcontroller platform used in this text is based on the Texas Instruments Tiva C series launchpad evaluation board. The development board is based on the TM4C123 microcontroller and provides expansion headers for peripheral interfacing as can be seen from Figure 8.6. This platform is equipped with an on-board integrated In-Circuit Debugging Interface (ICDI), which allows programming and debugging of the TM4C123 microcontroller. There are two USB interfaces available on the board. One of the USB interfaces is used for programming and debugging purposes, while the second USB interface allows the user to develop USB applications in USB device mode. There is a selector switch to select the board power source, which can be supplied from either of the two USB connectors.

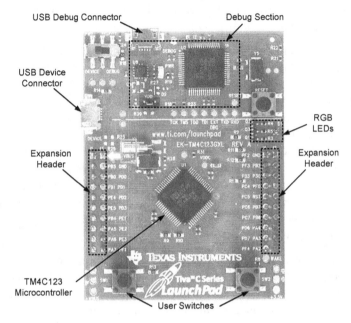

Figure 8.6: The TM4C123 microcontroller based evaluation platform.

The ICDI is implemented using a dedicated microcontroller. The debugging interface on one side connects to the host computer using a USB link, while it connects to the TM4C123G microcontroller using JTAG/SWD interface on the other side. The interface with the external devices, using the available peripherals, can be implemented using the two expansion headers provided on the board. The mapping of the microcontroller pins to the header pins is shown in Figure 8.7. It should be noted that not all the GPIOs are available on the header pins.

	J1	J3				J4	J2		
3.3V	1	1	VBUS		PF2	1	1	GND	
PB5	2	2	GND		PF3	2	2	PB2	
PB0	3	3	PD0		PB3	3	3	PE0	
PB1	4	4	PD1		PC4	4	4	PF0	
PE4	5	5	PD2		PC5	5	5	RESET	
PE5	6	6	PD3		PC6	6	6	PB7/PD1	
PB4	7	7	PE1		PC7	7	7	PB6/PD0	
PA5	8	8	PE2		PD6	8	8	PA4	
PA6	9	9	PE3		PD7	9	9	PA3	
PA7	10	10	PF1		PF4	10	10	PA2	

Figure 8.7: Port pin mappings for the expansion headers available on the TM4C123 hardware platform.

8.3   TM4C123 Microcontroller Peripherals

The TM4C123 microcontroller is equipped with a variety of peripherals, which are integrated with the ARM Cortex-M core. The TM4C123 on chip peripherals can be categorized into the following four groups.

- System integration peripherals: Different system integration peripherals include GPIOs, system clock management, direct memory access as well as hibernation module. The GPIOs can be used for parallel interfacing. For example, external memory can be interfaced using parallel interface. In addition, the GPIO pins on TM4C123 can also be configured as interrupt inputs for external binary events. The system clock management module is responsible for generating clock for system as well as for other peripherals. It allows the user to run the system at different operating frequencies. In many battery operated systems, the power conservation is of extreme importance. For that purpose, TM4C123G microcontroller has an integrated battery-backed hibernation module to efficiently keep the microcontroller in power down mode for reducing the power usage during intervals of inactivity.

- Timing interfaces: The TM4C123 has 12 general purpose timers, two watchdog timers, and one systick timer. Systick timer is integrated as part of the ARM processor core, while all other timers are integrated as peripheral modules. There are six 32-bit general purpose timers and each of these timers can be split into two 16-bit timers. Similarly, there are six 64-bit timers and each of these timers is made by concatenating two 32-bit timers. In addition to counter mode, a timer can be configured in input capture or output compare modes. Input capture and output compare will be used to create periodic interrupts and measure period, pulse width, phase as well as frequency as will be explained in Chapter 10. In addition, there are two 32-bit watchdog timers that can be used for regaining system control.

- Communication interfaces: The communication interfacing peripherals primarily consist of serial interfaces. The TM4C123 has eight Universal Asynchronous Receiver/Transmitter (UART) interfaces, which can be used for asynchronous point to point serial communication between two devices. The UART interface allows simultaneous communication in both directions making its communication full-duplex. The TM4C123 has four Synchronous Serial Interfaces (SSI), which is also called Serial Peripheral Interface (SPI). The SPI interface is also full-duplex. The microcontroller also supports four I2C interfaces, which is a serial bus interface that can be used to connect multiple devices. The I2C bus is half-duplex. There are two CAN (controller area network) interfaces and one USB device interface available as well on the TM4C123 microcontroller. The CAN bus is commonly used in automotive as well as distributed control systems.

- Analog interfacing peripherals: There are two 12-bit analog-to-digital converters (ADCs) available in TM4C123. The ADC is used for converting the analog signals to their digital equivalent, which is required for many data acquisition applications. In addition to ADCs, there are analog comparators available on the microcontroller. These analog comparators can be used for converting an

Table 8.1: Summary of peripherals integrated on TM4C123.

Peripheral	Description
GPIOs	43 configurable pins for this purpose.
System Clock	80 MHz maximum, can be configured arbitrarily using PLL
Hibernate	Operates in lower power mode using backup battery
UART	Eight UART modules with maximum baud rate of 10 Mbps
SPI	Four SPI modules with transmit & receive FIFOs
I2C	Four I2C bus interfaces configurable as master or slave
CAN	Two CAN modules supporting protocol versions 2.0 A/B
USB	One USB device interface
32-bit Timers	Six 32-bit timers with each timer constructed from two 16-bit timers concatenated. Can be used as 16-bit timer
64-bit Timers	Six 64-bit timers with each timer constructed from two 32-bit timers concatenated. Can be used as 32-bit timer
Watchdog Timers	Two watchdog 32-bit timer modules
ADC	Two analog to digital converters with 12-bit resolution
Comparators	Two analog comparators
QEI	Two quadrature encoder inputs
EEPROM	2 KB on chip EEPROM for storing configuration and other data

analog signal to binary logical signal based on the thresholding principle. In addition, it can also be used for comparing two analog input signals. The TM4C123 peripherals discussed above are summarized in Table 8.1.

Some of the GPIO pins are multiplexed for one or more alternate peripheral functionalities. These GPIO pins can be configured for one of the available alternate functions. In addition, the GPIO pin can be reconfigured to a different functionality at run time as well. Details of alternate pin functionalities are summarized in Tables 8.2 and 8.3. Alternate functionality on a pin can be configured using the associated registers for this purpose and will be discussed in this as well as subsequent chapters.

8.3.1 Peripherals on the Memory Map

The TM4C123 microcontroller memory map is described in this section. Since ARM Cortex-M architecture uses memory mapped peripherals, the address allocations to different peripherals will be discussed. The microcontroller has a fixed memory map that provides up to 4 GB of addressable memory. The memory map for the TM4C123 controller is provided in Table 8.4.

Since TM4C123 microcontroller has 256 KB on-chip Flash (code) memory, the corresponding addressable range in code memory region is 0x00000000 to 0x0003FFFF. Similarly the data memory (RAM) address range is 0x20000000 to 0x20007FFF, which shows that TM4C123 has 32 KB on-chip RAM. The first two entries in Table 8.4 correspond to code and data memories. From the third entry onward in Table 8.4, the address ranges belong to the peripheral address space. The address range assigned to a peripheral is allocated to different configuration, control, status, and data registers associated with that peripheral.

Table 8.2: Alternate functionality selection for ports A to D.

Port Pin	Analog	Port control register GPIO_PCTL_R, configuration values for alternate function.							
		1	2	3	4	5	6	7	8
PA0		U0Rx							CAN1Rx
PA1		U0Tx							CAN1Tx
PA2			SPI0Clk						
PA3			SPI0CS						
PA4			SPI0Rx						
PA5			SPI0Tx						
PA6				I2C1SCL		M1PWM2			
PA7				I2C1SDA		M1PWM3			
PB0	USB0ID	U1Rx						T2CC0	
PB1	USB0VBUS	U1Tx						T2CC1	
PB2				I2C0SCL				T3CC0	
PB3				I2C0SCL				T3CC1	
PB4	AIN10		SPI2Clk		M0PWM2			T1CC0	CAN0Rx
PB5	AIN11		SPI2CS		M0PWM3			T1CC1	CAN0Tx
PB6			SPI2Rx		M0PWM0			T0CC0	
PB7			SPI2Tx		M0PWM1			T0CC1	
PC0		TCK/ SWCLK						T4CC0	
PC1		TMS/ SWDIO						T4CC1	
PC2		TDI						T5CC0	
PC3		TDO/ SWO						T5CC1	
PC4	C1-	U4Rx	U1Rx		M0PWM6		IDX1	WT0CC0	U1RTS
PC5	C1+	U4Tx	U1Tx		M0PWM7		PhA1	WT0CC1	U1CTS
PC6	C0+	U3Rx					PhB1	WT1CC0	USB0EPEN
PC7	C0-	U3Tx						WT1CC1	USB0PFLT
PD0	AIN7	SPI3Clk	SPI3Clk	I2C3SCL	M0PWM6	M1PWM0		WT2CC0	
PD1	AIN6	SPI3CS	SPI1CS	I2C3SDA	M0PWM7	M1PWM1		WT2CC0	
PD2	AIN5	SPI3Rx	SPI1Rx		M0FAULT0			WT3CC0	USB0EPEN
PD3	AIN4	SPI3Tx	SPI1Tx				IDX0	WT3CC0	USB0PFLT
PD4	USB0DM	U6Rx						WT4CC0	
PD5	USB0DP	U6Tx						WT4CC1	
PD6		U2Rx			M0FAULT0		PhA0	WT5CC0	
PD7		U2Tx					PhB0	WT5CC0	NMI

8.4   Configuring Microcontroller Pins as GPIOs

The key initialization steps required to configure a microcontroller pin as GPIO pin will be outlined in this section. For some GPIO pins, a few additional steps might be required. For instance, some of the GPIO pins are assigned special functionality by default and to configure them as general purpose IO pins requires unlocking or unmasking. It is possible that a GPIO pin has some alternate functionalities multiplexed on it. The configuration of a GPIO for an alternate functionality will be explained in subsequent chapters. Following are the basic steps required for the configuration of a microcontroller pin as a GPIO.

- Clock and bus configuration
- Data control configuration

Table 8.3: Alternate functionality selection for ports E and F.

Port Pin	Analog	Port control register GPIO_PCTL_R, configuration values for alternate function.									
		1	2	3	4	5	6	7	8	9	14
PE0	AIN3	U7Rx									
PE1	AIN2	U7Tx									
PE2	AIN1										
PE3	AIN0										
PE4	AIN9	U5Rx		I2C2SCL	M0PWM4	M1PWM2			CAN0Rx		
PE5	AIN8	U5Tx		I2C2SDA	M0PWM5	M1PWM3			CAN0Tx		
PF0		U1RTS	SPI1Rx	CAN0Rx		M1PWM4	PhA0	T0CC0	NMI	C0o	
PF1		U1CTS	SPI1Tx			M1PWM5	PhB0	T0CC1		C1o	TRD1
PF2			SPI1Clk		M0FAULT0	M1PWM6		T1CC0			TRD0
PF3			SPI1CS	CAN0Tx		M1PWM7		T1CC1			TRClk
PF4						M1FAULT0	IDX0	T2CC0	USB0EPEN		

- Mode control configuration

- Pad control configuration

It is important to realize that the above-mentioned GPIO configuration steps are specific for TM4C123 microcontroller. Some of the above-mentioned configuration steps might be used for other microcontrollers, but in general there are differences, which need to be taken care of for proper configuration. Each configuration step requires setting or clearing a bit field in one or more registers. Effectively the configuration of a microcontroller for a certain functionality is equivalent to writing the specific values to the appropriate registers.

The process of configuration will involve setting and clearing bits in the associated registers to obtain the desired functionality. By setting a bit, we mean that the value of the bit is equal to 1. Similarly, clearing a bit is equivalent to make the bit value equal to 0. This generic terminology will be followed throughout the book.

8.4.1 Clock and Bus Configuration

The first step in the configuration of a peripheral is to enable the clock for that peripheral. This is also true when an entire port or few of its pins are to be configured for GPIO functionality. To enable the clock for a particular GPIO port it is required to set the appropriate bit field, for the required GPIO port, in the RCGC_GPIO_R register. The RCGC_GPIO_R register is memory mapped to address 0x400FE608. The bit 0 to bit 5 of RCGC_GPIO_R register can be set to enable the clock for PortA to PortF, respectively. After enabling the clock to a GPIO module, there must be a 3 clock cycle delay before accessing the GPIO registers.

Referring to Figure 8.4, it can be seen that the GPIO module or GPIO port registers can be accessed using two different buses. One of them is the legacy bus called Advanced Peripheral Bus (APB), while the other bus is the Advanced High-Performance Bus (AHB). All the registers associated with each port are accessible from both the buses, but they are mapped to different address spaces in the peripheral address space. For example, the registers associated with PortF on TM4C123 microcontroller are ac-

Table 8.4: Detailed memory map for some of the peripherals.

Start Address	End Address	Description
0x0000.0000	0x0003.FFFF	On-chip Flash Memory
0x2000.0000	0x2000.7FFF	On-chip SRAM
0x4000.0000	0x4000.0FFF	Watchdog timer 0
0x4000.1000	0x4000.1FFF	Watchdog timer 1
0x4000.4000	0x4000.4FFF	GPIO Port A Registers APB bus
0x4000.5000	0x4000.5FFF	GPIO Port B Registers APB bus
0x4000.6000	0x4000.6FFF	GPIO Port C Registers APB bus
0x4000.7000	0x4000.7FFF	GPIO Port D Registers APB bus
0x4000.8000	0x4000.8FFF	SSI 0 or SPI 0 Interface
0x4000.9000	0x4000.9FFF	SSI1 or SPI 1 Interface
0x4000.A000	0x4000.AFFF	SSI 2 or SPI 2 Interface
0x4000.B000	0x4000.BFFF	SSI 3 or SPI 3 Interface
0x4000.C000	0x4000.CFFF	UART 0 Communication Interface
0x4000.D000	0x4000.DFFF	UART 1 Communication Interface
$\vdots$	$\vdots$	$\vdots$
0x4001.3000	0x4001.3FFF	UART 7 Communication Interface
0x4002.0000	0x4002.0FFF	I2C 0 Communication Bus
0x4002.1000	0x4002.1FFF	I2C 1 Communication Bus
0x4002.2000	0x4002.2FFF	I2C 2 Communication Bus
0x4002.3000	0x4002.3FFF	I2C 3 Communication Bus
0x4002.4000	0x4002.4FFF	GPIO Port E Registers APB bus
0x4002.5000	0x4002.5FFF	GPIO Port F Registers APB bus
0x4002.8000	0x4002.8FFF	PWM 0 module registers
0x4002.9000	0x4002.9FFF	PWM 1 module registers
0x4003.0000	0x4003.0FFF	16/32-bit Timer 0
$\vdots$	$\vdots$	$\vdots$
0x4003.5000	0x4003.5FFF	16/32-bit Timer 5
0x4003.6000	0x4003.6FFF	32/64-bit Timer 0
0x4003.7000	0x4003.7FFF	32/64-bit Timer 1
0x4003.8000	0x4003.8FFF	ADC 0
0x4003.9000	0x4003.9FFF	ADC 1
0x4003.C000	0x4003.CFFF	Analog Comparators
0x4004.0000	0x4004.0FFF	CAN 0 Controller
0x4005.8000	0x4005.8FFF	GPIO Port A AHB bus
$\vdots$	$\vdots$	$\vdots$
0x4005.D000	0x4005.DFFF	GPIO Port F AHB bus

cessible from APB bus using the address space 0x40025000-0x40025FFF or they can be accessed from AHB bus using the address space 0x4005D000-0x4005DFFF.

8.4.2   Mode Control Configuration

The GPIO port pins in a fairly advanced microcontroller can be configured for different functionalities or modes. When the port pins are configured for GPIO mode, they are accessed directly by the software. When an alternate functionality is configured, in that case the port pins are controlled by the hardware of alternate functionality module. In case of GPIO mode, the GPIO_DATA_R register is used to read or write the corresponding port pins for data exchange.

The GPIO Alternate Function Select (GPIO_AFSEL_R) register is configured to allow one of the alternate peripheral modules to control the pin state. The corresponding bits in the GPIO_AFSEL_R register are set to enable the alternate functionality. To configure a specific alternate functionality, the GPIO Port Control (GPIO_PCTL_R) register is used, which selects one of several available peripheral functions multiplexed for the specific microcontroller pin. It is also possible to use some of the pins for analog input functionality. When a port pin is to be used as an analog to digital converter (ADC) input, the appropriate bit in the GPIO Analog Mode Select (GPIO_AMSEL_R) register must be set to disable the analog isolation circuit. This chapter discusses the GPIO function of the port pins and further details related to alternate (digital) or analog functioning of the pins will be provided in subsequent chapters.

8.4.3   Pad Control Configuration

The pad control registers allow software to configure the GPIO pads based on the application requirements. There are multiple pad control registers, which can be used to configure the following GPIO physical pad related attributes:

- GPIO pin current drive strength

- Open drain configuration

- Pull-up and pull-down configuration

- Slew rate configuration

- Digital enable

The registers assigned to the above-mentioned functionalities allow to configure each GPIO pin individually. The individual configuration capability of each GPIO pin is also valid for all other registers associated with the GPIO port pins. In order to use the pin as a digital input or output (either GPIO or alternate function), the corresponding bit in the GPIO_DEN_R register must be set. By default, on reset the GPIO_DEN_R register is cleared, with the exception of JTAG interface pins and NMI pin. Clearing the GPIO_DEN_R bit field is equivalent to GPIO pins configured to be in high-impedance or tristate. In tristate, the digital functionality is disabled, no logic value is applied to the pin and any voltage applied to the pin externally does not affect the GPIO data receiver register.

Before using the GPIO port pins, it is necessary to configure that GPIO for the desired operation. There could be a different number of steps required for different types of configurations. A generalized flow graph showing the basic steps of GPIO configuration is shown in Figure 8.8. The first step in the configuration of a peripheral (either GPIO or alternate function) is to enable the clock by setting the appropriate bit field in the clock configuration register. Once the clock is configured, then other configuration steps can be performed. However, a delay of 3 system clock cycles after enabling the clock is recommended before accessing the other registers for further configurations. In addition, when the clock for a specific peripheral is not enabled (this is done to save power), any accesses to the registers associated with that peripheral will result in a bus fault.

The sequence of configuration steps followed after the clock configuration, as depicted in Figure 8.8, are just an illustration and this order can be different. For instance, the order of digital enable and alternate function select steps can be reversed without any issues.

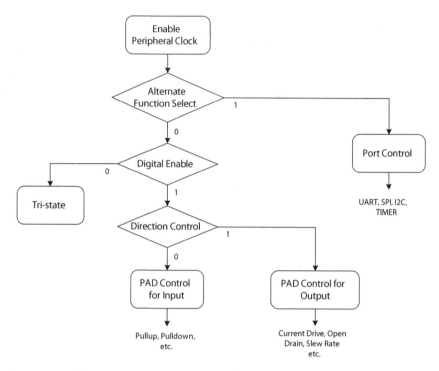

Figure 8.8: Flow chart explaining the steps involved in GPIO configuration.

8.4.4   Data Control Configuration

The data control configuration involves two aspects, namely the GPIO pin direction control and GPIO masking. The GPIO Direction (GPIO_DIR_R) register is responsible for configuring each individual port pin either as an input or an output. Clearing a data direction bit in the GPIODIR register configures the corresponding GPIO port pin as an input. When configured as an input, the value on the GPIO port pin is captured and stored in the corresponding bit of the GPIO Data (GPIO_DATA_R) register. The GPIO port pin is configured as an output when the data direction bit is set. In this case the corresponding GPIO_DATA_R register bit value is driven on the GPIO port pin. It is possible to configure some of the port pins as inputs while others as outputs. In this case, the same GPIO_DATA_R register is used for both read (data input) and write (data output) operations.

It is quite possible that different pins of the same GPIO port are used by different independent tasks. In that case, it becomes a desirable feature to have the capability to access individual GPIO port pins for read and write purposes. Conventionally, access to individual port pins is achieved indirectly in the software. In the case of selective read operation, the complete GPIO_DATA_R register value is first read and then it is masked by a desired mask value to eliminate the effect of unwanted bits. In case of selective write operation, read-modify-write sequence is followed in the software. Specifically, to change the state of an individual GPIO port pin, the GPIO_DATA_R register is read first then its value is updated by modifying the value of the desired bits and finally written back to the GPIO_DATA_R. By simply writing the value to the GPIO_DATA_R register (by skipping the preceding read and modify operations) may lead to an unwanted change in the state of some other GPIO port pins configured as outputs as well. We can conclude that the selective write operation described above, can be implemented using three instructions, which in turn require three instruction cycles to execute. This executional inefficiency can be taken care of by using the GPIO_DATA_R register masking capability provided by the TM4C123 microcontroller. It should be noted that this masking capability is a special feature of TM4C123 microcontroller and may not be available on many other microcontrollers. On a different note, some of the low-end microcontrollers have the feature of bit addressable GPIO port pins (e.g., see C51 based 8-bit microcontrollers).

Data Register Masking

To improve the software execution efficiency, the GPIO ports on TM4C123 allow access to individual bits in the GPIO_DATA_R register. This allows the software to perform read or write operation for individual (or multiple) GPIO pins in a single instruction cycle, without affecting the current state of other GPIO port pins. For this purpose, the [9:2] bits of the address bus are used to construct the mask. The underlying reason to choose address bits [9:2] will be clarified in the explanation given below. To implement this masking feature, GPIO_DATA_R register is assigned 256 different 32-bit addresses in the memory map. Using one of those addresses provides access to an individual or a selective group of port pins. To address the 256 different locations, we use the starting address as the base address and the address to an arbitrary location within those 256 addressable locations is constructed by adding an offset to the base address. For example, the first address that can be assigned to GPIO_DATA_R register of Port F is 0x40025000 (when using APB bus). Recall that it is same as the starting address of Port F. This is due to the fact that the offset of the first 32-bit address assigned to GPIO_DATA_R register in the memory map is at an offset of 0x0 and the GPIO_DATA_R register in the register list is also at an offset of 0x0 from the Port F base address. The next address location assigned to Port F GPIO_DATA_R register is 0x40025004, which is obtained by adding the offset to the base address i.e., 0x40025000 + 0x004. If GPIO_DATA_R register of Port F is accessed using 0x40025004 address then only bit 0, corresponding to Port F pin 0, is accessed for read and write operations. Similarly, using address equal to 0x40025008 allows an access to port pin 1 while 0x4002500C address grants access to both pins 0 and 1 simultaneously, for read and write operations. The address corresponding to accessing all the port pins is 0x400253FC. From the above discussion we can conclude that the addresses used are word aligned with the offset values, to generate the address, ranging from 0x000 to 0x3FC and correspondingly the address bus bits used for masking are [9:2]. This fact is illustrated in Table 8.5. Now let us illustrate the above-mentioned masking concept for both read and write operations.

Table 8.5: Offset address and corresponding GPIO_DATA_R masking.

Offset address	Accessible bits
0x000	no bit accessible
0x004	bit 0 is accessible
0x008	bit 1 is accessible
0x00C	bit 0 and 1 are accessible
⋮	⋮
0x098	bit 1, 2 and 5 are accessible
0x09C	bit 0, 1, 2 and 5 are accessible
⋮	⋮
0x3F8	bit 1 to 7 are accessible
0x3FC	all the bits, 0 to 7 are accessible

Example 8.1 (Illustration of GPIO_DATA_R register masking for write and read operations.). We illustrate the case for an arbitrary port. This is done by assuming that GPIO_DATA_R is defined as the port base address for an arbitrary port. From the GPIO_DATA_R masking viewpoint only the value of the offset is of importance. Now consider the example of writing a value of 0xA5 to the address GPIO_DATA_R + 0x0A4 (GPIO_DATA_R represent the base and 0x0A4 is the offset). The result of this write operation is illustrated in the figure below, where 'x' indicates that data is unchanged by the write operation.

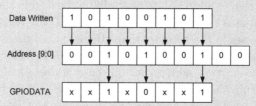

Similarly in case of read operation, the value of those data bits is read for which the corresponding address bits, in the offset address, are set. If the address bit is cleared then associated data bit is read as a zero, irrespective of actual value of the data bit. For example, a read operation performed using an address GPIO_DATA_R + 0x18C gives the result as illustrated in figure below. Actual value of port data is 0xAD, however due to address mask the read value is 0x21.

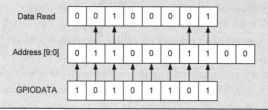

8.4.5 GPIO Configuration for Alternate Functionality

Each GPIO port has an associated set of registers that are used for its configuration. A selected subset of GPIO registers is summarized in Table 8.6, which lists the register labels and provides brief descriptions of different register groups. Table 8.6 also provides the GPIO register address offsets from the port base addresses. All the registers in Table 8.6 are 32-bits. With the exception of GPIO_PCTL_R register, these registers utilize only least significant 8-bits in case of Ports A, B, C, D, 6-bits in case of Port E and 5-bits in case of Port F. Each GPIO port pin has a corresponding bit in these registers, which is used to configure the required functionality.

Table 8.6: Register labels, their offset addresses and a brief description.

Register label	Offset	Brief description
Mode control registers		
GPIO_AFSEL_R	0x420	Alternate function selection register.
GPIO_AMSEL_R	0x528	Analog mode selection register.
GPIO_PCTL_R	0x52C	Port control register to select alternate function.
Pad control registers		
GPIO_DR2_R	0x500	2-mA drive selection register.
GPIO_DR4_R	0x504	4-mA drive selection register.
GPIO_DR8_R	0x508	8-mA drive selection register.
GPIO_OD_R	0x50C	Open drain selection register.
GPIO_PU_R	0x510	Pull-up selection register.
GPIO_PD_R	0x514	Pull-down selection register.
GPIO_SL_R	0x518	Slew rate control register.
GPIO_DEN_R	0x51C	Digital enable register.
Data control registers		
GPIO_DATA_R	0x000-x3FC	Data register.
GPIO_DIR_R	0x400	Direction register.

The bit field allocations for the registers in Table 8.6 are illustrated in Figure 8.9. All the registers in Table 8.6, with the exception of port control register (GPIO_PCTL_R), use the least significant 8 bits for GPIO port configuration as shown in Figure 8.9(a). Each of least significant 8 bits in these registers is associated with a corresponding GPIO pin.

When a GPIO pin is chosen for an alternate digital functionality, two registers, namely, GPIO_AFSEL_R and GPIO_PCTL_R should be configured to enable the desired alternate function. When a bit field in register GPIO_AFSEL_R is set to 1, the corresponding GPIO pins are enabled for alternate function. Since many of the GPIO pins have more than one alternate functionality available for configuration, the GPIO_PCTL_R register is used to select the desired alternate function. In GPIO_PCTL_R register, each 4 bit field, labeled as Port Mux Control (PMCx), is allocated for x^{th} port pin to select one of the alternate functions. This bit field allocation, for different port pins, in the GPIO_PCTL_R register is shown in Figure 8.9(b). Effectively, GPIO_PCTL_R register multiplexes different alternate functionalities for configuration. Different PMCx bit field values that can be configured to select the required alternate functionality are listed in Tables 8.2 and 8.3. For instance, assigning PMC6 a value of 7 corresponding to PB6 (port B pin 6) configures

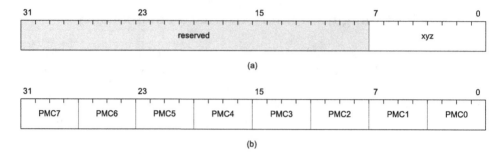

Figure 8.9: GPIO register bit field allocations. (a) The label 'xyz' refers to any of the registers in Table 8.6, except the port control register. (b) The port control register (GPIO_PCTL_R) bit field allocations. PMC0 refers to port pin 0 mux control to select the desired alternate functionality.

PB6 as timer 0 capture compare pin. Similarly, configuring PMC0, corresponding to port A, with a value 1 makes the PA0 to act as UART serial interface, while assigning PMC0 a value of 8 configures PA0 to act as CAN bus.

8.4.6   Configuring KEIL Tools for Hardware Debugging

In addition to the configuration of GPIO registers for input-output interfacing, we also need to configure KEIL μVision5 for downloading the program to the microcontroller flash memory and performing hardware debugging. For this purpose some simple configuration steps are required to be performed, which are detailed briefly in Appendix C.

8.5   Input-Output Interfacing for LED and Switch

The TM4C123 microcontroller platform (TI's ARM Cortex-M4F based TivaC) has three on board color LEDs (red, green and blue) and two switches for simple input-output interfacing. Both LEDs and switches are interfaced as digital devices. In the following, we will learn to interface each of these two different types of devices. LEDs will be interfaced using GPIO pins as digital outputs and switches will be connected by configuring the GPIO pins as digital inputs. Figure 8.10 shows the actual connections for three LEDs and two user switches (labeled as SW1 and SW2) to the microcontroller. From Figure 8.10, we observe that the LEDs are connected using digital NPN transistors. A digital transistor is the one that integrates the base (bias) resistors inside the transistor package. Unlike an ordinary transistor, a digital transistor operate in only two states, off and saturation.

The use of transistor in the above mentioned situation is to avoid any current overloading condition, which might occur if the GPIO pin is not capable of sinking/sourcing enough current. The magnitude of the current that GPIO pin has to sink in this scenario depends on both the value of diode current limiting series resistor as well as the supply voltage used.

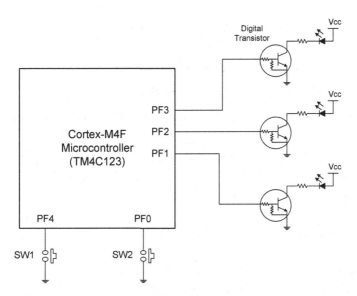

Figure 8.10: Connection diagram for TM4C123 on board LEDs and switches.

The switch interfacing shown in Figure 8.10 is not complete. This is true because, when the switch is pressed, logic low will be applied to the GPIO port pin connected to the switch. However, when the switch is released, there is no source (or supply) connection to ensure logic high appears at the GPIO port pin. With simple microcontrollers, the switch interface of Figure 8.10 might not work properly. However, in case of TM4C123 the above-mentioned problem is resolved by using the internal pull-up resistor. When the internal pull-up resistor is enabled, it pulls the GPIO pin to logic high when the switch is in released state. The internal pull-up resistor values are in general larger than 10kΩ. Since the current drawn by the GPIO port pin, when configured as input, is so low (due to its very high input impedance) that the voltage drop across the pull-up resistor is almost negligible and voltage level, approximately equal to the supply voltage, appears at the GPIO pin, which is translated as logic high by the internal logic.

A second option, different from the one shown in Figure 8.10, is to change the switch terminal connection from ground to supply (Vcc). For this case we need to enable internal pull-down resistor for proper switch functioning. For this case, logic high will be read by the user program, when the switch is pressed. Similar to the switch interfacing, a second option for LED interfacing is to reverse the LED direction, which will require to use a PNP transistor and the driving logic will be reversed.

8.5.1   Output Interfacing for LED

For LED interfacing we need to configure Port F pins 1, 2 and 3 as outputs. The basic steps to configure the required Port F pins are given below.

- Enable the clock to the GPIO port.

- Select the general purpose IO functionality by clearing the alternate function select register (GPIOAFSEL).

- Configure the GPIO direction for digital output function.

- Configure the PAD for the digital operation.

The assembly programming based LED interfacing is illustrated in Example 8.2.

Example 8.2 (Assembly program for toggling LED.). This assembly program toggles the on board green LED connected to port F pin 3.

```
    THUMB                   ; Marks the THUMB mode of operation

; Constants are declared in CODE AREA
    AREA    Constdata, DATA, READONLY

; Defining register addresses and constants
SYSCTL_RCGCGPIO_R         EQU    0x400FE608
GPIO_PORTF_DATA_R         EQU    0x40025020
GPIO_PORTF_DIR_R          EQU    0x40025400
GPIO_PORTF_AFSEL_R        EQU    0x40025420
GPIO_PORTF_DEN_R          EQU    0x4002551C

GPIO_PORTF_CLK_EN         EQU    0x20
GPIO_PORTF_PIN3_EN        EQU    0x08
SYSTEM_CLOCK_FREQUENCY    EQU    16000000    ; Default clock
DELAY_VALUE               EQU    SYSTEM_CLOCK_FREQUENCY/80
GPIO_PORTF_PIN3           EQU    0x08        ; For green LED

; The user code (program) is placed in CODE AREA
    AREA    |.text|, CODE, READONLY, ALIGN=2
    ENTRY               ;ENTRY marks the starting point of
    EXPORT __main

; User Code Starts from the next line
__main
    LDR    R1, =SYSCTL_RCGCGPIO_R    ; Enable clock for PORT F
    LDR    R0, [R1]
    ORR    R0, R0, #GPIO_PORTF_CLK_EN
    STR    R0, [R1]
    NOP                              ; Delay for clock settling
    NOP

    LDR    R1, =GPIO_PORTF_DIR_R     ; Set direction for PORT F
    LDR    R0, [R1]
    ORR    R0, R0, #GPIO_PORTF_PIN3
    STR    R0, [R1]

    LDR    R1, =GPIO_PORTF_DEN_R     ; Digital enable for PORT F
    LDR    R0, [R1]
    ORR    R0, r0, #GPIO_PORTF_PIN3
    STR    R0, [R1]

    LDR    R1, =GPIO_PORTF_DATA_R    ; Digital enable for PORT F

; Set the data for PORT F to toggle LED
LED_Flash
```

```
    LDR    R0 , [R1]
    EOR    R0 , R0 , #GPIO_PORTF_PIN3
    STR    R0 , [R1]

; Delay loop
    LDR    R5 , =DELAY_VALUE
Delay
    SUBS   R5,#1
    BNE    Delay
    B    LED_Flash

    ALIGN
    END ; End of the program, matched with ENTRY keyword
```

The executable for the program in Example 8.2 should be built using a project environment, which includes both the startup.s file as well as the user assembly code (given in the example). The user code can be added to the project by saving it in a separate assembly file and including that file to the project.

The assembly program in Example 8.2 uses two area segments. The first segment is used to define constants. The second segment contain the user program. In the first segment, the top five constants define the GPIO Port F related register addresses that will be used for the port configuration and LED toggling. The next group comprise of user defined constants.

The second segment contains the user program. The label __main defines the entry point of the user program. First part of this code segment configures the Port F pin 3 by enabling the clock using SYSCTL_RCGCGPIO_R register, setting the pin direction as output using GPIO_PORTF_DIR_R register and setting digital enable to ensure digital functioning using GPIO_PORTF_DEN_R. It should be noted that we did not configure the alternate function register (GPIOAFSEL) because its default value is 0x0 out of reset, which allows us to omit this configuration step. However, it is a good practice to include the configuration of GPIOAFSEL register as well, which will ensure proper working of the program in all scenarios. The above mentioned configuration steps are implemented using read-modify-write procedure, to ensure that other GPIO pins on Port F are not affected.

It is important to notice that the GPIO_PORTF_DATA_R register is defined as 0x40025020, which gives access to Port F pin 3 only. This allows us to perform write operation directly in one step to toggle the LED. The LED toggling is implemented in a loop with delay. The delay itself is implemented using a loop operation. Next we give the C programming based implementation for the LED toggling in Example 8.3. The C program precisely implements the same functionality.

Example 8.3 (C program for toggling LED.). This is the C programming based implementation that toggles the on board green LED connected to port F pin 3.

```
#define SYSCTL_RCGC2_R      *((volatile unsigned long *)0x400FE108)
#define SYSCTL_RCGC2_GPIOF 0x00000020 //Port F clock gating control
```

```
#define GPIO_PORTF_DATA_R  *((volatile unsigned long *)0x40025020)
#define GPIO_PORTF_DIR_R   *((volatile unsigned long *)0x40025400)
#define GPIO_PORTF_DEN_R   *((volatile unsigned long *)0x4002551C)

#define SYSTEM_CLOCK_FREQUENCY 16000000  // Default clock frequency
#define DELAY_VALUE             SYSTEM_CLOCK_FREQUENCY/80
#define GPIO_PORTF_PIN3         0x08      // For green LED

unsigned long j=0;


// User main program
int main()
{
  // Enable the clock for port F
  SYSCTL_RCGC2_R |= SYSCTL_RCGC2_GPIOF;

  // Configure port F pin 3 as digital output
  GPIO_PORTF_DEN_R |= GPIO_PORTF_PIN3;   // Digital enable for PF3
  GPIO_PORTF_DIR_R |= GPIO_PORTF_PIN3;   // PF3 as output

  while(1)
  {
    // Toggle the green LED
    GPIO_PORTF_DATA_R ^= GPIO_PORTF_PIN3;

    // delay loop
    for(j=0; j< DELAY_VALUE; j++);
  }
}
```

In the above C program, we can observe that the configuration steps are implemented using read-modify-write procedure. Same startup.s file is also required to build the executable for the C program based implementation. It is important to notice one key difference between the Assembly and C implementations in Examples 8.2 and 8.3. The assembly program in Example 8.2 uses a function __main, while the C program in Example 8.3 uses function main, which are not same. Their difference is explained in detail in Appendix A.3.

8.5.2   Input Interfacing for Switch

Mechanical switches are commonly used to feed any parameters to the digital systems. The switches can be interfaced to a microcontroller using digital inputs. The software program for switch interfacing can be implemented using one of the following two methods.

- Polling Based: In case of polling based method the GPIO pin connected with the switch is polled frequently enough, in the software, to avoid missing any key presses. By frequently enough, it is meant that the time interval between two consecutive read operations of the GPIO port pin connected with the switch and determining whether the switch is pressed or not, is smaller than the minimum time the switch is kept pressed by the user.

- Interrupt Based: In this method the GPIO pin is configured as an external interrupt and any key press leads to an interrupt. The edge triggering of the interrupt can be configured on rising or falling edge, depending on the switch hardware connectivity.

We will discuss polling based switch interfacing in this chapter. Further details, related to interrupt based switch interfacing, will be discussed in Chapter 9. Before we proceed further, it is important to first familiarize ourself with the switch physical behavior. Next we describe the switch bouncing, which is one of the critical attributes of its physical behavior.

Switch Bouncing

Electrical switches that use mechanical contacts to close or open a circuit are subject to bouncing of the contacts. Switch inputs are asynchronous and are not electrically clean. Switch contacts are usually made of springy metals. When the contacts strike together, their momentum and elasticity act together to cause them to bounce one or more times before making steady contact. It results in a pulsating electric signal instead of a clean transition. For a microcontroller GPIO pin configured as input, this switch bouncing can be interpreted as multiple switch presses. Figure 8.11 shows the bouncing of a switch with normally open contact and is captured using digital storage oscilloscope.

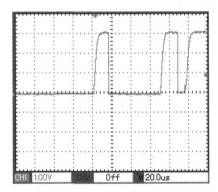

Figure 8.11: The switch bouncing when the switch is released.

Both hardware as well as software solutions exist to get around the switch bouncing problem. Analog filtering using an RC delay to filter out the rapid changes in switch output can be used as hardware solution, as shown in Figure 8.12. The design task is to choose R and C such that the input threshold is not crossed while bouncing is still occurring. In case of software solution to the switch bouncing problem, at the beginning of polling the switch is checked for being pressed. If the switch is detected as pressed, then a delay (corresponding to switch debouncing interval) is inserted and the switch is checked again for being pressed. If the switch is detected as pressed again, we declare that the switch is pressed. The rate at which the switch should be polled has to be fast enough that no switch press is missed, i.e., the time gap between two consecutive switch presses is larger than the scan cycle duration.

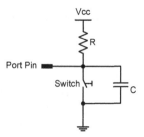

Figure 8.12: The hardware debouncing of the switch.

Switch Interfacing Illustration

There are two user switches on TM4C123 evaluation board. User switch labeled SW1 is connected to Port F pin 4 (abbreviated as PF4), while the second switch SW2 is connected to Port F pin 0 (PF0). In the following, we will describe the configuration steps required for interfacing switch SW1. The basic configuration steps are similar to the LED interfacing, with the exception of port pin direction configuration, as listed below:

- Enable the clock to the GPIO port.

- Select the general purpose IO functionality by clearing the alternate function select register (GPIOAFSEL).

- Configure the GPIO direction for digital input.

- Configure the PAD for digital operation. In addition, enable the internal pull-up resistor by setting the corresponding bit of GPIOPUR register. This is required for proper functioning of the switch, as no external pull-up resistor is placed on the board.

The C program illustration for interfacing the switch using polling method is provided in Example 8.4. The implementation in Example 8.4 is such that on every switch press an on board LED (green color) connected to PF3 will be toggled, which helps to visualize the proper functioning of the switch. The LED toggling on switch press can be implemented in two different ways. One possible approach is to toggle the LED, as soon as the switch is pressed and the next toggling of the LED only happens once the switch is first released and then pressed again. This approach can be termed as edge based LED toggling. In the second approach, if the switch is pressed then the LED is toggled and the next LED toggling happens after some predetermined delay if the switch is still pressed. This approach can be termed as level based LED toggling.

Example 8.4 (Switch interfacing using polling based method.).

This C program interfaces the on board user switch SW1 connected to PF4 and toggles the LED connected to PF3 pin, on each switch press.

```
#define   SYSCTL_RCGC2_R      *((volatile unsigned long *)0x400FE108)
#define   SYSCTL_RCGC2_GPIOF 0x0020 //Port F clock gating control
```

```c
#define   GPIO_PORTF_DATA_RD *((volatile unsigned long *)0x40025040)
#define   GPIO_PORTF_DATA_WR *((volatile unsigned long *)0x40025020)
#define   GPIO_PORTF_DIR_R   *((volatile unsigned long *)0x40025400)
#define   GPIO_PORTF_DEN_R   *((volatile unsigned long *)0x4002551C)
#define   GPIO_PORTF_PUR_R   *((volatile unsigned long *)0x40025510)


#define   GPIO_PORTF_PIN3    0x08
#define   GPIO_PORTF_PIN4    0x10

// Default clock frequency
#define   SYSTEM_CLOCK_FREQUENCY   16000000
#define   DEALY_DEBOUNCE           SYSTEM_CLOCK_FREQUENCY/1000

// Function prototypes
void Delay(unsigned long counter);

// User main program
int main()
{
  static char flag = 0;

  // Enable the clock for port F
  SYSCTL_RCGC2_R |= SYSCTL_RCGC2_GPIOF;

  // Configure PF3 as digital output and PF4 as digital input
  GPIO_PORTF_DEN_R |= GPIO_PORTF_PIN3 + GPIO_PORTF_PIN4;
  GPIO_PORTF_DIR_R |= GPIO_PORTF_PIN3;        // PF3 as output
  GPIO_PORTF_DIR_R &= (~GPIO_PORTF_PIN4);    // PF4 as input
  GPIO_PORTF_PUR_R |= 0x10;                   // Enable pull-up on PF4

  while(1)
  {
    if(GPIO_PORTF_DATA_RD == 0)
    {
      Delay(DEALY_DEBOUNCE);

      if((flag == 0) && (GPIO_PORTF_DATA_RD == 0))
      {
        // Toggle the green LED
        GPIO_PORTF_DATA_WR ^= GPIO_PORTF_PIN3;
        flag = 1;
      }
    }

    else  // clear the flag only when the switch is released
    {
      flag = 0;
    }
  }
}

/* This function implements the delay */
void Delay(unsigned long counter)
{
  unsigned long i = 0;
  for(i=0; i< counter; i++);
}
```

In Example 8.2 the value of the constant DEALY_DEBOUNCE depends on the switch characteristics. For many switches a debouncing delay of few milliseconds would be sufficient. The first `if` statement after `while(1)` loop determines if the switch is pressed. If it is pressed, then a delay sufficient for debouncing is inserted and the switch is checked for being pressed again by the second `if` statement, which also uses a `flag` variable to implement the edge based LED toggling as discussed above. If the `flag` variable is not used by the program, then the resulting implementation will be equivalent to level based LED toggling.

8.6   Seven-Segment LED Interfacing

The seven-segment display is one of the widely used displays that can display decimal, hexadecimal and in some cases special characters. Each of the seven LEDs in the display is called a segment, since when it is illuminated, it constructs a segment of the digit, to be displayed. The eighth LED is used for the indication of a decimal point, (DP) and is used when two or more seven-segment displays are put together to display larger numbers.

Each of the seven LEDs in the segment is assigned a label from *a* through *g* based on its position in the segment. In addition, one of the LED pins is brought out of the package for external connection, while all other terminals of the LEDs are connected together internally and brought out as single common pin. The common pin is also used to identify the type of the seven-segment display as *common anode* or *common cathode*.

- Common cathode: In this display type, all cathode terminals of the LED segments are joined together and are connected to logic low or ground. While based on the number to be displayed a collection of individual segments are illuminated by applying logic high though a current limiting resistor.

- Common anode: In this display type, the connections are reversed. The common terminal is of anode type and is connected with logic high, while individual segment pins are connected to logic low selectively based on the desired digit to be displayed.

The connection labels corresponding to both configurations are shown in Figure 8.13. Different characters can be displayed by selectively turning on the required LED segments.

Turning ON a specific combination of individual LEDs, one can drive the desired digit pattern on the seven segment LED display. To display the desired digit, an ON/OFF combination of different LEDs is generated. For example, to display the number 2, LED segments *a*, *b*, *d*, *e*, and *g* are turned ON. Table 8.7 provides the display pattern for numbers (0-9) and characters A, b, C, d, E and F for both common anode as well as common cathode type displays.

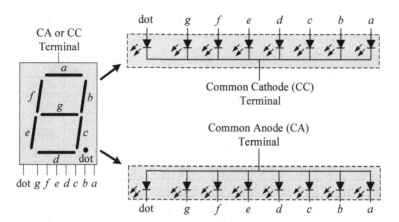

Figure 8.13: Common anode and common cathode configurations for seven segment display.

8.6.1 Time Multiplexing

Although it is possible to drive each segment of a multi-digit seven-segment display individually with its own dedicated GPIO pins, but the number of port pins required, becomes impractical when more than a few seven-segment displays are involved. To reduce the required number of GPIO pins, it is quite common to use a multiplexed (specifically time multiplexed) display architecture. In a time multiplexed display, each seven segment element is driven individually at a time, but due to the persistence of vision, it appears to the viewer that the entire display is continuously active. It is also important to realize that the implementation for multiplexed seven-segment display is quite straight forward due to the fact that each LED functions as a diode electrically, which conducts in one direction only. If the LED is replaced with a simple light source, implementation for time multiplexed display will become quite difficult.

In an n-digit time multiplexed display, n GPIO pins for digit selection and seven GPIO pins for each of the seven segments (or eight GPIO pins in case, dot point is also used) are required. The resulting total number of GPIO pin connections required is equal to $n + 7$ (or $n + 8$). If every segment is interfaced separately to the microcontroller then the number of GPIO pins required is $7n$ (or $8n$). For $n = 4$, the number of GPIO pins required for individually separate segment connections are 28, while we require only 11 GPIO pins for time multiplexed interfacing. It should be realized that for time multiplexed display each segment is only activated for $\frac{1}{n}$ time fraction of the total display refresh cycle duration. To avoid any flickering effect, the refresh frequency used is approximately 50 Hz in practice. The circuit diagram illustrating the interfacing of four digit seven segment display is shown in Figure 8.14. The PNP transistors are used as a switch with the common anode terminals of the display for turning ON one digit at a time. Additionally, the use of these transistors allows higher current sourcing for the common anode terminal.

Table 8.7: Seven-segment decoding for common anode and common cathode type displays.

Character	Common Anode		Common Cathode	
	dot g f e d c b a	Hex	dot g f e d c b a	Hex
0	1 1 0 0 0 0 0 0	0xC0	0 0 1 1 1 1 1 1	0x3F
1	1 1 1 1 1 0 0 1	0xF9	0 0 0 0 0 1 1 0	0x06
2	1 0 1 0 0 1 0 0	0xA4	0 1 0 1 1 0 1 1	0x5B
3	1 0 1 1 0 0 0 0	0xB0	0 1 0 0 1 1 1 1	0x4F
4	1 0 0 1 1 0 0 1	0x99	0 1 1 0 1 1 1 1	0x66
5	1 0 0 1 0 0 1 0	0x92	0 1 1 0 1 1 0 1	0x6D
6	1 0 0 0 0 0 1 0	0x82	0 1 1 1 1 1 0 1	0x7D
7	1 1 1 1 1 0 0 0	0xF8	0 0 0 0 0 1 1 1	0x07
8	1 0 0 0 0 0 0 0	0x80	0 1 1 1 1 1 1 1	0x7F
9	1 0 0 1 0 0 0 0	0x90	0 1 1 0 1 1 1 1	0x6F
A	1 0 0 0 1 0 0 0	0x88	0 1 1 1 0 1 1 1	0x77
b	1 0 0 0 0 0 1 1	0x83	0 1 1 1 1 1 0 0	0x7C
C	1 1 0 0 0 1 1 0	0xC6	0 0 1 1 1 0 0 1	0x39
d	1 0 1 0 0 0 0 1	0xA1	0 1 0 1 1 1 1 0	0x5E
E	1 0 0 0 0 1 1 0	0x86	0 1 1 1 1 0 0 1	0x79
F	1 0 0 0 1 1 1 0	0x8E	0 1 1 1 0 0 0 1	0x71

Example 8.5 (C program for multi-digit seven-segment display.). This program implements time-multiplexed four-digit seven-segment display interfacing of common anode type. PORT A pins 2 to 5 (i.e., PA2 to PA5) are used for digit selection, while entire PORT B is interfaced with the eight pins corresponding to seven-segment connections and one dot point connection. Both Port A and B are configured are output ports. Port pin PF4 is configured as input and is used for user switch SW1 input.

The switch is continuously scanned. On each switch press a counter variable labeled 'count' is incremented. The binary value of this counter is then converted to the equivalent BCD (binary coded decimal) value, which is displayed on the four digit seven segment display. The counter is reset to 0 when its value exceeds the maximum four digit value i.e., 9999.

```
#define   SYSCTL_RCGC2_R     *((volatile unsigned long *)0x400FE108)
#define   GPIO_PORTA_DATA_R *((volatile unsigned long *)0x400043FC)
#define   GPIO_PORTB_DATA_R *((volatile unsigned long *)0x400053FC)
#define   GPIO_PORTF_DATA_R *((volatile unsigned long *)0x40025040)

#define   GPIO_PORTA_DIR_R   *((volatile unsigned long *)0x40004400)
#define   GPIO_PORTB_DIR_R   *((volatile unsigned long *)0x40005400)
#define   GPIO_PORTA_DEN_R   *((volatile unsigned long *)0x4000451C)
#define   GPIO_PORTB_DEN_R   *((volatile unsigned long *)0x4000551C)
#define   GPIO_PORTF_DEN_R   *((volatile unsigned long *)0x4002551C)
#define   GPIO_PORTF_PUR_R   *((volatile unsigned long *)0x40025510)

#define SYSCTL_RCGC2_GPIOABF 0x00000023 //Port A,B,F clock enable
#define PORT_A_PINS          0x3C       // Port A pins 2,3,4 and 5
```

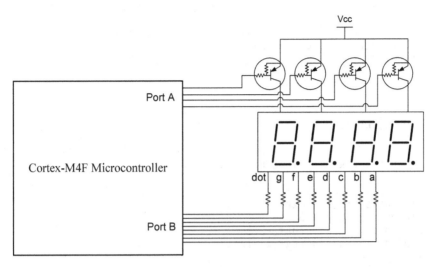

Figure 8.14: Four-digit seven-segment display connections using time multiplexing. The display used is of common anode type.

```
// Default clock frequency
#define SYSTEM_CLOCK_FREQUENCY  16000000
#define DEALY_DEBOUNCE           SYSTEM_CLOCK_FREQUENCY/1000
#define DEALY_DIGIT_REFRESH   SYSTEM_CLOCK_FREQUENCY/200

// Function prototypes
void Port_Init(void);
void Scan_Switch(void);
void Binary2BCD(unsigned int binary_val);
void Delay(unsigned long counter);


// Global variable declarations
unsigned char Lookup_7Seg_Disp[10] = {0xC0, 0xF9, 0xA4, 0xB0, 0x99,
                                      0x92, 0x82, 0xF8, 0x80, 0x90
                                      };
unsigned char Bcd_Val[4] = {0, 0, 0 ,0};


int main()
{
  unsigned char i, bcd_digit_value=0;

  // Initialize the ports
  Port_Init();

  while(1)
  {
    Scan_Switch();

    for(i = 1;  i <= 4;  i++)
    {
      bcd_digit_value = Bcd_Val[4-i];
      GPIO_PORTB_DATA_R = Lookup_7Seg_Disp[bcd_digit_value];
```

```c
        // Select the digit on Port A
        GPIO_PORTA_DATA_R = (PORT_A_PINS - (1 << (i+1)));
        Delay(DEALY_DIGIT_REFRESH);
    }
  }
}

/* Intialization routine for setting up the required ports. */
void Port_Init()
{
  // Enable the clock for port A, B and F
  SYSCTL_RCGC2_R |= SYSCTL_RCGC2_GPIOABF;

  // Configure Port A and B direction and digital enable functions
  GPIO_PORTB_DIR_R = 0xFF;
  GPIO_PORTB_DEN_R = 0xFF;
  GPIO_PORTA_DIR_R = PORT_A_PINS;
  GPIO_PORTA_DEN_R = PORT_A_PINS;

  // Configure port F as digital input with pullup
  GPIO_PORTF_DEN_R |= 0x10;      //      enable digital I/O on PF4
  GPIO_PORTF_PUR_R |= 0x10;      //      enable weak pull-up on PF4

  // Make port A and B logic high so all the LEDs are OFF.
  GPIO_PORTA_DATA_R = PORT_A_PINS;
  GPIO_PORTB_DATA_R = 0xFF;

}

/* This function scans switch SW1 based on edge transition and
   increments a local (static) counter everytime the switch is
   pressed. The counter binary value is converted to BCD every
   time the switch is pressed. */
void Scan_Switch(void)
{
static unsigned int count = 0;
static unsigned char flag = 0;

    if(GPIO_PORTF_DATA_R == 0)
    {
      // delay for switch debouncing
      Delay(DEALY_DEBOUNCE);

      if((flag == 0) && (GPIO_PORTF_DATA_R == 0))
      {
        count++;

        /* Since we have four digit display, the count value is
           reset, if it exceeds 9999. */
        if(count >= 9999)
          count = 0;

        Binary2BCD(count);
        flag = 1;
      }
    }

    /* reset the 'flag' to detect another switch press after it
       is released */
```

```
      else
      {
         flag = 0;
      }
}

/* The following function converts binary value to four digit BCD.
   Maximum permitted binary value is 9999. */
void Binary2BCD(unsigned int binary_val)
{
unsigned int temp;

   temp = binary_val;
   Bcd_Val[3] = temp/1000;
   temp  = temp - (Bcd_Val[3] * 1000);

   Bcd_Val[2] = temp/100;
   temp  = temp - (Bcd_Val[2] * 100);

   Bcd_Val[1] = temp/10;
   temp  = temp - (Bcd_Val[1] * 10);

   Bcd_Val[0] = temp;
}

/* This function implements the delay */
void Delay(unsigned long counter)
{
   unsigned long i = 0;

   for(i=0; i< counter; i++);

}
```

8.7 Keypad Interfacing

Use of keypad is a common mechanism for user input to the system. User input, in an embedded system application, is required in many different situations. For instance, the system operating parameters may require reconfiguration for performance or system requires user input under certain normal as well as abnormal situations. The user input from the keypad may be used to input data following a request from the microcontroller or simply to trigger the microcontroller to initiate a particular operation. There are two perspectives of keypad interfacing. From the software perspective a keypad interfacing can be either *polling based* or it can be *interrupt driven*. In polling based method the microcontroller spends most of its time by continuously scanning the keypad and checking for a possible key press. On the other hand, in interrupt driven method the microcontroller is free to execute any other tasks it is required to perform and scans the keypad only when an interrupt occurs due to a key press.

From the hardware perspective a keypad can be interfaced with a microcontroller using any of the three possible schemes discussed in the following. For each of the possible hardware keypad interfaces the corresponding software implementation can be either polling based or interrupt driven. The choice between polling based and interrupt driven approaches depends on the application as well as the hardware capability of the microcontroller. Next we discuss different possible solutions for hardware interfacing of the keypad.

- *Direct interface*: In direct interface each keypad switch (e.g., push button, slide switch, toggle switch, etc.) is interfaced directly to a dedicated digital GPIO pin of the microcontroller. This type of interfacing requires n number of GPIO pins for n switches. One key advantage of this interfacing approach is its capability to detect all possible combinations of multiple simultaneous switch presses. For n switches, the possible combinations are 2^n. This method can be used for specific switch combination that needs to be configured or when multiple simultaneous key presses are required to be recognized. The negative aspect of this approach is the fact that it is prohibitively demanding in terms of GPIO pins. Some of the possible uses of this approach are modifier keys (e.g., Shift, Ctrl, Alt, Func keys on a computer keyboard) as well as in the music keyboards. Figure 8.15 illustrates the direct hardware interfacing of an eight switch keypad.

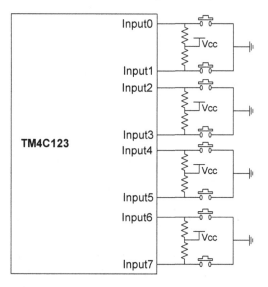

Figure 8.15: Direct keypad interfacing.

- *Scanned interface*: In scanned keypad interfacing the switches are arranged in such a manner that they form an $m \times n$ matrix. Assuming the software uses m rows as inputs and n columns as outputs, the scanning is performed by driving the digital output corresponding to each column one at a time and reading all the rows for each column drive to find out which key is pressed. Figure 8.16 shows one possible implementation of the scanned interfacing approach. The pull-down resistors connected to the rows might be omitted from the circuit if internal pull-down capability is available in the microcontroller used. Since TM4C123 GPIO module has an integrated internal pull-down capability, these external resistors can be eliminated from the circuit.

The key advantage of this approach is the reduced number of GPIO pins required, where $m \times n$ keys can be interfaced using only $m + n$ GPIO pins of the microcontroller. However, this approach is limited in its capability when multiple keys are pressed simultaneously. The scanned approach can only detect up to 2 simultaneous key presses. When three keys are pressed simultaneously, for instance, in 'L' shape then the fourth key forming the square will also appear pressed. For example, when keys '1-4-5' are pressed simultaneously, then key '2' also appears pressed irrespective of it being pressed or not physically. So it is not possible to differentiate between the three simultaneous key press scenario corresponding to keys '1-4-5' and '1-2-5'.

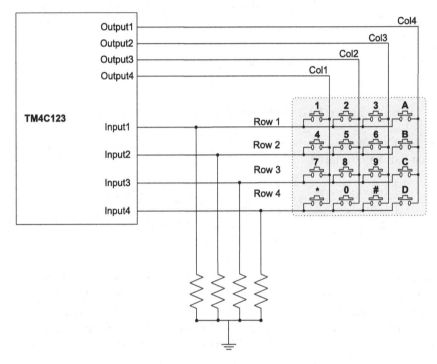

Figure 8.16: Matrix type keypad interfacing.

- *Multiplexed interface*: In a multiplexed keypad interface, the n microcontroller GPIO output pins corresponding to n columns drive the $n \times 2^n$ demultiplexer, while m GPIO input pins corresponding to m rows are interfaced with the $2^m \times m$ multiplexer. For $n = m = 4$, 4 GPIO outputs are connected to the 4 to 16 demultiplexer that drives 16 columns, while 16 rows of the keypad are connected, through 16 to 4 multiplexer, to the 4 GPIO inputs. This scenario is shown in Figure 8.17, providing the illustration of multiplexed keypad interfacing. It should be noted that using $m + n$ GPIO pins, multiplexed interfacing allows to interface a keypad having $2^m \times 2^n$ keys. This is the maximum possible number of keys or switches that can interfaced for the given number of GPIO pins. Since a large number of keys are connected in multiplexed interface, it takes significant processor execution time to scan the entire keypad once.

The relationship between the number of GPIO pins used and the corresponding pos-

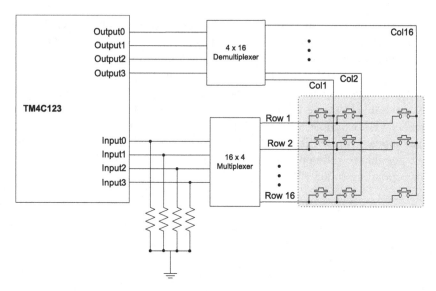

Figure 8.17: Multiplexed type keypad interfacing.

Table 8.8: The relationship between GPIO pins and the possible number of keypad keys that can be interfaced.

Interface type	GPIO pins	Number of keypad keys
Direct interface	$m + n$	$m + n$
Scanned interface	$m + n$	$m \times n$
Multiplexed interface	$m + n$	$2^m \times 2^n$

sible number of keys in the interfaced keypad is summarized in Table 8.8. Direct interfacing of keypad is simply equivalent to interfacing multiple switches individually. Individual switch interfacing has already been discussed in detail previously and extending it for multiple switches is straight forward. In the following, we will discuss scanned keypad interfacing further and illustrate it using an example program.

The scanned keypad interfacing discussed here neither handles the two key rollover situation nor it can handle key presses for long durations to result in repetitive key presses (for the same key). In case of two key rollover the first key is pressed and then the second key is pressed before the release of the first key. The keypad interfacing solution illustrated here requires the user to press the first key and then release it before pressing the second key. We will use a 16 button keypad for illustrating the scanned keypad interfacing method. To interface a 16 button keypad arranged as a 4×4 matrix, eight GPIO pins are required. Four of those GPIO pins are used as outputs while the other four are used as inputs as has been mentioned above.

Example 8.6 (C program for key pad interfacing.). This program implements the interfacing of 4×4 key pad interfacing. The four rows (Row 1 to Row 4) are connected to GPIO Port C pins 4 to 7 respectively. The four columns (Col 1 to Col 4) are connected to PF1, PE2, PA7 and PA6, respectively. This arrangement is due to the limited number of GPIOs available on the Tiva C.

```
// Default clock frequency and delay definition
#define   SYSTEM_CLOCK_FREQUENCY   16000000
#define   DEALY_COUNT                     SYSTEM_CLOCK_FREQUENCY/10000

// System clock register
#define SYSCTL_RCGCGPIO_R (*((volatile unsigned long *)0x400FE608))

// GPIO register definitions
#define GPIO_PORTA_DEN_R  (*((volatile unsigned long *)0x4000451C))
#define GPIO_PORTA_DIR_R  (*((volatile unsigned long *)0x40004400))
#define GPIO_PORTA_DATA_R (*((volatile unsigned long *)0x400043FC))
#define GPIO_PORTA_PIN6   (*((volatile unsigned long *)0x40004100))
#define GPIO_PORTA_PIN7   (*((volatile unsigned long *)0x40004200))

#define GPIO_PORTC_DEN_R  (*((volatile unsigned long *)0x4000651C))
#define GPIO_PORTC_DIR_R  (*((volatile unsigned long *)0x40006400))
#define GPIO_PORTC_DATA_R (*((volatile unsigned long *)0x400063FC))

#define GPIO_PORTE_DEN_R  (*((volatile unsigned long *)0x4002451C))
#define GPIO_PORTE_DIR_R  (*((volatile unsigned long *)0x40024400))
#define GPIO_PORTE_DATA_R (*((volatile unsigned long *)0x400243FC))
#define GPIO_PORTE_PIN2   (*((volatile unsigned long *)0x40024010))

#define GPIO_PORTF_DEN_R  (*((volatile unsigned long *)0x4002551C))
#define GPIO_PORTF_DIR_R  (*((volatile unsigned long *)0x40025400))
#define GPIO_PORTF_DATA_R (*((volatile unsigned long *)0x400253FC))
#define GPIO_PORTF_PIN1   (*((volatile unsigned long *)0x40025008))

#define row3                      (GPIO_PORTC_DATA_R & 0x10)
#define row2                      (GPIO_PORTC_DATA_R & 0x20)
#define row1                      (GPIO_PORTC_DATA_R & 0x40)
#define row0                      (GPIO_PORTC_DATA_R & 0x80)

// Function prototypes
void Delay(unsigned long counter);
void Init_Keypad_GPIOs(void);

unsigned char key_press = 0;

/* This function initialized the GPIOs for key pad interface. */
void Init_Keypad_GPIOs(void){
  volatile unsigned long delay;

  SYSCTL_RCGCGPIO_R |= 0x35;  // enable clock for GPIOPORT A,C,F,E
  delay = SYSCTL_RCGCGPIO_R;

  // enable digital function for rows and columns
  GPIO_PORTC_DEN_R |= 0xF0;
  GPIO_PORTA_DEN_R |= 0xC0;
  GPIO_PORTE_DEN_R |= 0x04;
  GPIO_PORTF_DEN_R |= 0x02;

  GPIO_PORTC_DIR_R &= ~0xF0; // configure columns as inputs
  GPIO_PORTA_DIR_R |= 0xC0;  // configure rows 0,1 as outputs
  GPIO_PORTE_DIR_R |= 0x04;
  GPIO_PORTF_DIR_R |= 0x02;  // configure rows 2,3 as outputs
}

/* This function scans the keypad for any key press and returns
```

```c
         the ASCII value for the pressed key. */
unsigned char Scan_Keypad(int *key_num){
   unsigned int i, cols, key = 0;
   const unsigned char key_val[4][4] =
                    { // col0, col1, col2, col3
                      {'1',   '2',   '3',   'A'}, // row0
                      {'4',   '5',   '6',   'B'}, // row1
                      {'7',   '8',   '9',   'C'}, // row2
                      {'*',   '0',   '#',   'D'}  // row3
                    };

   for(i = 0; i < 4; i++){          // loop for 4 rows
     cols = 0x01 << i;              // make columns high one by one
     GPIO_PORTF_PIN1 = cols << 1;
     GPIO_PORTE_PIN2 = cols << 1;
     GPIO_PORTA_PIN7 = cols << 5;
     GPIO_PORTA_PIN6 = cols << 3;

     if(row0 & 0x80){               // check row0
       *key_num = (i * 4) + 1;
       while(row0 & 0x80);          // wait for release
       return key_val[key][i];      // return value of key pressed
     }

     if(row1 & 0x40){               // check row1
       key += 1;                    // next key pressed
       *key_num = (i * 4) + 2;
       while(row1 & 0x40);          // wait for release
       return key_val[key][i];      // return value of key pressed
     }

     if(row2 & 0x20){               // check row2
       key += 2;                    // next key pressed
       *key_num = (i * 4) + 3;
       while(row2 & 0x20);          // wait for release
       return key_val[key][i];      // return value of key pressed
     }

     if(row3 & 0x10){               // check row3
       key += 3;                    // next key pressed
       *key_num = (i * 4) + 4;
       while(row3 & 0x10);          // wait for release
       return key_val[key][i];      // return value of key pressed
     }
     key = 0;
   }
   return 0;                        // no key pressed
}

int main ( void )
{
Init_Keypad_GPIOs();

// Loop forever and display Hello World message
while(1)
  {
    int val=0;
  // Insert delay for switch debouncing
    Delay(DEALY_COUNT);
```

```
      key_press = Scan_Keypad(&val);
   }
}

// This function implements the delay
void Delay(unsigned long counter)
{
   unsigned long i = 0;

   for(i=0; i< counter; i++);

}
```

8.8 Interfacing an LCD Module

The liquid crystal display (LCD) is replacing the seven-segment display due to its cost reductions and being more versatile for displaying alphanumeric characters. More advanced graphics displays are also available now at nominal prices. However, we will restrict our discussion to alpha-numeric displays with an integrated Hitachi's HD44780 display controller. The HD44780 LCD controller is one of the most widely used dot matrix controllers. The LCD displays with different standard screen configurations are available with common sizes of 8×1 (which is single row of eight characters), 16×1, 16×2, 20×2, 20×4, etc. We will illustrate LCD interfacing using 16×2 display, which will be equally applicable for other display sizes as well.

The HD44780 display controller can be interfaced with a microcontroller using GPIO pins allowing its configuration and use for information display. The first step toward LCD interfacing is to familiarize ourself with its pin functions, for proper interfacing, which is discussed next.

8.8.1 LCD Module Pin Functions

The LCD module pin functions are discussed here. The pins can be grouped as power (V_{ss}, V_{dd}, V_{ee}), control (RS, R/$\overline{W}$, E), and data (D7-D0). Their functionality is explained below.

1. **V_{ss}, V_{dd}, V_{ee}**: These pins are used for power supply (V_{dd} is +5V and V_{ss} is ground) and contrast control (V_{ee} pin). An adjustable voltage between +5V and 0V is applied to V_{ee} pin for adjusting the contrast.

2. **R/$\overline{W}$ (Read or Write) pin**: This is a control pin and is used to select between a Read or Write operation. When this pin is made logic 0 or low, a write operation is performed. In write operation, data pins D0-D7 are used to send data or command from the microcontroller to the LCD module. If R/$\overline{W}$ pin is made logic 1 or high, a read operation is performed. For read operation, D0-D7 are used to send data or status from the LCD module to the microcontroller.

3. **RS (Register Select) signal**: There are two types of read and write operations that can be performed. Any control or configuration of the LCD module is

performed using the instruction register. For instance, moving the cursor to a specific location, selection among different lines of the display, the cursor style, etc. are performed using the instruction register. On the other hand the data displayed is sent to the data register. The selection between instruction and data registers of the LCD module is performed using RS control pin. When RS pin is logic 0 or low, it selects the instruction register for a read or write operation (depending on the state of the $R/\overline{W}$ pin). For RS=0 and $R/\overline{W}$=0, a write operation to the instruction register is performed by sending data on D0-D7 data lines. The RS=0 and $R/\overline{W}$=1 performs read operation and obtains the busy flag status corresponding to previous control operation. Setting RS=1 selects the data register for read and write operation. Any data to be displayed on the LCD module should be written to this data register.

4. **E (enable) pin**: The enable pin is used for both read and write operations. In case of write operation, the $R/\overline{W}$ is cleared to logic 0, followed by the data written to D0-D7. Finally, a pulse is produced on the E pin following the sequence low-high-low. The data is latched by the LCD module on the high to low transition on the E pin. In case of read operation, the $R/\overline{W}$ is set to logic 1, followed by low-high-low transition on the E pin. The data is written by the LCD module to data bus (D0-D7) on the rising edge on E pin.

5. **D0-D7 (data) pins**: The D0 to D7 pin connections are used as an 8-bit data bus to perform read and write operations, for both commands as well as data. Pin D7 of the data bus serves as busy flag when performing a command register read operation. The LCD module also allows a 4-bit data bus interface to save GPIOs of the microcontroller for other purposes. In that case, an 8-bit transfer is performed using two 4-bit transfers. It is important to note that the default setting of LCD module for data bus width is 8-bit.

Different commands can be sent to an LCD module for its configuration. Table 8.9 provides a brief list of useful commands, which are used frequently while working with LCD modules.

8.8.2   Interfacing LCD Module with TM4C123

The microcontroller interfacing to LCD module requires connecting the control and data pins of the LCD module with appropriate GPIO pins of the microcontroller. In addition, LCD module should be powered using appropriate voltage source.

Pins 1 to 3 of the LCD module, respectively, are V_{ss} (GND), V_{dd} (+5V) and V_{ee} (contrast adjust). A 10 kΩ to 100 kΩ potentiometer is connected to this pin to adjust the voltage. Most character LCDs can achieve good display contrast with a voltage between 5V and 0V on pin 3. For programming and illustration purpose, the three control pins of the LCD module are connected to GPIO port pins PE3, PE4 and PE5, while the 8-bit data bus is connected to GPIO Port B. These connections are illustrated in Figure 8.18.

Table 8.9: LCD command codes with brief description.

Code	Brief command description
0x01	Clear display
0x02	Return home
0x04	Shift cursor left or decrement cursor
0x06	Shift cursor right or increment cursor
0x05	Shift display right
0x07	Shift display left
0x08	Display OFF, cursor OFF
0x0C	Display ON, cursor OFF
0x0E	Display ON, cursor ON
0x0F	Display ON, cursor ON and blinking
0x10	Shift cursor position to left
0x14	Shift cursor position to right
0x18	Shift entire display to left
0x1C	Shift entire display to right
0x20	2 lines, D4-D7 4-bit data interface, 5 × 11 character dot matrix
0x28	2 lines, D4-D7 4-bit data interface, 5 × 8 character dot matrix
0x30	2 lines, D4-D7 8-bit data interface, 5 × 11 character dot matrix
0x38	2 lines, D0-D7 8-bit data interface, 5 × 8 character dot matrix
0x80	Move cursor to the start of first line
0xC0	Move cursor to the start of second line

LCD Initialization

LCD module should be configured and initialized before the information can be displayed on it. Different configurations are available for an LCD display, of which the following four are important and are performed during initialization of the LCD module. Each of the following configurations is performed by writing an appropriate one byte command value using command write operation.

1. Function setting: This configuration is responsible for data bus width selection (8-bit or 4-bit), number of display lines (2 or 1) and the display font type (5×11 dots or 5×8 dots).
2. Display and cursor control: The display and cursor ON/OFF control is performed by setting or resetting appropriate bit fields using this command. In addition, this command also allows cursor blinking.
3. Entry mode setting: The entry mode setting allows configuration for cursor moving direction and entire display shift direction.
4. Clearing of display: This configuration allows to clear the display by writing 0x01 command byte to LCD module as can be seen from Table 8.9.

LCD Write Operation

The write operation is performed for sending data as well as commands to the LCD module. Assuming 8-bit data bus width, the write operation is performed by first writing data to data bus, followed by $R/\overline{W}$ and RS pins configuration. For write

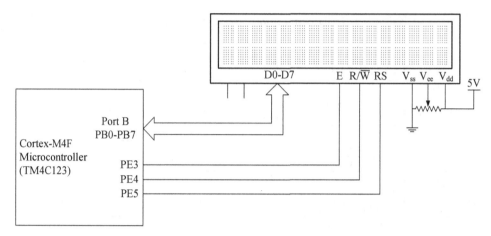

Figure 8.18: LCD interfacing with microcontroller using 8-bit data bus.

operation, the $R/\overline{W}$ pin should be held at logic low. Finally, a pulse (low to high followed by high to low) is produced on the E pin to latch data or command to LCD module. The timing diagram for data write operation is shown in Figure 8.19. From Figure 8.19, it can be observed that logic high to low transition on E pin is used for write operation.

The data or command write operation using 4-bit data bus width requires writing two 4-bit nibbles. High 4-bits are sent first followed by low 4-bits. Further details regarding 4-bit data bus interfacing are provided as part of Exercise 8.5.

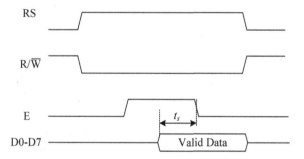

Figure 8.19: LCD write operation timing. The variable t_s represents the data setup delay. This delay is required before the falling edge on E pin for write operation.

LCD Read Operation

In case of read operation, the $R/\overline{W}$ and RS pins are configured first. For read operation, the $R/\overline{W}$ pin should be held at logic high. This is followed by low to high transition on the E pin to initiate data read operation. The timing diagram for data read operation is shown in Figure 8.20, which shows that low to high transition on E pin initiates read operation. Figure 8.20 also shows that data is available on the data bus to be read by microcontroller after time t_d, which represents the data output delay.

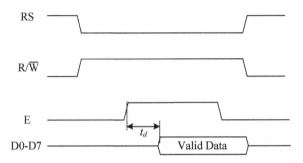

Figure 8.20: LCD read operation timing. The variable t_d represents the data output delay, once read operation is initiated by the rising edge on E pin.

Example 8.7 (C program for LCD module interfacing using 8-bit data bus.). This program illustrates the initialization of LCD module. Additionally, functions for data string display and moving the cursor to any arbitrary position on the display are also implemented.

```c
#define SYSCTL_RCGCGPIO_R   *((volatile unsigned long *)0x400FE608)

// register definitions for Port B, which is used as LCD data bus
#define GPIO_PORTB_DATA_R   *((volatile unsigned long *)0x400053FC)
#define GPIO_PORTB_DIR_R    *((volatile unsigned long *)0x40005400)
#define GPIO_PORTB_DEN_R    *((volatile unsigned long *)0x4000551C)
#define GPIO_PORTB_AFSEL_R  *((volatile unsigned long *)0x40005420)

/* register definitions for Port E (pins 3, 4 and 5), which is
used as LCD control bus */
#define GPIO_PORTE_DATA_R   *((volatile unsigned long *)0x400240E0)
#define GPIO_PORTE_DIR_R    *((volatile unsigned long *)0x40024400)
#define GPIO_PORTE_DEN_R    *((volatile unsigned long *)0x4002451C)
#define GPIO_PORTE_AFSEL_R  *((volatile unsigned long *)0x40024420)

// LCD module control pins and data bus definitions
#define LCD_E_PIN           0x08
#define LCR_RW_PIN          0x10
#define LCD_RS_PIN          0x20
#define LCD_DATA_BUS        GPIO_PORTB_DATA_R
#define LCD_CONTROL_BUS     GPIO_PORTE_DATA_R

// Function prototype definitions
void GPIO_Port_Init(void);
void Delay(volatile unsigned int delay);

void LCD_Init(void);
void LCD_Clear(void);
void LCD_Send_Command(unsigned char command);
void LCD_Send_Data(unsigned char data);
void LCD_Goto_XY(unsigned char x, unsigned char y);
void LCD_Send_String(char *ptr);
```

```c
/* This function initializes Port B and pins 3,4 and 5 of Port E
   as digital outputs. */
void GPIO_Port_Init(void)
{
 //Clock Gating Control
  SYSCTL_RCGCGPIO_R |= 0x12;      //enable clock for Port B & E

  //GPIO configuration for Port B
  GPIO_PORTB_DIR_R |= 0xFF;       // Port B as output
  GPIO_PORTB_DEN_R |= 0xFF;       // digital enable
  GPIO_PORTB_AFSEL_R &= ~0xFF;    // regular port function

  //GPIO Configuration for Port E pins 3, 4 and 5
  GPIO_PORTE_DIR_R |= 0x38;       // Port E pins as output
  GPIO_PORTE_DEN_R |= 0x38;       // digital enable
  GPIO_PORTE_AFSEL_R &= ~0x38;    // regular port function
}

/* This function writes user command to the LCD module. */
void LCD_Send_Command(unsigned char command)
{
  LCD_DATA_BUS = command;
  LCD_CONTROL_BUS = 0;

  // Generate a pulse on the LCD Enable pin
  Delay(1);
  LCD_CONTROL_BUS |= LCD_E_PIN;
  Delay(1);
  LCD_CONTROL_BUS &= ~(LCD_E_PIN);
  Delay(4);
}

/* This function writes user data to the LCD module. */
void LCD_Send_Data(unsigned char data)
{
  LCD_DATA_BUS = data;
  LCD_CONTROL_BUS = LCD_RS_PIN;

  // Generate a pulse on the LCD Enable pin
  Delay(1);
  LCD_CONTROL_BUS |= LCD_E_PIN;
  Delay(1);
  LCD_CONTROL_BUS &= ~(LCD_E_PIN);
  Delay(4);
}

/* This function initializes the LCD module for 8-bit data
   interface, turns ON the display and cursor. */
void LCD_Init(void)
{
  LCD_CONTROL_BUS = 0;
  Delay(1500);                    // Wait for approx. 15ms

  // Configure LCD for required functionality
  LCD_Send_Command(0x38);         // Function Set: 8 bit, 2 line
  LCD_Send_Command(0x10);         // Set cursor
  LCD_Send_Command(0x0E);         // Display ON, cursor ON
  LCD_Send_Command(0x06);         // Entry mode
```

```
}

/* This function clears LCD module and brings cursor to home
   position on the display. */
void LCD_Clear(void)
{
  LCD_Send_Command(0x01);      // Clear display
  Delay(170);                  // Delay for approx. 1.6ms
  LCD_Send_Command(0x02);      // Move cursor to home
  Delay(170);                  // Delay for approx. 1.6ms
}

/* This function writes character string to LCD display. */
void LCD_Send_String(char *ptr){
  while(*ptr){
    LCD_Send_Data(*ptr);
    ptr++;
  }
}

/* This function moves the cursor to (x,y) coordinates on
   the LCD display. */
void LCD_Goto_XY(unsigned char x, unsigned char y)
{
  unsigned char row_start_address[] = {0x80, 0xC0};

  // Move cursor to (x,y) location on display
  LCD_Send_Command(row_start_address[y-1] + x - 1);
  Delay(170);                  // Delay for approx. 1.6ms
}

/* This function generates the delay. */
void Delay(volatile unsigned int delay)
{
    volatile unsigned int i, j;
    for(i = 0; i < delay; i++)
    {  //introduces a delay of about 10us at 16MHz
       for(j = 0; j < 12; j++);
    }
}
```

Now we use LCD module interfacing explained in Example 8.7 and write a simple user program to display 'Hello' and 'World' at the center of first and second lines of the display. The C program implementing this functionality is given in Example 8.8, which uses the functions of Example 8.7.

Example 8.8 (C program for illustrating data display on LCD module.). This program uses the LCD driver in Example 8.7 to display user data string on the LCD module.

```
int main(void)
{
  GPIO_Port_Init();
  LCD_Init();
  LCD_Clear();
```

```
// Loop forever and display Hello World message
while(1)
{
  LCD_Goto_XY(5, 1);
  LCD_Send_String("Hello");
  LCD_Goto_XY(5, 2);
  LCD_Send_String("World");
}
}
```

8.9   Summary of Key Concepts

The focus of this chapter has been a broad introduction to the GPIO interfacing. Specifically, the following key concepts have been introduced.

- The key attributes related to a GPIO pin hardware include bidirectional capability, open drain, current drive strength, slew rate, etc.

- To use a microcontroller for a specific task, a minimum connectivity is required that covers power, clock, reset and programming/debugging wiring.

- In modern microcontrollers, each GPIO pin is multiplexed for many different functionalities. Configuration of a GPIO pin for a specific function is performed during initialization and can be switched to an arbitrary functionality at run time as well.

- The basic steps to configure a microcontroller pin for GPIO purpose include enabling clock to the GPIO module as well as its mode, data and pad related configurations.

- In Cortex-M based microcontroller the peripherals are memory mapped. To use a microcontroller pin as GPIO or as part of another peripheral interface, the understanding of the peripheral space memory map is highly crucial.

- The TM4C123 microcontroller provides a special data register masking. This masking allows direct access to selected single GPIO pin or combination of pins without requiring read-modify-write sequence.

- When using a GPIO pin as an output, its current drive strength and slew rate should be configured to appropriate values.

- To use a GPIO pin as input it is important to ensure that the voltage levels of the external signal do not exceed the GPIO pin tolerable voltage levels.

- The parallel port interfacing is illustrated by interfacing GPIO port pins with seven segment LED and LCD display devices and matrix type keypad.

- Time multiplexing is used to reduce the required number of GPIO pins, when interfacing to LED display and keypad. When using time-multiplexing, the GPIO pins requirement reduces from $m \times n$ to $m + n$, where m and n represent the number of display digits and the number of LEDs per digit in case of seven-segment LED display, while they represent the number of rows and columns in case of matrix keypad.

Review Questions

Question 8.1. What are the uses of open-drain functionality, when it is available on a GPIO pin?

Question 8.2. Name different communication and timing peripherals that are available on TM4C123.

Question 8.3. What debug/programming interfaces are available on TM4C123?

Question 8.4. What are four basic hardware connections that are essential for proper functioning of a microcontroller?

Question 8.5. How much RAM and flash (on chip) memories are integrated in TM4C123 and what are their address ranges?

Question 8.6. List four configuration steps required to use a microcontroller pin for GPIO purpose.

Question 8.7. Why does the GPIO data register have more than one address assigned to it? This is in contrast to other registers, which are assigned only one address.

Question 8.8. What is the disadvantage of special data register masking supported by TM4C123 microcontroller?

Question 8.9. When a GPIO pin is to be used for an alternate functionality, which two registers should be configured for this purpose?

Question 8.10. The LEDs in Figure 8.10 are turned on when a logic high signal is applied on the corresponding GPIO pin. Now we are interested in turning on the LED when a logic low signal is applied. For that purpose, redraw Figure 8.10 for LED wiring so that a logic low signal on the corresponding GPIO turns on the LED.

Question 8.11. While defining register labels in a C program, why do we use the 'volatile' key word?

Question 8.12. When the GPIO pins have limited current sourcing but large current sinking capability, which seven-segment display (common anode or common cathode) would be preferred?

Question 8.13. We are interested in using a keypad interface with the option of multiple simultaneous key presses. Among the three different types of keypad interfaces discussed in this chapter, which one is more suitable for this case?

Exercises

Exercise 8.1. Among the four basic configuration steps for a GPIO, the clock and bus configuration is the first one that needs to be performed. What happens if we perform any of the other three steps before clock configuration?

Exercise 8.2. Consider using PORT C pins for both input as well as output purposes. Specifically, we want to use PORT C pins 0, 2, and 7 for output, and pins 1 and 4 for input. The PORT C base address is 0x40006000 and the offset for the DATA register ranges from 0x000 to 0x3FC. Assume that the direction register is configured accordingly. We are interested in generating logic high/low only on pins 0, 2, and 7 without affecting any other pins configured as outputs and it is not permitted to use Read-Modify-Write sequence. Similarly, reading data from input pins 1 and 4 should be done without using Read-Modify (which is equivalent to software masking).

(a) What should be the PORT C DATA register address, when used as an output port to update the logic levels at pins 0, 2, and 7?

(b) What should be the PORT C DATA register address, when used as an input port to read the data from pins 1 and 4?

Exercise 8.3. Recall that we used 12 GPIO lines to interface four-digit seven-segment display (eight lines were used for seven-segments and dot, while four lines were used to select the desired digit). Now assume we have run short of GPIOs and we would like to interface the same four digit seven-segment display using 10 GPIOs. Redraw the four-digit seven-segment interfacing block diagram using 10 GPIOs and show the necessary details.

Exercise 8.4. In the standard algorithm to scan a matrix keypad, each column (or row) is driven by a GPIO pin one at a time and correspondingly all the GPIO pins connected to the rows (or columns) are scanned. This method enables the detection of multiple (specifically two) simultaneous key presses. Using this approach, the entire keypad scanning takes significant execution time of the processor. When only single key presses are permitted, then more efficient algorithm described below can be used instead. The key steps for this efficient algorithm are given next.

1. First, drive all columns high simultaneously and read all the rows.

2. Next, drive all the rows high simultaneously and read all the columns.

3. Finally, combine the read results from the above two steps to find which key or switch was pressed.

Write a C program that implements this algorithm.

Exercise 8.5. This problem discusses 4-bit data bus based LCD module interfacing. To enable 4-bit data interface, command 0x2U (where U is user parameter and can assume values of 0x0, 0x4, 0x8 or 0xC) is used when configuring for function setting. Now consider 4-bit write operations. The timing diagram for 4-bit write operation is shown in Figure 8.21. Write a C program that configures LCD module with 4-bit data bus interface and displays 'Hello' on line 2 of the display.

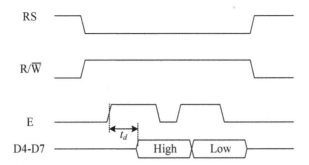

Figure 8.21: LCD write operation timing for 4-bit data bus interface.

Exercise 8.6. Two switches SW1 and SW2 are connected to port E pins PE1 and PE2, respectively as shown in Figure 8.22. Pressing switch SW1 each time toggles red LED connected to PF1, while pressing SW2 toggles blue LED connected to PF2. Write a program that performs the desired operation. Configure all the necessary GPIO registers. You are free to write your program using either polling or interrupt based approach.

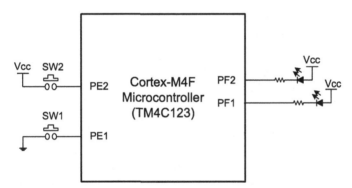

Figure 8.22: LCD write operation timing for 4-bit data bus interface.

Exercise 8.7. Consider the case where a multi-digit seven-segment display is constructed using single digit seven segment displays. Assuming we have two common anode and two common cathode digits to construct a four digit display. Redraw the connection diagram, similar to Figure 8.14, for this case.

Exercise 8.8. Consider an n digit, time-multiplexed, seven-segment display. Since one digit is turned on at a time in this case, what is the minimum possible number of GPIO pins required to interface this display?

Chapter 9

I/O Synchronization and Interrupt Programming

Overview

Speed of input and output (I/O) devices mismatches that of the processor. We need to synchronize the activities of processor and I/O devices to have an efficient embedded system. In the absence of synchronization, one possibility is that the processor is highly underutilized when the I/O is too slow, while it can potentially lead to data loss in the case of too fast I/O device. There is an inherent need to synchronize the processor with the I/O devices. This chapter first describes different methods that can be employed for synchronization of input and output devices with that of processor execution. Later part of the chapter is dedicated to interrupt based synchronization, where we specifically present how Cortex-M processor uses interrupts for this job.

9.1 Why Synchronization is Needed

One of the distinguishing features of an embedded system, compared to a conventional computer, is its strong coupling between the processor execution and external interfaces. Microcontrollers have a variety of integrated peripherals, which serve as interfaces between the physical phenomenon of interest and the processor. These interfaces play a key role to realize this coupling between processor and external world. A large proportion of these interfaces run quite slow (on the order of few Hz to kHz) compared to the execution speed of processor (which is on the order of tens to hundreds of MHz). Many input-output devices, e.g., displays, sensors, switch inputs, control actuators, etc. operate quite slow compared to the execution speed of the processor. For instance, consider the example of an autonomous robot with obstacle avoidance capability. The data from obstacle detection sensor is checked continuously for the presence of an obstacle. The processor might be kept busy in waiting for an obstacle before it can take an action by controlling the servo motors of the robot. If there are other tasks to be executed by the processor then keeping the processor busy in waiting results in wastage of thousands of processor clock cycles, which otherwise could have been used to execute other tasks. Even if the processor does not have any other task to be performed, busy waiting should still be avoided to reduce system power consumption, which is possible by changing the processor state from busy to low power idle state. To deal with the above mentioned speed mismatch between software execution and the hardware interfaces, I/O synchronization is used. Different approaches for I/O synchronization are available. Some of the approaches are simple (e.g., busy wait, blind cycle) but are inefficient, while others including periodic poling, interrupts, DMA, etc. are more complex to implement but improve execution performance.

Among the different approaches available, the use of interrupts to achieve the above mentioned synchronization is one of the fundamental and widely used approach. Due to this reason main focus of this chapter will be on the use of interrupts for I/O synchronization. Specifically, the following topics will be discussed in detail in this chapter.

- Introduction to I/O synchronization

- Methods for I/O synchronization

- Interrupts and their types

- Interrupt configuration for Cortex-M microcontroller

- Configuring multiple priority interrupts

9.2 Introduction to Input-Output Synchronization

A common practice in microcontroller architecture is to integrate many different peripherals with the processor and memory. These peripherals are primarily used for data handling. Broadly speaking data handing can be either in the form of data acquisition/generation or data transmission/reception. A peripheral operates as an input device in case of data acquisition as well as data reception, while its role is changed to an output device for data generation and data transmission. For example, analog to digital converters are used for data acquisition from different information sources (e.g., sensors and transducers) and digital to analog converters are used for signal generation purpose. Similarly most of the communication interfaces (e.g., UART, USB, Ethernet, etc.) are used for data transmission as well as reception among different devices. Based on this description, we can observe that some peripherals are either input or output devices, while others are both input and output devices.

From an operational perspective, the hardware device can be in one of the two possible states, namely, *active state* or *inactive state*. No input-output operation can be performed by the device in an inactive state. When the device is active, it can be either in *busy state* or in *ready state*. This hierarchy of device states is shown in Figure 9.1. A transition from inactive to active state can be generated by the software (i.e., user application program) and the device is in ready state when activated by the user as can be seen from Figure 9.1(b). A transition from ready to busy state occurs either due to user task assignment or due to an event. The device remains in the busy state when performing the assigned task. When the device is finished with the current task a transition from busy to ready state occurs. The device remains in ready state, waiting for another task assignment. A transition from ready to inactive state can be triggered either by the user program or it can happen due to expiration of predefined inactivity time interval. A state transition from busy state to inactive state can be triggered by the software (user program), which can lead to data loss. The state transitions shown in Figure 9.1(b) are equally valid for both input as well as output modes of operation of a peripheral device.

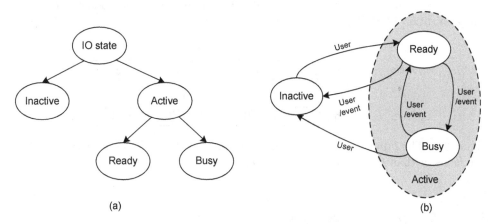

Figure 9.1: (a) Peripheral device operating states. (b) Device state transition diagram that is valid for both input as well as output modes of operation.

For efficient data handling, the peripheral hardware and the system software should work in synchronization, (also termed as I/O synchronization) to avoid delays in data handling as well as any possibility of losing the data. If the software execution is faster compared to the peripheral hardware speed then software needs to wait for the completion of the activity being performed by the hardware before it can request for the next activity. In the absence of synchronization, the software might wait longer than necessary, incurring further delays. On the other hand, in case of input devices, data loss becomes a possibility if the hardware generates data faster than the rate at which the software can process it. Different methods for I/O synchronization are primarily responsible for minimizing the delays in data handling as well as avoiding any data losses. These methods differentiate in terms of their implementation complexity as well as their capability of efficiently using the processing resource.

9.2.1 Input Device Synchronization

As mentioned above, synchronization requires the peripheral hardware and the associated software to wait for each other for proper data handling. In case of an input device, hardware is the data producer, while software is responsible for receiving and processing data and is labeled as data consumer. Once the hardware has produced data, the software reads it and allows the hardware device to start producing new data, while the software is busy in processing that data. If the time required by the hardware to produce data is higher than the time it takes the software to consume data then software needs to wait for the hardware. This scenario is depicted in Figure 9.2, which shows the timing diagram for the device software and hardware states. The arrows from hardware to the software mark the synchronizing instances. In case of second possible scenario, if the hardware produces data faster than the software can consume it then there is always a possibility of data loss. Different mechanisms, including buffering, hand shaking, etc. are used to avoid any data loss and will be discussed laster in this chapter as well as subsequent chapters. In an ideal scenario, the hardware and software data producing and consuming speeds should match exactly, which is highly unlikely in practical situation and some methodology for their synchronization is required.

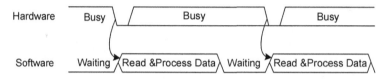

Figure 9.2: The input device synchronization showing the case where software execution is faster than the associated hardware.

9.2.2 Output Device Synchronization

For an output device synchronization the above mentioned two scenarios are also applicable. In case of an output device, hardware is the data consumer, while software is responsible for producing data. If the time taken by the software to produce data is less than the time required by the hardware to consume that data (i.e., to transmit or generate corresponding signal for that data) then software needs to wait for synchronization. For the opposite case, the hardware is faster and needs to wait for the software. Both of these scenarios are illustrated in Figure 9.3 by depicting device software and hardware states as a function of time. Figure 9.3(a) illustrates the case where software is faster than the hardware and needs to wait, while the case of faster device hardware is shown in Figure 9.3(b). The synchronizing instances are marked by the arrows from the software to the hardware. From the above discussion, we can observe that in case of an output device synchronization, the mismatch in the hardware and software speeds results in delay only and unlike an input device, there is no possibility of data loss.

9.3 Methods for Input-Output Synchronization

As discussed in the previous section I/O synchronization is important for efficient use of processing capability as well as data integrity. Different methods are available for synchronization with corresponding merits and demerits. Before discussing these

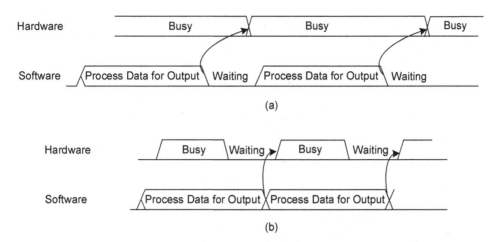

Figure 9.3: The output device synchronization for the case of (a) faster device software waiting for the hardware and (b) faster device hardware waiting for the software.

methods, it is beneficial to define some of the important parameters that can be used to quantify the performance of different synchronization approaches. For that purpose, the following three performance parameters are critical and will be described next.

- Latency

- Bandwidth or throughput

- Priority

The delay between time instance when the I/O device indicates that it needs to be processed till the time instance when the required processing is initiated is termed *latency*. Latency includes any delays incurred due to the hardware as well as software delays due to processor occupancy by another task. For an input device, the latency can be considered as the time elapsed between new data available till the time when software reads the data for processing. Similarly from an output device perspective, latency can be defined in terms of time interval from the instance the output device becomes idle till the software (processor) provides new data to the output device. A real-time system is one that can guarantee a worst-case latency. It is required that the latency is bounded and should be as small as possible. The term bounded implies that it is always below an already specified threshold value, which is acceptable to the consumers/users of the real-time system.

Bandwidth or throughput is defined as the maximum data processing/transfer rate (measured in terms of bytes/bits per second) that can be handled by the system. In some cases the system bandwidth is limited by the I/O device, while in other scenarios it is limited by the execution speed of the processor. *Priority* determines the order in which the processor processes the requests from different I/O devices, when two or more devices have made requests simultaneously. In addition, priority also determines whether a high-priority request is allowed to suspend the execution of a lower priority device request, currently being executed.

Different methods for I/O synchronization can be grouped into the following three main categories.

1. Blind cycle

2. Polling based methods

   - Continuous polling or busy waiting
   - Periodic polling

3. Interrupt-driven methods

   - I/O synchronization using Interrupts
   - Direct memory access

In the following each of these methods will be discussed and the pros and cons of different methods will be highlighted.

9.3.1 Blind Cycle

When an I/O operation is performed, the device hardware incurs some delay in responding to the software request. This delay corresponds to the time interval starting from the time instance when software makes the I/O operation request, till the device hardware has finished the operation. If this delay is highly predictable due to the fact that any variations in this delay are relatively small then it is possible that the software can initiate a new I/O request after fixed delay. This is precisely what is done in *blind cycle* synchronization, where the software waits for a fixed time interval assuming that the I/O device hardware has completed the operation within this time interval. The choice of fixed delay depends on the maximum possible time that the device hardware may require to complete the operation. To ensure proper functioning, the fixed delay used by blind cycle cannot be less than the maximum time required by the device hardware. If the mean time for the device hardware to complete the operation is significantly smaller than the corresponding maximum time, then blind cycle method incurs unnecessarily large delays resulting in wastage of processing time, which is the main disadvantage of this method. The software and hardware operations are performed sequentially in case of blind cycle method resulting in poor system performance by wasting processing resource. The key advantage of blind cycle method is its implementation simplicity.

In case of an input device using blind cycle synchronization, once the device hardware is triggered for and input operation, the device software waits for fixed time interval before reading the data from device hardware. On the other hand, for an output device, the software triggers the hardware after writing data to the device and then waits for a fixed time interval. The sequence of operations using blind cycle synchronization are shown in Figure 9.4(a) and (b) for input and output devices, respectively. This method is called blind cycle, since no status information is obtained by the software to know about the current state of the device hardware. As mentioned above, this method is useful for those I/O operations where the device hardware speed is highly predictable. Recall the LCD interfacing discussed in Chapter

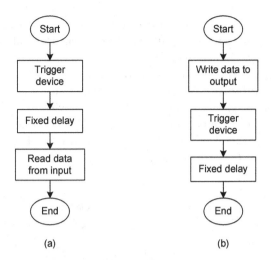

Figure 9.4: Blind cycle synchronization for (a) an input device and (b) an output device.

8, where different fixed delays were used during initialization as well as other display related operations. In other words, the LCD interfacing was performed using blind cycle synchronization.

9.3.2 Polling Based Methods

The unnecessarily large delays incurred by blind cycle method can be reduced by checking the status of the device hardware either continuously or periodically. Using this status information the software can eliminate any excessive delays and waits only for the required duration. Status checking of the device hardware, by the software, is also termed as *polling*.

Continuous Polling or Busy Wait

The *continuous polling* is implemented by the software using a loop, where the status of the device hardware is continuously checked for operation completion. As soon as the device hardware is done with the operation, the corresponding status is changed and the software immediately knows about it and can initiate new operation. Continuous polling is also called *busy waiting*. One major drawback of busy waiting is the possibility of blocking software execution. This can happen in case of some hardware malfunctioning, which halts the corresponding status information updating and as a result the software execution is blocked.

From the perspective of an input device, the software triggers the device hardware then waits until the device has new data available, which is read and processed by the software. This is shown in Figure 9.5(a). In case of an output device, there are two possible implementations for busy waiting. In the first case the software writes the data to the device followed by device triggering and then waiting for the operation completion by checking the status continuously. These busy waiting implementations, for output device as well as the input device, perform the software and hardware operations sequentially. In the second possible implementation for output device, the software first checks the device status for being ready then writes the data and triggers the device. In this case, the software does not wait for the device to finish the operation and can start preparing new data for the device in the mean time, while the device hardware is busy in performing the operation. This second implementation has the advantage of parallelizing (partially) the software and hardware activities and can improve the overall performance of the system.

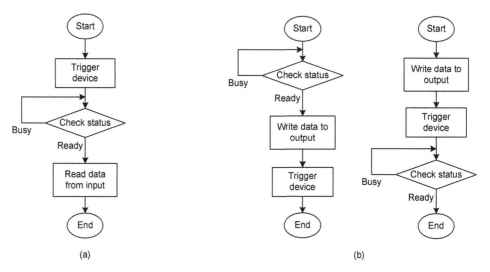

Figure 9.5: Busy waiting based device synchronization for (a) an input device and (b) an output device. Two different methods for output device synchronization are shown.

Periodic Polling

The key disadvantage of continuous polling is the possibility of blocking the software execution. This problem is resolved by using periodic polling. In periodic polling a timer is used to generate periodic interrupts. The device status is checked at each timer interrupt and appropriate action is performed. In case of an input device, when the data is ready the device will set the status flag and the software will read this data at next timer interrupt. This is shown in the form of a block diagram in Figure 9.6(a). For an output device the status flag is set by the hardware when the device is done with the previous operation. At the next timer interrupt the software will check the status flag and can request for next output operation as depicted in Figure 9.6(b). Polling based switch interfacing has been discussed in Section 8.5, which effectively implements periodic polling. The periodicity of switch interfacing in Section 8.5 is not determined by a timer, rather it is equal to one execution cycle time of endless loop.

One of the limitations of periodic polling is an extra delay, compared to continuous polling, that might occur in the I/O operation. In the worst case this extra delay can be equal to the time period of timer interrupt. This delay can be reduced by increasing the frequency of timer interrupts. However, doing so will require more number of executions of timer interrupt service routine (ISR). So in conclusion there exists a trade-off between performance improvement and the associated additional overhead.

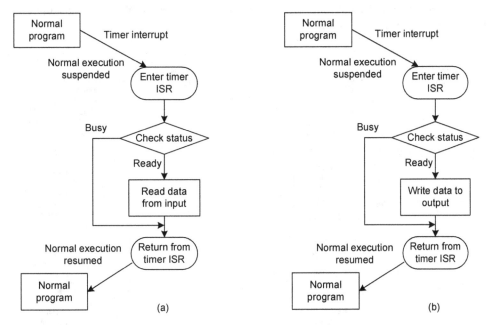

Figure 9.6: Periodic polling based device synchronization for (a) an input device and (b) an output device.

9.3.3 Interrupt-Driven Methods

I/O Synchronization Using Interrupts

Some of the key limitations associated with the above-mentioned I/O synchronization approaches are software execution blocking and serial execution of software and hardware operations. When the software is waiting for the completion of device hardware operation, the processing time is being wasted, which otherwise could have been utilized for performing another useful task. This is precisely what is achieved by interrupt based I/O synchronization, which overcomes the limitations of the above mentioned synchronization approaches.

For an input device, the hardware will generate an interrupt when the data is ready and the corresponding software ISR will read the data as shown in Figure 9.7(a). In case of an output device, the hardware will generate an interrupt on the completion of assigned operation to inform the software. The software then can initiate another output operation while executing the corresponding ISR. This scenario is shown in Figure 9.7(b).

When multiple different peripheral devices are synchronized using interrupts, then it is possible to assign different priorities as well as masking to their corresponding interrupts. In contrast, it is not possible to assign priorities when synchronizing devices using polling based approach. Interrupt based synchronization is quite useful for real-time applications as well as for event driven systems. The key advantage of interrupt based synchronization is the parallelization of software and hardware operations with minimal overhead. It also avoids any software blocking and improves the systems overall performance.

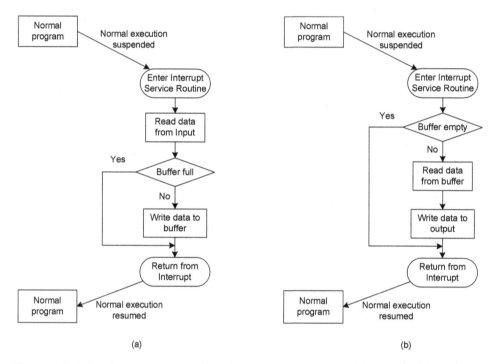

Figure 9.7: The device interrupt based synchronization for (a) an input device and (b) an output device.

Direct Memory Access

Direct memory access (DMA) is a useful attribute of the microcontroller architecture, which allows data transfers among different peripheral devices and system memory without involving the main processor. Specifically, Cortex-M based microcontroller uses a DMA controller as a peripheral device that can transfer data among the supported peripherals and main memory. In addition, memory to memory transfers are also possible using a DMA controller. The DMA controller is used for transferring large data volumes by offloading the main processor. The main processor is interrupted at the end of the transfer to inform about the completion of the operation. DMA is primarily used for high speed data transfers, where efficient bandwidth utilization and low latency are essential requirements. It is important to realize that in DMA based data transfer, the peripheral and system memory are directly accessed by the DMA controller, it is primarily a hardware based data transfer approach. Since the main processor is not involved in the data transfer, correspondingly there is not any associated software execution.

For an input device, the device hardware will generate a request to the DMA controller for data transfer when it has new data available. The DMA controller will read the data from the input device and will put it to the assigned memory location without any software execution. For an output device, the peripheral device hardware will generate a request to the DMA controller for data transfer when the output device is ready. In this case the DMA controller will obtain data from the specified memory location and writes it to the output device.

9.4 Types of Exceptions or Interrupts

- Synchronous exceptions
- Asynchronous exceptions

9.4.1 Synchronous Exceptions

Synchronous exceptions or interrupts are generated by the processor while executing instructions and are requested only after the completion of execution of the instruction. Below is the list of synchronous exceptions

- Memory management (MemManage) exception
- BusFault exception
- UsageFault exception
- SVCall exception

The first three are system exceptions that are executed as a result of fault. The last exception is also called a supervisor call (SVC), which is initiated by SVC instruction. Applications executing in operating system environment can make use of SVC instruction for accessing operating system kernel functions as well as any device driver functions. In addition, when a computational error or an exceptional condition due to instruction execution is detected, it leads to the execution of SVC instruction, causing the processor exception.

9.4.2 Asynchronous Exceptions

An asynchronous exception is the one that can occur, at an arbitrary time, during the execution of an instruction. These exceptions are generated by events other than the processor execution. Below is the list of asynchronous exceptions

- Reset exception
- NMI exception

- PenSV exception

- Systick exception

- Interrupt

Reset exception is used by the system on power up or in response to warm reset for restarting the system and is the highest priority exception. The NMI is the second highest priority exception used by the system to handle a highly critical situation. PendSV is a software triggered exception and unlike other software interrupts is an asynchronous exception. The SysTick exception is generated by the system timer when its count has reached zero. It will be discussed in detail in Chapter 10. An interrupt is an exception triggered by a peripheral device or it can be generated by the associated software and is also called a *peripheral interrupt* or *external interrupt*. For asynchronous exceptions apart from reset exception, it is possible that the processor can complete the execution of currently being executed instruction before entering the exception handler.

The maximum number of external interrupts that are supported by Cortex-M microcontrollers are 240. They are assigned to different peripherals including timers, communication interfaces (UART, SPI, I2c, CAN bus etc.), data converter modules, general purpose I/Os etc. A specific microcontroller can use an arbitrary number of external interrupts depending on its peripherals. A total of 139 interrupts are supported by the NVIC controller of Texas Instrument's TM4C123 microcontroller and are labeled as IRQ0 to IRQ138. Table 9.1 lists the assignment of IRQ numbers to the peripheral devices. An interrupt from each peripheral device is assigned a specific IRQ number. This unique IRQ number is used to determine the address of corresponding interrupt service routine (ISR) in the vector table. Assuming interrupt vector table base address is 0x00000000, then n^{th} IRQ vector address is obtained as $4n + 64$. This address when defined in terms of exception number m, is obtained as $4m$, providing the relation between exception number and IRQ number as $m = n + 16$.

Table 9.1: IRQ assignment to peripheral devices.

IRQ no.	Peripheral	IRQ no.	Peripheral
IRQ0	GPIO port A	IRQ1	GPIO port B
IRQ2	GPIO port C	IRQ3	GPIO port D
IRQ4	GPIO port E	IRQ5	UART0
IRQ6	UART1	IRQ7	SPI0
IRQ8	I2C0	IRQ9	PWM0 Fault
IRQ10	PWM0 Generator 0	IRQ11	PWM0 Generator 1
IRQ12	PWM0 Generator 2	IRQ13	QEI0
IRQ14	ADC0 Sequencer 0	IRQ15	ADC0 Sequencer 1
IRQ16	ADC0 Sequencer 2	IRQ17	ADC0 Sequencer 3
IRQ18	WD Timer 0 & 1	IRQ19	16/32 Bit Timer 0A
IRQ20	16/32 Bit Timer 0B	IRQ21	16/32 Bit Timer 1A
IRQ22	16/32 Bit Timer 1B	IRQ23	16/32 Bit Timer 2A
IRQ24	16/32 Bit Timer 2B	IRQ25	Analog comparator 0
IRQ26	Analog comparator 1	IRQ27	Reserved
IRQ28	System control	IRQ29	Flash and EEPROM control

continued ...

...continued

IRQ no.	Peripheral	IRQ no.	Peripheral
IRQ30	GPIO port F	IRQ31	Reserved
IRQ32	Reserved	IRQ33	UART2
IRQ34	SPI1	IRQ35	16/32 Bit Timer 3A
IRQ36	16/32 Bit Timer 3B	IRQ37	I2C1
IRQ38	QEI1	IRQ39	CAN0
IRQ40	CAN1	IRQ41	Reserved
IRQ42	Reserved	IRQ43	Hibernation module
IRQ44	USB	IRQ45	PWM0 Generator 3
IRQ46	uDMA Software	IRQ47	uDMA error
IRQ48	ADC1 Sequence 0	IRQ49	ADC1 Sequence 1
IRQ50	ADC1 Sequence 2	IRQ51	ADC1 Sequence 3
IRQ52-56	Reserved	IRQ57	SPI2
IRQ58	SPI3	IRQ59	UART3
IRQ60	UART4	IRQ61	UART5
IRQ62	UART6	IRQ63	UART7
IRQ64-67	Reserved	IRQ68	I2C2
IRQ69	I2C3	IRQ70	16/32 Bit Timer 4A
IRQ71	16/32 Bit Timer 4B	IRQ72-91	Reserved
IRQ92	16/32 Bit Timer 5A	IRQ93	16/32 Bit Timer 5B
IRQ94	32/64 Bit Timer 0A	IRQ95	32/64 Bit Timer 0B
IRQ96	32/64 Bit Timer 1A	IRQ97	32/64 Bit Timer 1B
IRQ98	32/64 Bit Timer 2A	IRQ99	32/64 Bit Timer 2B
IRQ100	32/64 Bit Timer 3A	IRQ101	32/64 Bit Timer 3B
IRQ102	32/64 Bit Timer 4A	IRQ103	32/64 Bit Timer 4B
IRQ104	32/64 Bit Timer 5A	IRQ105	32/64 Bit Timer 5B
IRQ106	System Exception	IRQ107-133	Reserved
IRQ134	PWM1 Generator 0	IRQ135	PWM1 Generator 1
IRQ136	PWM1 Generator 2	IRQ137	PWM1 Generator 3
IRQ138	PWM1 Fault		

The use of GPIO pins as external interrupts will be discussed in detail in this chapter. External interrupts associated with other peripheral devices will be discussed in subsequent chapters.

9.5 Configuring Interrupts for Cortex-M Devices

Configuring different peripheral devices to use interrupts for synchronization is performed during their initialization and is considered as one aspect of device initialization. To enable interrupts for a peripheral device requires configuration at the following three different levels.

- Processor interrupt configuration

- NVIC configuration

- Device interrupt configuration

The NVIC and processor interrupt configurations are quite similar for different peripheral devices. However, the device interrupt configuration varies from device to device and the configuration for one device may not be used for a different peripheral device. In the following, we will discuss in detail the configuration of a GPIO port as a source of external interrupts. The interrupt configuration for other peripheral devices will be discussed in subsequent chapters.

The processor interrupt configuration provides global enabling and disabling of interrupts. The NVIC configuration allows device level interrupt configuration, priority assignment and pending behavior management. A separate interrupt is assigned to a device or device functionality, with an associated priority by the NVIC. For instance, a GPIO device is assigned one interrupt for all of its I/O pins when configured as source of external interrupts, while an ADC device has multiple interrupts assignments, one for each ADC sequencer functionality. The NVIC can be viewed as an interrupt multiplexer with multiple interrupt inputs from different devices and single interrupt output to the processor.

The device level interrupt configuration provides another level of interrupt multiplexing. For a given device, there can be multiple interrupt sources. For instance, when all the pins of an 8-bit GPIO port are configured as sources of external interrupts, then there can be eight different interrupts from that GPIO device. However, only one interrupt is assigned to this device by NVIC. In this case, the GPIO device functions as an 8×1 interrupt multiplexer. To differentiate among different sources of interrupts, the GPIO device has an interrupt status register that is used by the application program to implement the interrupt service routine. This hierarchical organization of external device interrupts is illustrated by Figure 9.8 showing multiplexing of interrupts at different levels.

9.5.1   Processor Interrupt Configuration on TM4C123

There are three processor registers, FAULTMASK, PRIMASK and BASEPRI that are used for interrupt masking as mentioned in Section 3.3.2. The default values of these registers are such that all interrupts are enabled. However, in case of critical code execution, that should not be interrupted, global interrupt enabling and disabling can be performed. For that purpose, either the PRIMASK or FAULTMASK register can be used. See Section 3.3.2 to find the difference in interrupt masking, when using PRIMASK or FAULTMASK. The BASEPRI register is used for enabling or disabling interrupts selectively, depending on their priorities.

The assembly program that implements interrupt enabling and disabling subroutines using PRIMASK register is provided in Listing 9.1, while the use of FAULTMASK register for this purpose is illustrated in Listing 9.2.

```
; Subroutine for enabling interrupts
EnableInterrupts
      CPSIE   I       ; clear PRIMASK to enable interrupts
      BX      LR

; Subroutine for disabling interrupts
DisableInterrupts
```

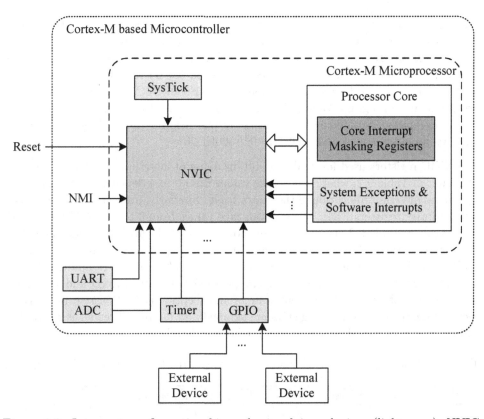

Figure 9.8: Interrupt configuration hierarchy involving, devices (light gray), NVIC (gray) and processor (dark gray).

```
        CPSID  I       ; set PRIMASK to disable interrupts
        BX     LR
```

Listing 9.1: Interrupt enable and disable subroutines using PRIMASK register.

```
; Subroutine for enabling interrupts
EnableInterrupts
        CPSIE  F       ; clear FAULTMASK to enable interrupts
        BX     LR

; Subroutine for disabling interrupts
DisableInterrupts
        CPSID  F       ; set FAULTMASK to disable interrupts
        BX     LR
```

Listing 9.2: Interrupt enable and disable subroutines using FAULTMASK register.

9.5.2 NVIC Configuration on TM4C123

The NVIC configuration has two main components, namely enabling/disabling device interrupts and assigning priority to the interrupts. In addition, interrupt pending behavior can also be configured using NVIC configuration. Separate NVIC registers are allocated for enabling and disabling of interrupts (IRQ0-IRQ138 or correspondingly exceptions 16 to 154). On the other hand system exceptions (i.e., exceptions 1 to 15) are configured using system control block (SCB) registers (see SCB register descriptions for address range 0xE000ED04-0xE000ED2C). Some of these system exceptions have programmable priority, as mentioned in Table 3.2. The priority for system exceptions can be configured using system priority (SYSPRIx) registers (see registers in address space 0xE000ED18-0xE000ED20).

The NVIC registers used for enabling and disabling of interrupts are labeled as ENx and DISx in Table 9.2, where x can take values from 0 to 4 depending on the specific peripheral interrupt allocation in the vector table. Similarly, priorities can be assigned using priority registers PRIy, where y can take values from 0 to 34 for TM4C123 microcontroller. Table 9.2 also provides memory address allocation for registers as well as their reset values for enable, disable and priority registers. Figure 9.9 shows the NVIC register bit field allocations for their proper configuration. In addition, NVIC has allocated registers for controlling the pending behavior (PEND and UNPEND registers) and observing the current state of an exception (ACTIVE registers). However, details related to these registers are skipped and interested reader is referred to TM4C123 device datasheet [TI(2014)] for further details.

Table 9.2: Selected NVIC register labels, addresses, reset value and brief description.

Register label	Address	Reset value	Brief description
Interrupt enable registers			
EN0	0xE000E100	0x00000000	Used to enable interrupts 0 to 31.
EN1	0xE000E104	0x00000000	Used to enable interrupts 32 to 63.
⋮	⋮	⋮	⋮
EN4	0xE000E110	0x00000000	Used to enable interrupts 128 to 138.
Interrupt disable registers			
DIS0	0xE000E180	0x00000000	Used to disable interrupts 0 to 31.
DIS1	0xE000E184	0x00000000	Used to disable interrupts 32 to 63.
⋮	⋮	⋮	⋮
DIS4	0xE000E190	0x00000000	Used to disable interrupts 128 to 138.
Priority configuration registers			
PRI0	0xE000E400	0x00000000	Priority for interrupts 0 to 3.
PRI1	0xE000E404	0x00000000	Priority for interrupts 4 to 7.
⋮	⋮	⋮	⋮
PRI34	0xE000E488	0x00000000	Priority for interrupts 136 to 138.

Writing '1' to specific bit location of the interrupt enable register, enables the corresponding device interrupt. Writing '1' to interrupt disable register, disables the interrupt and clears the corresponding bit in the interrupt enable register. Writing '0' to these registers does not have any effect. An eight level (i.e., three bit) priority is implemented for each interrupt source as can be seen from Figure 9.9(b) and can be configured using the appropriate priority register. Each priority register can configure priority for four interrupt sources. When i^{th} priority register, PRIi, is used then the corresponding interrupt sources, that can be configured are IntA$= 4i$, IntB$= 4i + 1$,

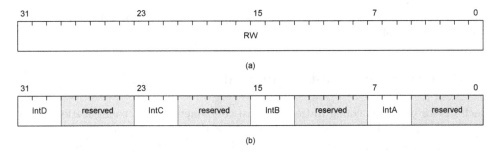

Figure 9.9: NVIC register bit field allocations for (a) enable and disable registers and (b) priority registers.

IntC= $4i + 2$ and IntD= $4i + 3$.

Next we provide a C code implementation for NVIC configuration. For this purpose, port F is configured as source of external interrupt. The NVIC configuration is independent of which port pin will be used as a source of external interrupt. In addition, NVIC configuration is not affected by the fact that one or more port pins are configured as external interrupt source. Listing 9.3 provides and illustration, configuring port F as source of external interrupt with priority 5.

```
// IRQ 0 to 31 enable and disable registers
#define NVIC_EN0_R          *((volatile unsigned long *)0xE000E100)
#define NVIC_DIS0_R         *((volatile unsigned long *)0xE000E180)

// IRQ 28 to 31 Priority Register
#define NVIC_PRI7_R         *((volatile unsigned long *)0xE000E41C)

// Defines for port F interrupt
#define NVIC_EN0_INT30      0x40000000  // Interrupt 30 enable

// Processor interrupt enable/disable functions
extern void DisableInterrupts(void);
extern void EnableInterrupts(void);

// User main program
int main()
{
   // NVIC configuration
   DisableInterrupts();

   NVIC_PRI7_R |= 0x00A00000;          // Priority is 5
   NVIC_EN0_R |= NVIC_EN0_INT30;       // Enable INT 30 in NVIC

   EnableInterrupts();

   // Other configurations and tasks to follow
   ...
   ...
}
```

Listing 9.3: NVIC configuration for enabling port F interrupt with priority 5.

The processor interrupt enable and disable subroutines, discussed in Section 9.5.1, are implemented as part of the system startup file 'startup.s', which is provided in Appendix A for reference.

9.5.3 Device Interrupt Configuration on TM4C123

Each device interrupt configuration has two components. One configuration is performed for the device using the corresponding registers, while the second configuration is performed for the interrupt vector table and implementation of associated ISR. Below we describe each of these configurations.

Interrupt Vector Table Configuration and ISR Implementation

The interrupt vector table was introduced in Section 3.3. It contains the addresses (labels or function names) of interrupt service routines corresponding to each source of interrupt. A vector table with just one vector entry (i.e., reset vector) was introduced in our first assembly program discussed in Section 4.3. The interrupt vector table can always be extended to include more vector entries. Listing 9.4 provides first few entries of the entire vector table, which is made part of the system startup file 'startup.s'.

```
; ************************************************************
; The interrupt vector table.
; ************************************************************
        EXPORT   __Vectors
__Vectors
        DCD      StackMem + Stack        ; Top of Stack
        DCD      Reset_Handler           ; Reset Handler
        DCD      NMI_Handler             ; NMI Handler
        DCD      HardFault_Handler       ; Hard Fault Handler
        DCD      MemManage_Handler       ; MPU Fault Handler
        DCD      BusFault_Handler        ; Bus Fault Handler
        DCD      UsageFault_Handler      ; Usage Fault Handler
        DCD      0                       ; Reserved
        DCD      0                       ; Reserved
        DCD      0                       ; Reserved
        DCD      0                       ; Reserved
        DCD      SVC_Handler             ; SVCall Handler
        DCD      DebugMon_Handler        ; Debug Monitor Handler
        DCD      0                       ; Reserved
        DCD      PendSV_Handler          ; PendSV Handler
        DCD      SysTick_Handler         ; SysTick Handler
        DCD      GPIOPortA_Handler       ; GPIO Port A
        DCD      GPIOPortB_Handler       ; GPIO Port B
        DCD      GPIOPortC_Handler       ; GPIO Port C
        DCD      GPIOPortD_Handler       ; GPIO Port D
        DCD      GPIOPortE_Handler       ; GPIO Port E
        DCD      UART0_Handler           ; UART0 Rx and Tx
        DCD      UART1_Handler           ; UART1 Rx and Tx
        DCD      SSI0_Handler            ; SSI0 Rx and Tx
        DCD      I2C0_Handler            ; I2C0 Master and Slave
        DCD      PWM0Fault_Handler       ; PWM 0 Fault
```

Listing 9.4: First few entries from the complete interrupt vector table for illustration.

To handle an interrupt from an arbitrary peripheral device, for instance say port F, we need to have either all the vector entries included in the table or reserve equivalent memory space before including port F ISR label in the vector table. This is due to the fact that fixed exception/IRQ numbers are assigned to the sources of interrupts and correspondingly the relative addresses (with respect to the base address of the vector table) of ISRs are also fixed. As a result, an interrupt vector table entry can not be skipped even if it is not required. This is the reason for vector entries even for reserved interrupts in the vector table, as can be seen from Listing 9.4.

Similar to the vector table entry for port A ISR in Listing 9.4, there is an entry for port F ISR as well at its appropriate location in the vector table. For Cortex-M, an ISR is implemented similar to an ordinary C function call and is illustrated in Listing 9.5 for port F interrupt.

```
// Interrupt service routine for Port F
void GPIOPortF_Handler(void){

    // Take the aapropriate action in response to interrupt
    ...
    ...
    // Clear the interrupt status flag before exiting ISR
}
```

Listing 9.5: Example ISR implementation.

Device Interrupt Configuration

Each device can have multiple possible sources of interrupts that can be enabled/disabled (unmasked/masked) individually using device interrupt related registers. The device level configuration of GPIO as a source of external interrupts is discussed here. Specifically, we will explain in detail, the configuration of port F pin 4 (also abbreviated as PF4) as source of external interrupt. The choice of this pin as source of external interrupt is motivated by the fact that an external user switch is connected with this pin on the TivaC evaluation board, as discussed previously in Section 8.5.

When a GPIO port pin is configured as an external interrupt, then interrupt triggering can be performed by one of the following two possible activations.

- Level triggered

- Edge triggered

Cortex-M devices support both level-triggered as well as edge-triggered interrupts. According to ARM terminology, edge-triggered interrupts are also termed as pulse interrupts. A level-triggered interrupt remains active until the peripheral deactivates it by deasserting the interrupt signal. In case of level triggered interrupt, the processor executes the ISR and on the exit if the interrupt condition is still valid, the processor immediately enters the ISR again, provided no higher priority interrupt occurs in the meantime. In case of an edge triggered interrupt, the interrupt signal is generated by the rising or falling edge (or both edges) on the corresponding GPIO port pin.

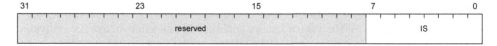

Figure 9.10: GPIO interrupt sense (IS) register bit field allocation.

There are seven 32-bit registers associated with GPIO port for interrupt configuration. Only least significant 8-bits are used by all these registers to configure each GPIO port pin individually. Below we briefly discuss the usage and configuration of each of these registers.

- *Interrupt Sense* (IS) register is used to configure interrupt triggering as edge or level. When a bit in this register is cleared to 0 the corresponding GPIO port pin is configured as edge triggered interrupt.

- *Interrupt Event* (IEV) register configures the type of the interrupt event. Setting to 1 configures the port pin for rising edge interrupt (when edge type interrupt is configured in IS register) or a logic level high triggers the interrupt when IS register is configured for level sensitive interrupt. Clearing it to 0 configures either the falling edge or logic low as the interrupt condition depending on the IS register.

- *Interrupt Both Edges* (IBE), when set to 1, configures both rising as well as falling edges as the source of interrupt. For this to be valid, IS register should be configured for edge type interrupt. Configuring IBE register with a value of 1 overrides the IEV register configuration.

- *Interrupt Mask* (IM) register is used to enable or disable individual port pin interrupts. Writing 1 to this register enables the corresponding port pin interrupt.

- *Raw Interrupt Status* (RIS) and *Masked Interrupt Status* (MIS) are the two registers, which indicate the occurrence of an interrupt condition. The RIS register indicates that a GPIO pin satisfies the requirements for an interrupt, but whether the interrupt is sent to NVIC or not depends on the IM register. On the other hand, the MIS register only indicates those GPIO pin interrupts that are enabled and sent to NVIC.

- Interrupt Clear (ICR) register is used to clear an interrupt flag in the status registers. To clear an interrupt status flag in RIS as well as MIS registers, it is required to write '1' to the corresponding bit in the ICR register. If an interrupt is not cleared using ICR before exiting the corresponding interrupt service routine, then the ISR is immediately entered again, provided no other interrupt of higher priority has occurred in the meantime.

Figure 9.10 shows the bit allocations for GPIO port F interrupt sense register. The addresses and reset values for all interrupt related registers, for port F, are tabulated in Table 9.3.

The address information in Table 9.3 is based on APB bus as discussed previously in Section 8.4. It is important to mention that each GPIO register is at a fixed offset

Table 9.3: GPIO device interrupt configuration registers for port F.

Register label	Address	Reset value	Brief description
GPIO_PORTF_IS_R	0x40025404	0x00000000	Interrupt sense register.
GPIO_PORTF_IBE_R	0x40025408	0x00000000	Interrupt on both edges register.
GPIO_PORTF_IEV_R	0x4002540C	0x00000000	Interrupt event register.
GPIO_PORTF_IM_R	0x40025410	0x00000000	Interrupt mask register.
GPIO_PORTF_RIS_R	0x40025414	0x00000000	Raw interrupt status register.
GPIO_PORTF_MIS_R	0x40025418	0x00000000	Masked interrupt status register.
GPIO_PORTF_ICR_R	0x4002541C	0x00000000	Interrupt clear register.

from a base address. When accessing different port registers, each GPIO port has a different base address while a common offset is applicable for all GPIO ports.

To configure PF4 as an external interrupt, it required to know what type (level or edge) of interrupt is generated by the external hardware to the GPIO pin. Assuming it is required that a falling edge on PF4 should generate an interrupt, the C code in Listing 9.6 illustrates PF4 configuration. In Listing 9.6, we have combined the NVIC configuration with device configuration for completeness as well as better understanding. Clock enabling for port F and configuration of PF4 as an input is required for proper operation and is included in Listing 9.6.

```
// Port clock control register
#define   SYSCTL_RCGC2_R      *(volatile unsigned long *)0x400FE108
#define   SYSCTL_RCGC2_GPIOF 0x00000020 //Port F clock control

// port I/O control registers
#define   GPIO_PORTF_DIR_R    *((volatile unsigned long *)0x40025400)
#define   GPIO_PORTF_DEN_R    *((volatile unsigned long *)0x4002551C)
#define   GPIO_PORTF_PUR_R    *((volatile unsigned long *)0x40025510)

// port F interrupt configuration registers
#define GPIO_PORTF_IS_R      *((volatile unsigned long *)0x40025404)
#define GPIO_PORTF_IBE_R     *((volatile unsigned long *)0x40025408)
#define GPIO_PORTF_IEV_R     *((volatile unsigned long *)0x4002540C)
#define GPIO_PORTF_IM_R      *((volatile unsigned long *)0x40025410)
#define GPIO_PORTF_ICR_R     *((volatile unsigned long *)0x4002541C)


// IRQ 0 to 31 enable and disable registers
#define NVIC_EN0_R           *((volatile unsigned long *)0xE000E100)
#define NVIC_DIS0_R          *((volatile unsigned long *)0xE000E180)

// IRQ 28 to 31 Priority Register
#define NVIC_PRI7_R          *((volatile unsigned long *)0xE000E41C)

// Defines for port F interrupt
#define NVIC_EN0_INT30       0x40000000  // Interrupt 30 enable

// Port pin definitions
#define   GPIO_PORTF_PIN4     0x10


// Processor interrupt enable/disable functions
extern void DisableInterrupts(void);
```

```
extern void EnableInterrupts(void);

// User main program
int main()
{
    DisableInterrupts();

    // NVIC configuration
    NVIC_PRI7_R |= 0x00A00000;              // Priority is 5
    NVIC_EN0_R = NVIC_EN0_INT30;            // Enable INT 30 in NVIC

    // Device configurations for PF4

    SYSCTL_RCGC2_R |= SYSCTL_RCGC2_GPIOF;   // Enable port F clock

    // Configure PF4 as digital input
    GPIO_PORTF_DEN_R |= GPIO_PORTF_PIN4;
    GPIO_PORTF_DIR_R &= (~GPIO_PORTF_PIN4); // PF4 as input
    GPIO_PORTF_PUR_R |= GPIO_PORTF_PIN4;    // Enable pull-up on PF4

    // Configure PF4 as interrupt source
    GPIO_PORTF_IS_R &= (~GPIO_PORTF_PIN4);  // PF4 is edge-sensitive
    GPIO_PORTF_IBE_R &= (~GPIO_PORTF_PIN4); // PF4 is not both edges
    GPIO_PORTF_IEV_R &= (~GPIO_PORTF_PIN4); // PF4 falling edge event
    GPIO_PORTF_ICR_R = GPIO_PORTF_PIN4;     // Clear flag for pin 4
    GPIO_PORTF_IM_R |= GPIO_PORTF_PIN4;     // Unmask interrupt for PF4

    EnableInterrupts();

    // Other tasks to follow
    ...
    ...
}
```

Listing 9.6: Complete interrupt configuration for PF4 including device and NVIC configurations.

9.6 Interrupt-Based Switch/Keypad Interfacing

The use of external interrupts will be illustrated through examples in this section. Specifically, switch and keypad interfacing will be discussed using interrupts. Polling based switch and keypad interfacing has been discussed in Chapter 8. In case of interrupt based implementation, the processor only responds to a switch or key press by executing the ISR and the processing overhead is zero in case no switch or key is pressed. This is in contrast to polling based method, where switch or keypad interfaces are scanned (polled) periodically.

9.6.1 Switch Interfacing Using Interrupts

Switch interfacing using interrupts does not require any hardware level modification. We will illustrate the use of SW1 as a source of external interrupt. Example 9.1 next explains the use of SW1 as source of external interrupt. The interrupt is generated on the falling edge, which corresponds to pressing of the switch. However, it is important to note that there is an overhead associated with the calling of an ISR.

Example 9.1 (Switch interfacing using interrupt based method.).

This C program interfaces the on board user switch SW1 connected to PF4 by configuring PF4 as an external interrupt. The interrupt is generated on the falling edge, corresponding to the pressing of the switch. On every key press an on board LED (green color) connected to PF3 is toggled. If the PF4 is configured for rising edge interrupt, then the interrupt occurs on the switch release action and the LED will toggle on the switch release rather on switch press.

```c
#define   SYSCTL_RCGC2_R       *((volatile unsigned long *)0x400FE108)
#define   SYSCTL_RCGC2_GPIOF 0x00000020 //Port F clock control

#define   GPIO_PORTF_DATA_RD *((volatile unsigned long *)0x40025040)
#define   GPIO_PORTF_DATA_WR *((volatile unsigned long *)0x40025020)
#define   GPIO_PORTF_DIR_R   *((volatile unsigned long *)0x40025400)
#define   GPIO_PORTF_DEN_R   *((volatile unsigned long *)0x4002551C)
#define   GPIO_PORTF_PUR_R   *((volatile unsigned long *)0x40025510)


#define NVIC_EN0_INT30      0x40000000  // Interrupt 30 enable
// IRQ 0 to 31 Set Enable Register
#define NVIC_EN0_R          *((volatile unsigned long *)0xE000E100)
// IRQ 28 to 31 Priority Register
#define NVIC_PRI7_R         *((volatile unsigned long *)0xE000E41C)
#define GPIO_PORTF_IS_R     *((volatile unsigned long *)0x40025404)
#define GPIO_PORTF_IBE_R    *((volatile unsigned long *)0x40025408)
#define GPIO_PORTF_IEV_R    *((volatile unsigned long *)0x4002540C)
#define GPIO_PORTF_IM_R     *((volatile unsigned long *)0x40025410)
#define GPIO_PORTF_ICR_R    *((volatile unsigned long *)0x4002541C)

// Port pin definitions
#define   GPIO_PORTF_PIN3      0x08
#define   GPIO_PORTF_PIN4      0x10

// Default clock frequency and delay definition
#define   SYSTEM_CLOCK_FREQUENCY   16000000
#define   DEALY_COUNT             SYSTEM_CLOCK_FREQUENCY/1000

// Function prototypes
void Delay(unsigned long counter);


// User main program
int main()
{

   // Enable the clock for port F
   SYSCTL_RCGC2_R |= SYSCTL_RCGC2_GPIOF;

   // Configure PF3 as digital output and PF4 as digital input
   GPIO_PORTF_DEN_R |= GPIO_PORTF_PIN3 + GPIO_PORTF_PIN4;
   GPIO_PORTF_DIR_R |= GPIO_PORTF_PIN3;    // PF3 as output
   GPIO_PORTF_DIR_R &= (~GPIO_PORTF_PIN4); // PF4 as input
```

```
    GPIO_PORTF_PUR_R |= GPIO_PORTF_PIN4;     // Enable pull-up on PF4

    GPIO_PORTF_IS_R &= (~GPIO_PORTF_PIN4);   // PF4 is edge-sensitive
    GPIO_PORTF_IBE_R &= (~GPIO_PORTF_PIN4);  // PF4 is not both edges
    GPIO_PORTF_IEV_R &= (~GPIO_PORTF_PIN4);  // PF4 falling edge
                                             // event
    GPIO_PORTF_ICR_R = GPIO_PORTF_PIN4;      // Clear flag for pin 4
    GPIO_PORTF_IM_R |= GPIO_PORTF_PIN4;      // Unmask interrupt for PF4
    NVIC_PRI7_R |= 0x00A00000;               // Priority is 5
    NVIC_EN0_R = NVIC_EN0_INT30;             // Enable INT 30 in NVIC

    while(1)
    {
    }
}


// Interrupt service routine for Port F
void GPIOPortF_Handler(void){

    // Insert delay for switch debouncing
    Delay(DEALY_COUNT);

    if((GPIO_PORTF_DATA_RD) == 0)
    {
        GPIO_PORTF_DATA_WR ^= GPIO_PORTF_PIN3;
    }

    GPIO_PORTF_ICR_R = 0x10;       // acknowledge flag4
}


// This function implements the delay
void Delay(unsigned long counter)
{
    unsigned long i = 0;

    for(i=0; i< counter; i++);

}
```

The ISR in Example 9.1 implements software switch debouncing. When switch is pressed a falling edge is detected and correspondingly the ISR is executed. However, due to switch bouncing, it is possible that multiple falling edges are produced. To ensure that ISR is executed only once in response to each switch press, the interrupt is acknowledged at the end of ISR. If the interrupt is acknowledged at the start of ISR, then due to switch bouncing it is quite possible that another falling edge occurs after the acknowledgment of the interrupt, which can result in the execution of ISR for the second time.

9.6.2 Interrupt-Driven Keypad Interfacing

The interrupt driven scanned keypad interface requires hardware modification. Figure 9.11 illustrates the required additional hardware for implementing interrupt driven interfacing of a matrix type keypad. For proper functioning of keypad interface, it is required that an interrupt is generated in response to each key press. To generate an interrupt in response to each key press, while still being able to identify the pressed key uniquely, four diodes are used as shown in Figure 9.11.

All four columns in Figure 9.11 are applied logic '1' during keypad interface initialization. When a key is pressed a low to high transition is generated at PF4 resulting in an interrupt condition. The key pad scanning is then performed as part of the port F ISR. It is important to realize that the key pad scan function is executed only once in response to each key press. Example 9.2 provides an implementation of interrupt driven keypad interfacing.

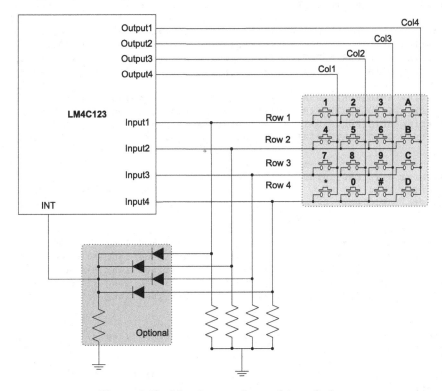

Figure 9.11: Matrix type keypad interfacing.

Example 9.2 (Keypad interfacing using interrupt based method.).

The C program in this example modifies the polling based keypad interfacing discussed in Example 8.6. The main difference is PF4 configuration as an external interrupt.

```
// Default clock frequency and delay definition
#define  SYSTEM_CLOCK_FREQUENCY  16000000
#define  DEALY_COUNT              SYSTEM_CLOCK_FREQUENCY/10000

// System clock register
#define SYSCTL_RCGCGPIO_R (*((volatile unsigned long *)0x400FE608))

// GPIO register definitions
```

```c
#define GPIO_PORTA_DEN_R    (*((volatile unsigned long *)0x4000451C))
#define GPIO_PORTA_DIR_R    (*((volatile unsigned long *)0x40004400))
#define GPIO_PORTA_DATA_R   (*((volatile unsigned long *)0x400043FC))
#define GPIO_PORTA_PIN6     (*((volatile unsigned long *)0x40004100))
#define GPIO_PORTA_PIN7     (*((volatile unsigned long *)0x40004200))

#define GPIO_PORTB_DEN_R    (*((volatile unsigned long *)0x4000551C))
#define GPIO_PORTB_DIR_R    (*((volatile unsigned long *)0x40005400))
#define GPIO_PORTB_DATA_R   (*((volatile unsigned long *)0x400053FC))

#define GPIO_PORTC_DEN_R    (*((volatile unsigned long *)0x4000651C))
#define GPIO_PORTC_DIR_R    (*((volatile unsigned long *)0x40006400))
#define GPIO_PORTC_DATA_R   (*((volatile unsigned long *)0x400063FC))

#define GPIO_PORTE_DEN_R    (*((volatile unsigned long *)0x4002451C))
#define GPIO_PORTE_DIR_R    (*((volatile unsigned long *)0x40024400))
#define GPIO_PORTE_DATA_R   (*((volatile unsigned long *)0x400243FC))
#define GPIO_PORTE_PIN2     (*((volatile unsigned long *)0x40024010))

#define GPIO_PORTF_DEN_R    (*((volatile unsigned long *)0x4002551C))
#define GPIO_PORTF_DIR_R    (*((volatile unsigned long *)0x40025400))
#define GPIO_PORTF_DATA_R   (*((volatile unsigned long *)0x400253FC))
#define GPIO_PORTF_PIN1     (*((volatile unsigned long *)0x40025008))

// NVIC register definitions
#define NVIC_EN0_INT30      0x40000000  // Interrupt 30
#define NVIC_EN0_R          *((volatile unsigned long *)0xE000E100)
#define NVIC_PRI7_R         *((volatile unsigned long *)0xE000E41C)

// Port F interrupt related register definitions
#define GPIO_PORTF_IS_R     *((volatile unsigned long *)0x40025404)
#define GPIO_PORTF_IBE_R    *((volatile unsigned long *)0x40025408)
#define GPIO_PORTF_IEV_R    *((volatile unsigned long *)0x4002540C)
#define GPIO_PORTF_IM_R     *((volatile unsigned long *)0x40025410)
#define GPIO_PORTF_ICR_R    *((volatile unsigned long *)0x4002541C)

// Port pin definitions
#define GPIO_PORTF_DATA_RD  *((volatile unsigned long *)0x40025040)
#define GPIO_PORTF_PIN4     0x10

#define ROW3                (GPIO_PORTC_DATA_R & 0x10)
#define ROW2                (GPIO_PORTC_DATA_R & 0x20)
#define ROW1                (GPIO_PORTC_DATA_R & 0x40)
#define ROW0                (GPIO_PORTC_DATA_R & 0x80)

// Function prototypes
void Delay(unsigned long counter);
extern void DisableInterrupts(void);
extern void EnableInterrupts(void);

unsigned char key_press = 0;

/* This function initialized the GPIOs for key pad interface. */
void Init_Keypad_GPIOs(void){

  SYSCTL_RCGCGPIO_R |= 0x35;  // enable clock for GPIOPORT A,C,F,E

  // Enable digital function for rows and columns
  GPIO_PORTC_DEN_R |= 0xF0;
```

```
    GPIO_PORTA_DEN_R  |= 0xC0;
    GPIO_PORTE_DEN_R  |= 0x04;
    GPIO_PORTF_DEN_R  |= 0x12;

    GPIO_PORTC_DIR_R &= ~0xF0; // configure rows as inputs
    GPIO_PORTA_DIR_R |= 0xC0;  // configure columns 0,1 as outputs
    GPIO_PORTE_DIR_R |= 0x04;
    GPIO_PORTF_DIR_R |= 0x02;  // configure columns 2,3 as outputs

    // Interrupt related configuration
    DisableInterrupts();

    GPIO_PORTF_IS_R  &= (~GPIO_PORTF_PIN4);   // PF4 is edge-sensitive
    GPIO_PORTF_IBE_R &= (~GPIO_PORTF_PIN4);   // PF4 is not both edges
    GPIO_PORTF_IEV_R |= (GPIO_PORTF_PIN4);    // PF4 rising edge event
    GPIO_PORTF_ICR_R = GPIO_PORTF_PIN4;       // Clear flag for pin 4
    GPIO_PORTF_IM_R  |= GPIO_PORTF_PIN4;      // Unmask interrupt for PF4
    NVIC_PRI7_R |= 0x00A00000;                // Priority is 5
    NVIC_EN0_R = NVIC_EN0_INT30;              // Enable INT 30 in NVIC

    EnableInterrupts();

    // Apply logic high on column outputs for key press detection
    GPIO_PORTF_PIN1 |= 2;
    GPIO_PORTE_PIN2 |= 0x04;
    GPIO_PORTA_PIN7 |= 0x80;
    GPIO_PORTA_PIN6 |= 0x40;
}

/* This function scans the keypad for any key press and returns
   the ASCII value for the pressed key. */
unsigned char Scan_Keypad(int *key_num){
  unsigned int i, cols, key = 0;
  const unsigned char key_val[4][4] =
      { // col0, col1, col2, col3
      {'1',  '2',  '3',  'A'}, // row0
      {'4',  '5',  '6',  'B'}, // row1
      {'7',  '8',  '9',  'C'}, // row2
      {'*',  '0',  '#',  'D'}  // row3
      };

  for(i = 0; i < 4; i++){        // loop for 4 rows
    cols = 0x01 << i;            // make columns high one by one
    GPIO_PORTF_PIN1 = cols << 1;
    GPIO_PORTE_PIN2 = cols << 1;
    GPIO_PORTA_PIN7 = cols << 5;
    GPIO_PORTA_PIN6 = cols << 3;

    if(ROW0 & 0x80){             // check row0
        *key_num = (i * 4) + 1;
        while(ROW0 & 0x80);       // wait for release
        return key_val[key][i];   // return value of key pressed
    }

    if(ROW1 & 0x40){             // check row1
        key += 1;                 // next key pressed
        *key_num = (i * 4) + 2;
        while(ROW1 & 0x40);       // wait for release
        return key_val[key][i];   // return value of key pressed
```

```
    }

              if(ROW2 & 0x20){                    // check row2
      key += 2;                        // next key pressed
      *key_num = (i * 4) + 3;
      while(ROW2 & 0x20);              // wait for release
      return key_val[key][i];          // return value of key pressed
    }

    if(ROW3 & 0x10){                   // check row3
      key += 3;                        // next key pressed
      *key_num = (i * 4) + 4;
      while(ROW3 & 0x10);              // wait for release
      return key_val[key][i];          // return value of key pressed
    }
    key = 0;
  }
  return 0;                            // no key pressed
}

int main ( void )
{
Init_Keypad_GPIOs ();

// Loop forever and display Hello World message
while(1)
{ }
}

// Interrupt service routine for Port F
void GPIOPortF_Handler(void){
   int val=0;
  // Insert delay for switch debouncing
    Delay(DEALY_COUNT);
    key_press = Scan_Keypad(&val);

  // Apply logic high on column outputs to detect next key press
    GPIO_PORTF_PIN1 |= 2;
    GPIO_PORTE_PIN2 |= 0x04;
    GPIO_PORTA_PIN7 |= 0x80;
    GPIO_PORTA_PIN6 |= 0x40;

    GPIO_PORTF_ICR_R = 0x10;        // acknowledge flag4
}

// This function implements the delay
void Delay(unsigned long counter)
{
  unsigned long i = 0;

  for(i=0; i< counter; i++);

}
```

Implementing an interrupt driven interface for multiplexed keypad interface will require significant additional hardware rendering it a nonviable approach. As a result, a polling based approach would be used for these types of interfaces.

9.7 Summary of Key Concepts

In this chapter, we have introduced the concepts related to input/output (I/O) synchronization and interrupt programming. Specifically, the following key concepts were introduced and discussed.

- Introduction to I/O synchronization was provided. The processor is faster as compared to the I/O devices. To overcome this mismatch, the system should be designed such that the processor executes software programs efficiently and paying attention to the hardware devices on time.

- A hardware device can be in one of the two possible states, namely, active state or inactive state. When the device is active, it can be either in busy state or in ready state. The states of the I/O device were presented with a help of state transition diagram.

- There are three main methods for I/O synchronization. Blind cycle, polling based methods, and interrupt driven methods. In blind cycle, the software waits for a fixed amount of time assuming that the I/O device hardware has completed the operation within this time. Status checking of the device hardware, by the software is the basis for the polling methods. In interrupt driven methods, the hardware seeks the attention of the software whenever it is ready for data transfer.

- The performance of synchronization methods is quantified using three parameters: latency, bandwidth, and priority.

- Exceptions can be asynchronous or synchronous. An asynchronous exception or interrupt is the one that can occur at an arbitrary time during the execution of an instruction. Synchronous exceptions are generated by the processor while executing instructions and are requested only after the completion of execution of the instruction.

- Configuration of Interrupts for Cortex-M can be performed at three levels. At the processor level, at NVIC level and at the device level.

- Interrupt based switch and keypad interfacing are presented towards the end of the chapter. These real-life programming examples were presented to illustrate the concepts one has learnt about interrupts.

Review Questions

Question 9.1. What is the key difference between blind cycle and busy wait synchronization?

Question 9.2. How does periodic polling based synchronization differ from continuous polling synchronization?

Question 9.3. Are external interrupts synchronous or asynchronous exceptions?

Question 9.4. Assume that the starting address for the interrupt vector table is 0x00000000. What will be the interrupt service routine address of IRQ30?

Question 9.5. What are different configuration levels to enable an interrupt from a peripheral device?

Question 9.6. How does the priority configuration for system exceptions (with programmable priority) differ from that of interrupt requests (IRQs)?

Question 9.7. Is it necessary to list the addresses of all interrupt service routines (ISRs) in the interrupt vector table?

Question 9.8. Show the interrupt vector table entries, if only ISRs for exceptions 1-3 and IRQ 2 are implemented.

Question 9.9. What different device level configurations are required to enable an interrupt from a peripheral device?

Question 9.10. When a GPIO is used as a source of an external interrupt, then how can it be configured as level or edge triggered?

Question 9.11. Why is it necessary to clear the interrupt flag before exiting the ISR?

Question 9.12. How can we mask interrupts from different port pins of port A using a single register configuration? How can it be achieved for multiple GPIO ports and again using a single register?

Exercises

Exercise 9.1. Illustrate the working of the peripheral device state transition diagram for two input and output devices.

Exercise 9.2. What can be the impact on an embedded system if we don't synchronize the hardware and software speeds? Provide real-life examples with timing diagrams as given in Figures 9.2 and 9.3.

Exercise 9.3. Consider a scenario, where state transitions (interrupts) from multiple devices are connected to port A pins. Is it possible to assign different priorities to these devices? If the answer is no, then suggest a solution suitable for this situation.

Exercise 9.4. Different IRQs are assigned to peripheral devices for their interrupt handling. Is it possible to assign higher priority to a (a) lower numbered IRQ or (b) higher numbered IRQ?

Exercise 9.5. Write a C program that configures priorities 2 and 6 for IRQ10 and IRQ15, respectively.

Exercise 9.6. Write a C program to configure a GPIO port pin (any port pin of your choice) as a source of external interrupt. The first interrupt should be triggered on a falling edge, while all subsequent interrupts should be triggered on rising edges.

Show all the required register configurations and any necessary steps that should be performed as part of ISR.

Chapter 10

Timing Interfaces

Overview

Many embedded applications require a microcontroller to operate at a different clock frequency depending on the mode of operation. In addition, the microcontroller is required to generate accurate timing intervals, time/count external events (timer as an input device) and generate digital signals of desired frequency and pulse duration (timer as an output device).

In this chapter, we will first discuss the issue of clocking a microcontroller. The configurable system clock frequency provides the application developer to operate the microcontroller at its desired frequency. When no clock configuration is performed, the microcontroller operates at its default frequency. The default configuration will be discussed to verify the default operating frequency. Then we will learn the key steps involved to configure the microcontroller to operate at a different frequency. Once done with the clocking of a microcontroller, the basic timer concepts will be introduced. Next we discuss the use of timers as input device for frequency and pulse duration measurement, which is followed by the discussion on timer as an output device for signal generation. Later we highlight the capabilities and features of the available timing interfaces on TM4C123. We also provide details regarding the use of TM4C123 timer modules for different applications.

10.1 Basics of Timing Interfaces

Each microcontroller has two types of timing interfaces. One timing interface is used to generate the system clock to operate the microcontroller at desired frequency. The second timing interface is based on a timer, which is primarily a peripheral module. One primary use of a timer module is to generate timing intervals of desired value. In addition, the timer modules can be configured to operate either as an input or an output device. The timer module when used as an input device, it can be configured to time or count external events. On the other hand, when the timer module is used as an output device, it can generate signals of varying duty cycle as well as frequency.

Timers can be used in variety of application areas. At the basic level, the timer can be used to generate accurate timing signals, to measure time intervals, to generate interrupts at specific time intervals as well as to count events of interest. However, many complex applications are build using these basic timer uses. Some of the key application areas, where timers act as fundamental building blocks, are listed below.

- Periodic data sampling

- Scheduling of different tasks

- Data communication rate control

- Motion control

- Navigation

10.2   Clocking a Microcontroller

With the technological advancements, microcontrollers have become more complex and provide many different options for its clocking. This section discusses different mechanisms used for appropriate provisioning of clock signal to a microcontroller. Generally, an oscillator is used to generate clock signal for the microcontroller. Two types of oscillators are most widely used for clocking a microcontroller, namely crystal oscillators and RC oscillators. Crystal oscillators typically provide high frequency accuracy and low frequency variation with respect to temperature. On the other hand RC oscillators are capable of providing fast startup and are relatively cheaper, but they suffer from poor frequency accuracy in the presence of variations in supply voltage as well as temperature.

A microcontroller can have an internal integrated oscillator or it can be clocked from an external oscillator. Majority of microcontrollers have one or more internal oscillators. In case of internal oscillator with fixed frequency, usually no external components are needed. However, when an internal oscillator is required to produce a range of frequencies, an external crystal or RC circuit is used to select the desired frequency. An external oscillator is used to obtain higher frequency accuracy by making the clock source independent of any temperature variations of the microcontroller chip.

Both internal as well as external oscillators produce fixed frequency, for the given hardware configuration. Fixed frequency configuration is perfectly acceptable for those applications which do not require the processor to operate at different frequencies for different operating states. However, for those applications requiring the processor to operate at different frequencies, there is an alternate solution to the problem. This alternate approach is based on using phased lock loop (PLL), which is used to multiply the fixed frequency from the oscillator by a configurable factor. Figure 10.1 summarizes different possibilities when selecting the clock source for the microcontroller. In addition, the use of PLL as a frequency multiplier is also depicted. In the following, we will describe the operation of PLL as well as its use as a frequency multiplier briefly.

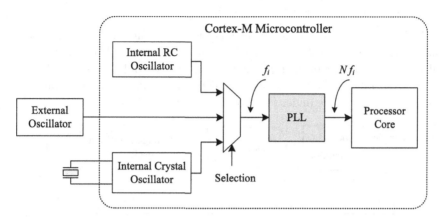

Figure 10.1: Illustration of different clock sources for microcontroller and use of internal PLL as frequency multiplier by factor N.

10.2.1 Phase Lock Loop

Phase locked loop is a device that is used to track the phase of an input signal for a range of frequencies. By tracking of phase we mean that the phase difference of PLL output and incoming signal is constant, provided the frequency of incoming signal does not change. When the PLL output tracks input signal perfectly, the corresponding frequency difference between incoming signal and the PLL output is zero. This can be justified by using the relationship, $\theta(t) = \int \omega(t)dt$, between phase $\theta(t)$ and frequency, $\omega(t)$. Let $\theta_i(t)$ and $\theta_o(t)$ denote the phases of input signal and PLL output, respectively, then the phase difference is given by

$$\theta_i(t) - \theta_o(t) = \int \left(\omega_i(t) - \omega_o(t)\right) dt \qquad (10.1)$$

From (10.1) it can be observed that for constant phase difference, the corresponding frequency difference is zero, which can be obtained by differentiating the above expression. This is a generic result and is applicable to PLL for a limited range of frequencies depending on the configuration of PLL building blocks. In general, PLL comprises of the following three basic building blocks.

- Phase detector (PD)

- Low pass filter (LPF)

- Voltage controlled oscillator (VCO)

The three building blocks are connected to form a feedback loop as shown in Figure 10.2, which is also the basic PLL configuration. The *phase detector* generates an error signal proportional to the phase difference by comparing the phases of two inputs (one of them in external input signal, while the second input is same as the PLL output).

The phase detector output is passed through a lowpass filter, which is also termed as *loop filter*. The loop filter output is connected to the input of the voltage controlled oscillator. VCO is a controlled oscillator and its output frequency f_o is proportional to the voltage applied at the input of VCO.

When PLL is tracking the phase (by keeping the phase difference constant), then PLL is considered in *lock condition*. Under lock condition, the PLL tracks the incoming signal and any variations in the input signal frequency are followed by the PLL output (i.e., VCO output). In addition, the signal frequencies at the two inputs of phase detector are exactly same under lock condition.

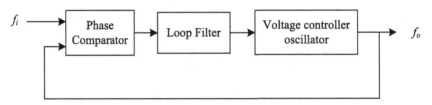

Figure 10.2: Basic block diagram of a phase locked loop.

10.2.2   PLL as Frequency Multiplier

As mentioned above, the frequency difference between two inputs of phase detector is zero under lock conditions. Now let us not feedback the VCO output signal directly to the PD, rather the signal from VCO is divided in its frequency by some integer N_1 before being fed to the PD. This can be accomplished by using a simple N_1 counter in the feedback path. For this scenario, the PLL under lock condition will ensure that the frequencies of the two PD input signals are exactly same and the VCO output frequency is N_1 times higher than the input signal frequency. In addition, using N_2 counter at the output of VCO, as shown in Figure 10.3, results in an output frequency, f_s given by

$$f_s = \frac{N_1}{N_2} f_i. \tag{10.2}$$

This method implements a non-integer frequency multiplication factor by using two integer frequency counters along with PLL.

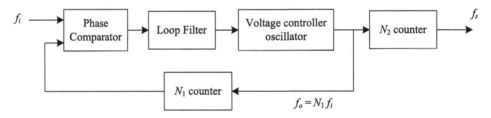

Figure 10.3: PLL as frequency multiplier.

10.3 TM4C123 Clock and Frequency Configuration

A user program can perform system configuration during initialization before executing the application task. The system configuration might have different components related to clock, power and reset control of the microcontroller. In addition, the microcontroller can be configured for its operating mode as running, low power, sleep or deep sleep, while executing the application task. However, we will only concentrate on clock configuration aspect here.

The clock configuration for Cortex-M microcontroller has two associated components. First component is the selection of the clock source and the second component is the clock frequency. Every microcontroller when powered, starts with the default clock source and frequency. For some applications the default frequency is sufficient and no clock configuration is required. If this is not the case, then clock configuration is performed to run the processor at the desired operating frequency. Furthermore, there are situations where the processor operating frequency is dependent on the current operating state and is reconfigured for power conservation as well as heat dissipation management. In this section, we will first discuss the default clock configuration for the Cortex-M microcontroller. This will be followed by the procedure to select a different clock source and configure the system clock frequency to a value, different from the default frequency.

10.3.1 Default Clock Source and Frequency Configuration

Even if the user program does not perform any clock configuration as part of system initialization, the microcontroller will still be running at a default frequency. We will verify the system default operating frequency using its default clock configuration. This will allow us to observe, what components and configurations are involved in the selection of default frequency. Based on this information we will be in a position to configure the microcontroller for the desired operating frequency.

The processor clock source and frequency configuration can be performed using appropriate clock configuration registers. The run-mode clock configuration (RCC) and run-mode clock configuration 2 (RCC2) registers are used to configure the system clock. Figure 10.4 shows the simplified block diagram for the system clock configuration tree. From Figure 10.4, we observe that there are multiple choices available for clock source selection. In addition, the clock form the selected source can either be fed directly for system clock purpose or it can be first multiplied using PLL. When enabled, the PLL automatically chooses its multiplier value based on the selected source frequency and provides a fixed 400 MHz signal at its output. The dashed line boxes in Figure 10.4 represent the configurable bit-fields in RCC and RCC2 registers. The default values of these configurable bit-fields, after reset, are responsible for selecting the clock source and default system clock frequency.

The RCC and RCC2 register addresses in the memory map and their default values after reset are provided in Table 10.1. The bit field allocations for registers RCC and RCC2 are shown in Figure 10.5, while descriptions for different bit fields are provided in Table 10.2. In Table 10.2, for each bit field label, its corresponding register

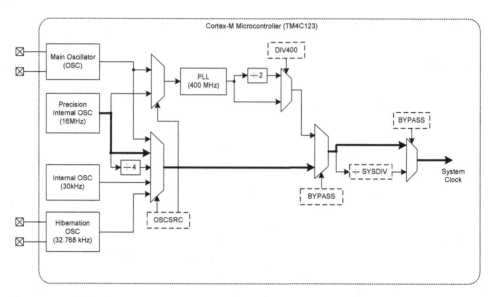

Figure 10.4: Block diagram illustrating the system oscillator and clock configuration possibilities.

Table 10.1: Run-mode clock configuration registers with their labels, addresses, reset value, and brief description.

Label	Address	Reset value	Brief description
RCC	0x400FE060	0x078E3AD1	Configure system clock and oscillators.
RCC2	0x400FE070	0x07C06810	Configure clock and overrides RCC equivalent fields.

information is also provided in parenthesis. In addition, the descriptions for common bit fields in registers RCC and RCC2 are provided together. It is important to mention here that once the use of register RCC2 is enabled by setting bit USERCC2, its configuration overrides the equivalent RCC register bit fields. Each RCC2 field that supersedes an RCC field is located at the same LSB bit position.

Based on the default values of RCC and RCC2 registers, we observe that the USERCC2 bit field is cleared, which disables the clock configuration using register RCC2. As a result only register RCC is involved in the clock configuration. The default values for OSCSRC and BYPASS bit fields, from register RCC, are important for system default clock frequency. The default value of bit field OSCSRC is 0x1, which selects the internal precision oscillator as the clock source. The default value of BYPASS is 0x1, which bypasses the use of PLL as well as the system clock divisor (i.e., use of SYSDIV). Based on these default values of OSCSRC and BYPASS the 16 MHz clock from the precision internal oscillator becomes the system clock frequency and is the default clock configuration.

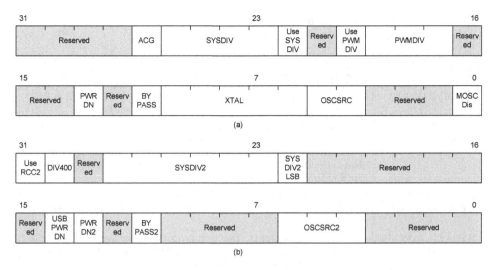

Figure 10.5: Bit field allocations for registers (a) RCC and (b) RCC2.

10.3.2 Configuring Clock Frequency

Many different clock frequencies can be configured for system clock, depending on the application requirements. We will illustrate the basic steps involved in clock configuration by changing the system clock frequency to 80 MHz, which is the maximum permissible operating frequency for TM4C123 microcontroller. To operate the system at 80 MHz, the use of PLL is essential. If we select 400 MHz as the PLL output by setting DIV400 bit, then using SYSDIV2 equal to 5 will allow us to obtain 80 MHz system clock frequency. It should be remembered that in this configuration, SYS-DIV2LSB is appended to SYSDIV2 as LSB to construct a 7 bit divisor. Following are the key steps performed for successful configuration of system clock.

- Bypass the PLL and system clock divider by setting the BYPASS bit and clearing the USESYSDIV bit in the RCC register. If USERCC2 bit is set in register RCC2, then BYPASS2 bit should also be set. It is important to write the RCC register prior to writing the RCC2 register. Doing so will allow us to configure the PLL and validate its operation, while running the microcontroller from an oscillator directly before switching it to PLL.

- Select the oscillator source by configuring OSCSRC (OSCSRC2), the crystal value using bit field XTAL and clear the PWRDN (PWRDN2) bit in RCC (RCC2). Setting the XTAL field automatically pulls a valid PLL configuration data for the selected crystal, while clearing the PWRDN (PWRDN2) bit powers the PLL and enables its output.

- Select desired system divider SYSDIV (SYSDIV2 and SYSDIV2LSB) in RCC (RCC2) and set USESYSDIV bit in RCC. The SYSDIV (SYSDIV2 and SYS-DIV2LSB) field determines system clock frequency for microcontroller.

- Wait for the PLL to lock by polling the PLLLRIS bit in the Raw Interrupt Status (RIS) register.

Table 10.2: Selected bit field descriptions for register RCC and RCC2 to perform clock configuration.

Bit field	Description
SYSDIV (RCC), SYSDIV2 (RCC2)	System clock divisor, used to specify divisor value to generate system clock. SYSDIV2 overrides SYSDIV when both USESYSDIV bit in RCC register and USERCC2 bit in RCC2 register are set to 1.
USESYSDIV (RCC)	This bit when set to 1, enables the use of system clock divider. The divider is forced to be used when PLL is selected as the source. If USERCC2 bit in RCC2 register is set, then the corresponding SYSDIV2 field in the RCC2 register is used as the system clock divider rather than the SYSDIV field in this register.
PWRDN (RCC), PWRDN2 (RCC2)	When this bit is set, the PLL is powered down. PWRDN2 overrides PWRDN when USERCC2 bit in RCC2 register is set. Clearing this bit turns on PLL.
BYPASS (RCC), BYPASS2 (RCC2)	When this bit set to 1, the system clock is derived from oscillator directly by bypassing the PLL. BYPASS2 overrides BYPASS when USERCC2 bit in RCC2 register is set.
XTAL	This bit field is used to specify crystal value that is attached to the *main* oscillator. The values can be configured from 0x05 to 0x1A with corresponding crystal frequencies ranging from 4MHz to 25MHz.
OSCSRC (RCC), OSCSRC2 (RCC2)	Oscillator source selection, with default value of 0x1 that selects precision internal oscillator. Setting to 0x0 selects main oscillator, while setting to 0x3 selects low frequency internal oscillator. OSCSRC2 overrides OSCSRC when USERCC2 bit in RCC2 register is set. For OSCSRC2 it is possible to configure a value of 0x7 to select a 32.768 kHz external oscillator.
MOSCDIS	This bit when set to 1, disables the main oscillator.
USERCC2	When this bit is set, the RCC2 register bit fields override corresponding the RCC register bit fields.
DIV400	This bit field selects between 400 MHz (when set to 1) or 200 MHz as the PLL output. When this bit is 0, SYSDIV2 is used as clock divisor and 200 MHz pre-divided PLL output is used.
SYSDIV2LSB	This bit becomes the LSB of SYSDIV2 when DIV400 is set to 1. By appending SYSDIV2LSB bit to the SYSDIV2 field provides a 7 bit divisor and uses 400 MHz PLL output.

- Enable use of PLL by clearing the BYPASS (BYPASS2) bit in RCC (RCC2).

Using the steps discussed above, clock configuration to operate TM4C123 at 80 MHz is illustrated in Example 10.1. To run the microcontroller at 80 MHz, PLL is used and is configured to operate at 400 MHz. The clock output from PLL is divided by a factor of 5 to get the desired system clock frequency.

Example 10.1 (System clock configuration.). For the divisor value of 5, the SYSDIV2 is configured with a value of 2 and SYSDIV2LSB is set to 1.

```
// Clock frequency configuration register definitions
#define SYSCTL_RCC_R    (*((volatile unsigned long *)0x400FE060))
#define SYSCTL_RCC2_R   (*((volatile unsigned long *)0x400FE070))
#define SYSCTL_RIS_R    (*((volatile unsigned long *)0x400FE050))

// Bit field definitions for RCC register
```

```
#define RCC_BYPASS        0x00000800   // PLL bypass
#define RCC_USESYSDIV     0x00400000   // Enable system clock divider
#define RCC_PWRDN         0x00002000   // PLL power down
#define RCC_BYPASS        0x00000800   // PLL bypass
#define RCC_XTAL_M        0x000007C0   // Crystal value
#define RCC_XTAL_16MHZ    0x00000540   // 16 MHz
#define RCC_OSCSRC        0x00000030   // Oscillator source
#define RCC_OSCSRC_MAIN   0x00000000   // MOSC

// Bit field definitions for RCC2 register
#define RCC2_USERCC2      0x80000000 // Use RCC2
#define RCC2_DIV400       0x40000000 // Divide PLL as 400/200 MHz
#define RCC2_SYSDIV2      0x1F800000 // System clock divisor 2
#define RCC2_PWRDN2       0x00002000 // Power-down PLL 2
#define RCC2_USBPWRDN     0x00004000 // Power-down USB PLL
#define RCC2_BYPASS2      0x00000800 // PLL bypass 2
#define RCC2_OSCSRC2      0x00000070 // Oscillator source 2
#define RCC2_OSCSRC2_MO   0x00000000 // MOSC

// For 80 MHz system clock, 400 MHz output from PLL is divided by
// 5. For that SYSDIV2 is set to 2 and SYSDIV2LSB to 1. The binary
// value for these bit fields combined looks like 101 (=5).
#define RCC2_SYSDIV2_80M 0x01000000
#define RCC2_SYSDIV2LSB  0x00400000


#define RIS_PLLLRIS       0x00000040 // PLL lock interrupt status

// Function prototype definition
void System_Clock_Freq_Config(void);

// Configure the system clock to 80 MHz using PLL
void System_Clock_Freq_Config(void){

  // Configure the system to use RCC2 for advanced features
  // such as 400 MHz PLL and non-integer System Clock Divisor
  SYSCTL_RCC2_R |= RCC2_USERCC2;

  // Bypass PLL and system clock divider while initializing
  SYSCTL_RCC2_R |= RCC2_BYPASS2;
  SYSCTL_RCC_R  &= ~(RCC_USESYSDIV);

  // Select the crystal value and oscillator source
  SYSCTL_RCC_R  &= ~(RCC_XTAL_M);    // clear XTAL field
  SYSCTL_RCC_R  |= RCC_XTAL_16MHZ;   // configure 16 MHz crystal
  SYSCTL_RCC2_R &= ~(RCC2_OSCSRC2);  // clear oscillator source
  SYSCTL_RCC2_R |= RCC2_OSCSRC2_MO;  // configure for main osc.
  SYSCTL_RCC2_R |= RCC2_USBPWRDN;    // disable USB PLL operation

  // Activate PLL by clearing PWRDN and use 400 MHz PLL
  SYSCTL_RCC2_R &= ~(RCC2_PWRDN2);
  SYSCTL_RCC2_R |= RCC2_DIV400;

  // Set the desired system divider and the least significant bit
  SYSCTL_RCC2_R &= ~(RCC2_SYSDIV2);  // Clear clock divider field
  SYSCTL_RCC2_R |= RCC2_SYSDIV2_80M; // Set SYSDIV2 for 80MHz
  SYSCTL_RCC2_R |= RCC2_SYSDIV2LSB;  // Set SYSDIV2 LSB bit

  // Wait for the PLL to acquire lock condition
```

```
    while((SYSCTL_RIS_R & RIS_PLLLRIS) == 0);

    // Enable use of PLL by clearing BYPASS2
    SYSCTL_RCC2_R &= ~(RCC2_BYPASS2);
}
```

10.4   Timer Basics

The basic timer module comprise of an up (down) counter with an associated counter
overflow (underflow) detection capability. Each timer can be enabled or disabled using
a control signal. When enabled, an input clock signal is applied to the counter, which
increments (decrements) counter value. The *timer size* or *timer width* is actually the
number of bits allocated for its counter. When a 16-bit timer is counting up, its
counter will overflow when it reaches 0xFFFF. After timer overflow, the counter rolls
over and is loaded either with 0x0000 or with a user specified value from the *timer
reload register* to start count up again. In addition, the overflow also sets a *timer
overflow flag*. The user can either poll this flag bit to determine the occurrence of
an overflow or a timer interrupt (when configured) can be generated by this overflow
flag. This flag bit should be cleared by the user software, so that the occurrence of
timer overflow next time can be known. The user can also measure time elapsed since
the occurrence of previous overflow, by reading current timer value.

If the timer is counting down, it underflows when the counter reaches 0x0000. After
underflow, the counter is either loaded with 0xFFFF or a user specified reload value
to start its next count down. The underflow condition also sets the timer overflow
flag. The basic timer block diagram using either a 16-bit or 32-bit counter is shown
in Figure 10.6. A register holding the counter reload value is also depicted in Figure
10.6. Example 10.2 describes the evaluation of timer overflow intervals for different
timer clock frequencies and reload values.

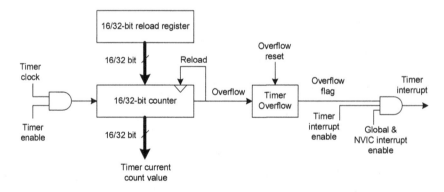

Figure 10.6: Basic timer block diagram with reload and timer interrupt capability.

Example 10.2 (Evaluation of timer overflow interval.)**.** In this example, we
evaluate the timer overflow interval for 16-bit timer. For that purpose it is

assumed that timer is counting down and the reload value is 0xFFFF. If the frequency of the clock input to the timer is 4 MHz then the timer overflow interval is

$$\frac{2^{16}}{4 \times 10^6} = 16.384 \text{ ms}.$$

Now the next question is, if we want to generate a timer interrupt every 10 ms, what do we need to do? There are two possible choices. One option is that we change the frequency of the input clock signal to 6.5536 MHz. A second possibility is that, the timer starts count down from 0x9C40 (40000) rather 0xFFFF. This second option is more viable and is discussed next to generate desired timer overflow intervals.

Consider the scenario, where we are interested in generating a timer interrupt every Δt seconds. If the timer width is w-bit and it is operating at a frequency, f Hz with count down configuration, then its reload value x_d, when the timer underflows after reaching 0x0000, is obtained as

$$x_d = f \times \Delta t. \tag{10.3}$$

However, when the timer is counting up, it overflows after reaching 0xFFFF and the corresponding reload value x_u is obtained using the following modified expression

$$x_u = 2^w - f \times \Delta t. \tag{10.4}$$

For the above expressions to be valid, both x_d and x_u are upper bounded by 2^w.

10.4.1 Timer Resolution and Range

As discussed above, the timer size is the number of counter bits, which determine the maximum value the counter can accumulate before an overflow occurs. A 16-bit timer can accumulate 2^{16} number of clock cycles, which also determines the *timer range* for a given timer clock frequency. *Timer resolution* is the smallest time interval that can be measured by the timer. Since counter value changes by +1 or −1, when time elapsed is equal to one clock cycle, hence timer resolution is simply equal to time period of input clock given by $\frac{1}{f}$, where f is the frequency of timer input clock. For a given resolution, timer range determines the maximum interval, a timer can measure without overflowing. Using Example 10.2, we can see that for a 16-bit timer with 1 μs resolution (i.e., timer clock is 1MHz), the maximum time interval (i.e., timer range) that can be measured is 65.536 ms. If 16-bit timer range is required to be 1 s the corresponding resolution is approximately 16 us. This tradeoff, between timer resolution and range, can be achieved using timer prescaler.

A prescaler divides the frequency of timer input clock signal before feeding it to the counter. This scaling increases the timer operating range or the maximum time interval that it can generate using timer overflow. However, using a prescaler reduces the timer resolution. Figure 10.7 illustrates the use of prescaler.

To extend timer range, while maintaining its resolution, a possible solution is to extend the counter size. For instance, using a 32-bit timer, provides a range of 4295 s at 1 μs resolution, or alternatively 42.95 s at 0.01 μs resolution.

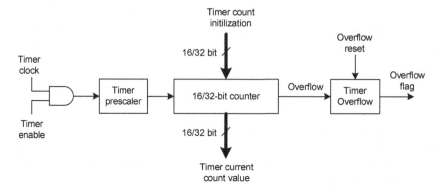

Figure 10.7: Timer block diagram with prescaler to increase timer operating range.

10.5   TM4C123 Timing Interfaces and Systick Timer

The TM4C123 microcontroller is equipped with different timer modules that have varying capabilities. An application uses either a timer as an independent device or for some complex task multiple timers (or timer modules) can be used in combination. It is also possible that a specific timer mode of operation is not available on each timer module requiring an intelligent assignment of timers. Below we list different timer modules that are integrated in TM4C123.

- Systick timer

- General purpose timer module

- PWM timer module

- Watchdog timer module

The *Systick* timer is integrated as part of Cortex-M processor and is referred as system timer. All other timer modules are integrated as peripheral devices and the memory space for these timers (except Systick timer) is allocated in peripheral address space. On the other hand the Systick timer is assigned memory space in the system address space. Next we discuss Systick and general purpose timer modules in fair detail and illustrate their use through practical applications.

10.5.1   Systick Timer

The Cortex-M processor integrates a system timer called Systick. Systick timer is based on a 24-bit down counter and reloads on counting to zero, which is also termed as wrap-on-zero. The counter current value is cleared by write operation, which is also referred as clear-on-write. The Systick timer has a simple control configuration and can be used for different purposes as listed below.

- One key usage of Systick timer is to generate timer ticks, of specified interval, for real-time operating system (RTOS). For instance, it can be used to generate 10 ms timer ticks for scheduling tasks.

- Systick can also be used as a simple counter for measuring time to finish or time elapsed while executing a task.

- It can be configured as high speed alarm timer based on the system clock.

The Systick timer has three registers for its proper functioning. These registers are listed in Table 10.3, which provides the memory address allocation, the reset values and brief description for these registers. The bit field allocations for Systick registers are shown in Figure 10.8.

Table 10.3: Systick register labels, memory addresses, reset value, and brief description.

Register label	Address	Reset value	Brief description
STCTRL	0xE000E010	0x00000004	Systick control and status register.
STRELOAD	0xE000E014	0x00000000	Systick reload value register.
STCURRENT	0xE000E018	0x00000000	Systick current value register.

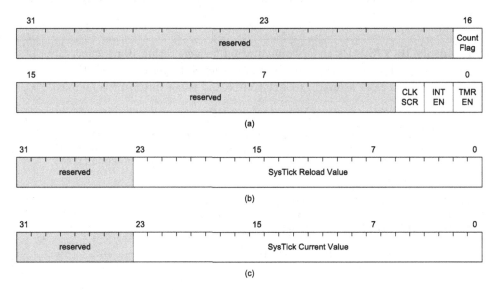

Figure 10.8: Bit field allocations for Systick timer registers. (a) Systick control (STCTRL) register, (b) reload value register (STRELOAD), and (c) current value register (STCURRENT).

The Systick reload and current value registers use least significant 24 bits for holding their corresponding values. If the Systick timer has to generate timer ticks every N clock cycles, then the reload register should be initialized with $N - 1$. The maximum reload value that can be configured to the reload register is 0x00FFFFFF. Different bit field descriptions for Systick control and status register are provided in Table 10.4.

Systick current value register can be read at any time for the timer current count. However, this register is a clear on write, implying that writing any arbitrary value to this register, clears this register. Clearing this register also clears the 'Count Flag' bit in STCTRL register. Example 10.3 details the configuration of Systick timer to

Table 10.4: Bit field descriptions for Systick control and status (STCTRL) register.

Bit field	Description
Count Flag	When this flag bit is read as 1, it indicates that the timer has counted to 0 since the last time of reading this bit. This flag bit is cleared either by performing a read of the STCTRL register or an arbitrary value is written to STCURRENT register.
CLK SCR	This bit signifies the selection of clock source. When set to 1, system clock is used as clock source. When cleared to 0, precision internal oscillator output divided by 4 becomes the Systick clock source.
INT EN	Timer interrupt can be enabled or disabled using this bit. When set to 1, an interrupt is generated, on counting to 0, to the NVIC. An interrupt service routine should be implemented before enabling the interrupt.
TMR EN	This bit controls enabling of timer. When set to 1, Systick timer is enabled and counts down.

generate 10 ms timer interrupts, which can be used as timer ticks by the system software.

Example 10.3 (Systick timer configuration for generating 10 ms timer ticks.). This example illustrates the use of systick to generate periodic timer ticks every 10 ms. System clock is used as clock source and timer interrupt is enabled.

```
// Definitions for Systick registers
#define ST_CTRL_R       (*((volatile unsigned long *)0xE000E010))
#define ST_RELOAD_R     (*((volatile unsigned long *)0xE000E014))
#define ST_CURRENT_R    (*((volatile unsigned long *)0xE000E018))

#define ST_CTRL_COUNT    0x00010000   // Count underflow flag
#define ST_CTRL_ENABLE   0x00000001   // Timer enable control
#define ST_CTRL_INTEN    0x00000002   // Interrupt enable control
#define ST_CTRL_CLK_SRC  0x00000004   // Clock source is system
                                      // clock
#define ST_RELOAD_VALUE  0x0027100    // Counter reload value for
                                      // 10ms interval when system
                                      // clock frequency is 16MHz
                                      // (1/100)*16,000,000
                                      // = 160,000 = 0x27100

// Initialize SysTick running at bus clock.
void SysTick_Init(void)
{
  ST_CTRL_R &= (~ST_CTRL_ENABLE);   // Disable SysTick
  ST_RELOAD_R = ST_RELOAD_VALUE;    // Reload initialization
  ST_CURRENT_R = 0;                 // Any write to current value
                                    // register clears it

  // Enable SysTick, its interrupt and system clock as its
  // clock source
  ST_CTRL_R = ST_CTRL_ENABLE + ST_CTRL_CLK_SRC + ST_CTRL_INTEN;
}

// Systick interrupt handler
void SysTick_Handler(void)
```

```
{
    ST_CTRL_R &= (~ST_CTRL_COUNT);    // Clear the count flag

    // Other activites scheduled every 10ms
}

// main function
int main(void)
{
    SysTick_Init();                   // initialize Systick timer

    while(1){                         // Endless loop
    }
}
```

From Example 10.3, we can observe that there is no mandatory NVIC configuration required for enabling the Systick timer interrupt. However, the optional priority configuration for Systick can be performed using system priority (SYSPRIx) registers available in memory address space 0xE000ED18-0xE000ED20.

The Systick timer is a system timer with limited capability. On the other hand peripheral timer modules integrated in TM4C123 microcontroller are highly flexible with powerful features. This flexibility along with feature set makes the use these timer modules more complex. In the following, we will first discuss further capabilities of timer modules and their use as an input an output device. The use of general purpose timer modules and their configurations will be discussed later in this chapter.

10.6   Timer as Input Device

One basic use of timers, as discussed in the previous section, is to generate periodic intervals. This is implemented by triggering an interrupt on each timer overflow. Apart from this basic timer use there are numerous other uses of timers. As mentioned before, a timer can be configured as an input as well as output device. The use of timer as an input device is discussed in this section, while timers as output devices will be covered in the next section.

There are two possible choices for timer input signal. One possibility is to use an internal periodic clock signal, operating at known frequency, as timer input. This scenario has been discussed when the timer was used to generate periodic intervals. A second possibility is to use an external signal (either periodic or aperiodic but digital) as timer input using a GPIO pin configured for timer alternate functionality, which is discussed in detail in this section. When an external signal applied to the timer input is aperiodic, we might be interested in its accumulated count. The use of timer for this purpose is termed as external event counter. Another possibility is that the external signal is periodic but of unknown frequency. The timer operating as an input capture device can be used to measure this unknown frequency of the external signal. It is also possible to measure pulse duration or width, when the timer is operating as an input capture device as will be explained later. To summarize, when the input to the timer is an external signal, following tasks can be performed using one or more timers.

- External event counting

- Pulse duration measurement

- Signal frequency measurement

In the following, we will discuss each of these uses of timers.

10.6.1   Timer as External Event Counter

Enumerating external events is required by many different applications. For instance, timer as an event counter can be used to enumerate the number of objects passing a fixed point on an assembly line. An external event can be marked by a rising, a falling or both edges. The external events can be either periodic or aperiodic. A timer can be easily configured to count external events by connecting the counter input to the external signal source. The functional diagram of timer as an external event counter is shown in Figure 10.9. From Figure 10.9, we can observe that there is a selector switch, which can be configured for timer input selection as internal or external. When an external signal is used as an input to the timer, the timer counter value shows the number of external events. If the event count reaches the maximum count value (e.g., $2^{16} - 1$ for 16 bit timer), then an overflow interrupt can be generated by the timer.

It is also possible to configure the timer as down counter, while enumerating external events. In this case an underflow interrupt can be generated after a predefined number of external events, depending on the counter initial value.

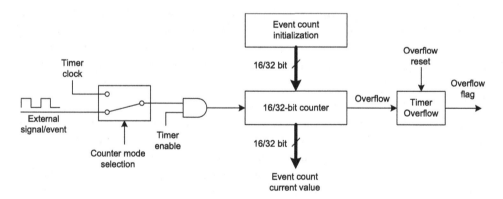

Figure 10.9: Timer configuration as external event counter.

10.6.2   Timer as Input Capture Device

For many different applications, the occurrence time of external events is important. When the timer is configured as an input capture device, it allows to measure external event times accurately. For this purpose, an external signal is connected to the timer input pin. When an event occurs on the timer input pin, the timer current value is captured and is stored in an associated register (latch). The user program can read this value as capture register contents, for further processing. The functioning of timer as input capture device is illustrated in Figure 10.10. Depending on the timer configuration, a rising, falling or even both edges of an external signal can be used for capturing the corresponding event time.

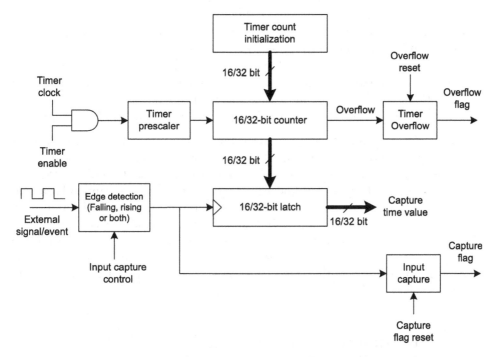

Figure 10.10: Timer configuration to capture the time of external event.

Input capture can be used for different applications. Below we list two important uses of input capture mechanism.

- Pulse duration measurement

- Time period measurement

The use of input capture mechanism for pulse duration measurement is discussed below, while its use for time period measurement is illustrated in the next section.

Pulse Duration Measurement

The timer input capture mode can be used for measuring pulse durations. In addition, we define the *pulse duration* as the time interval between a rising (low to high transition) edge followed by a falling edge. This type of pulse is called a positive pulse and the corresponding counterpart is a negative pulse. The pulse duration measurement solution discussed here for positive pulses can be easily modified for negative pulses. To keep the discussion simple, we require the pulse duration smaller than the maximum interval of the timer. To measure the duration of positive pulses we can use either of the following two methods.

- *Method 1*: The timer is configured to capture event time on both the edges. In addition, a timer interrupt is generated on each event. While processing the first edge interrupt, it is determined whether the event was due to a rising or a falling edge. If the event is a rising edge, the edge time is recorded. To determine the edge type, the corresponding GPIO pin state is checked after the edge. A logic high on the GPIO pin indicates that the previous edge was indeed a rising edge. The event time is captured again on the next event, which corresponds to a falling edge. Subtracting the event time corresponding to first edge from that of second edge, we can determine the pulse duration. The minimum pulse duration that can be measured using this method depends on the time elapsed from the occurrence of event until it is processed by the timer ISR.

- *Method 2*: In this method, timer is configured to capture event time on a rising edge and generate an interrupt. While processing the timer interrupt for rising edge event, the timer is reconfigured for falling edge event. In addition, the edge time is recorded for further processing. Now the event time is captured again on the occurrence of a falling edge. The pulse duration then can be determined as mentioned in Method 1 above. It is important to recognize that the recording of event time as well as timer reconfiguration should be completed before the occurrence of a falling edge. This requirement also determines the lower limit on the pulse duration that can be measured reliably. We do not require GPIO status checking for this method.

10.7   Frequency Measurement Using Timers

One important application of timers is frequency measurement. The unknown frequency signal can be either analog or digital. For frequency measurement of a digital signal no extra circuitry is required, provided the digital signal has appropriate logic levels. Otherwise a level shifter can be used, to translate the input signal logic levels, before connecting it to the timer input. For a bipolar analog signal, a DC offset is added to the signal before its digitization using comparator. For unipolar analog signal, only a comparator is sufficient. Once we have a digital signal, its frequency can be measured using one of the following two fundamental approaches. These two approaches are discussed next.

- Frequency evaluation using period measurement

- Frequency measurement by cycle counting

10.7.1   Period Measurement

Period measurement can be performed by using timer as an input capture device. We have already seen the use of timers for capturing event times in case of pulse duration measurement. For period measurement, the timer is configured to capture event times either on the rising or falling edges but not both. Capturing the event times for two consecutive rising or falling edges allows us to measure the signal time period, which can be evaluated by subtracting the event time corresponding to first edge from that of the second edge.

Next we evaluate the frequency measurement accuracy of this method. Consider an extreme scenario, where the external signal frequency is comparable to timer clock frequency. In a microcontroller the timer clock is derived from the system clock. For simplicity, we assume that the timer clock frequency is same as that of system clock. At this point, it is important to mention that there is no synchronization between the external signal (of unknown frequency) and the system clock used for timer input. This independence of two clock sources can also be verified from Figure 10.10. This lack of synchronization can always lead to well known *one-count error*. To clarify this, let us consider a high frequency external signal, such that the system clock frequency is 1.5 times the external signal frequency. For illustration, let us choose system clock frequency of 15 MHz and the corresponding external signal frequency is 10 MHz. To measure the time period of external signal, the timer value is captured at two consecutive rising (or falling) edges of the external signal. Since the two signals are not synchronized, it is always possible to have an arbitrary phase delay between the corresponding cycles of the two signals. Depending on the actual phase delay between the two signals, it is possible that the counter value increments by either 1 or 2. This fact is illustrated in Figure 10.11, where frequency of the system clock is selected 1.5 times higher than that of external signal. For the case shown in Figure 10.11(a) the counter will increment by 2 while for the scenario in Figure 10.11(b) the counter value will increment by 1. The measured external signal frequency is 7.5 MHz for the case in Figure 10.11(a) while it is 15 MHz for the scenario shown in Figure 10.11(b). The maximum frequency measurement error in this case is either -50% or +25 %. Now consider an external signal of 2 MHz frequency, while the system clock frequency is kept at 15 MHz. The timer captured values, for this case, will be either 7 or 8 due to one count error as illustrated in Figure 10.12. For this case, the measured external signal frequency can be either 2.143 MHz or 1.875 MHz. The frequency measurement error in this case is either -7.14% or +6.25%.

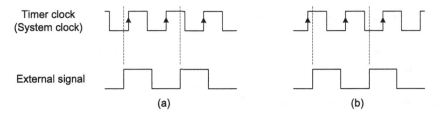

Figure 10.11: Illustration of one count error for high frequency (10 MHz) external signal. It is assumed that the counter increments on the rising edges of the system clock (15 MHz). The timer count value is captured on the two consecutive rising edges of the external signal. During one time period of external signal, (a) two system clock cycles are counted and (b) one system clock cycle is counted.

From the above discussion, we observe that the frequency measurement accuracy improves if the timer clock frequency to external signal frequency ratio increases. For instance, if this frequency ratio is 100 the frequency measurement error is approximately 1%. This in general, is valid for time period based frequency measurement, where the time period is evaluated using timer as an input event time capture device. In addition, it should be observed that, for the two cases illustrated above, the timer clock frequency to external signal frequency ratio was not a whole number. When this frequency ratio is not a whole number, it is always possible to have one count error. On the other hand, if this ratio is a whole number, then ideally there will not be one count error. But due to frequency drift/deviation, which is inherent to clock sources, one count error is always inevitable.

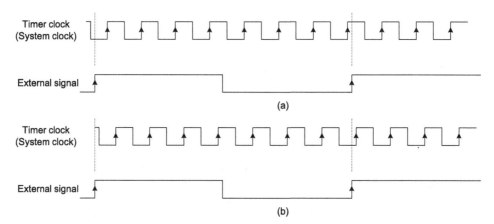

Figure 10.12: Reduced significance of one count error for relatively lower frequency (2 MHz) external signal. During one time period of external signal, (a) eight system clock cycles are counted and (b) seven system clock cycles are counted.

Based on the above discussion, we observe that time period based frequency measurement is more accurate for low frequency input signals. For high frequency signals, it is more appropriate to measure the signal frequency directly by cycle counting as explained next.

10.7.2   Cycle Counting

The high frequency measurement limitation of time period based method can be addressed by using direct frequency measurement approach, which employees *cycle counting*. In this method, we count the cycles of an unknown (but relatively high) frequency external signal, for a predefined time interval. To count the cycles of an external signal, the timer is configured as an external event counter. A second timer is required to generate a predefined time interval. For instance, we can evaluate the signal frequency by counting the cycles for $\frac{1}{4}$s. In this case, the signal frequency is obtained by multiplying the timer count value by a factor of 4, which gives a 4 Hz *frequency resolution*. The frequency measurement resolution is the minimum frequency that can be measured reliably. For a given frequency of an external signal, the frequency measurement resolution can be improved by increasing the counter size used by the timer.

The selection of predefined time interval, primarily depends on the external signal frequency as well as the size of the counter used by the timer. For instance, assume that the frequency of an external signal is 10MHz and the counter is 16-bit. Then the question is, what is the maximum predefined time interval that can be used for this case, without timer overflow. For a 16-bit timer, the overflow occurs, when the timer has counted $2^{16} - 1$ events. Using an external signal frequency of 10 MHz, the maximum permissible value of predefined time interval, that does not result in timer overflow, is evaluated as $\frac{2^{16}-1}{10^7} = 6.55ms$. For general case, assume that the external signal frequency is f_i (Hz) and the timer width is w (bits). Then the maximum value of predefined time interval t_c, which can be used for cycle counting, without incurring timer overflow, is bounded as

$$t_c \leq \frac{2^w - 1}{f_i}. \tag{10.5}$$

The accuracy of cycle counting method reduces for lower frequency of external signals. Consider an extreme scenario, where the signal frequency is 1.5 Hz. Using a predefined time interval of 1s, the measured frequency will be either 1 Hz or 2 Hz, due to one count error. The measurement error in this case is either -50% or 50%. The frequency measurement accuracy for this method can be improved increasing the predefined time interval to a larger value. However, it is not practical to wait for tens of seconds to measure the frequency for low frequency signals. To summarize, we have the following conclusions.

- The time period based frequency measurement should be used for relatively low frequency signals. An increase in the external signal frequency, reduces the frequency measurement accuracy due to the fact that one count error becomes more significant.

- The cycle counting based frequency measurement should be used for high frequency signals. The measurement accuracy of this method reduces for relatively lower frequencies. This is attributed to an increase in the significance of one count error.

10.8 Timer as Output Device

We have explained the functioning and use of timers as input devices in the previous sections. Now we concentrate on the use of timers as output devices. When a timer is used as an output device, its main purpose is to generate signals with desired attributes. For instance, we can use timer to generate periodic signal (square wave shape) of certain frequency as well as to generate pulses of varying width. It can also be used to generate time delays. We will discuss the following two main uses of timers in this section.

- Pulse width modulation

- Generating variable frequency signals

10.8.1   Pulse Width Modulation

Pulse width modulation (PWM) is a technique to generate variable width pulses, while maintaining a constant signal frequency. The time period, t_p and the pulse width t_w are the two parameters that must be specified to generate the desired PWM signal, such that $t_w < t_p$. An important attribute of PWM signal is its *duty cycle, d,* which is defined as a percentage value using the following expression

$$d = \frac{t_w}{t_p} \times 100. \tag{10.6}$$

Figure 10.13 shows a typical PWM signal, for three different duty cycle values. It can be observed from Figure 10.13 that in a PWM signal the frequency remains the same and it is only the duty cycle that changes. PWM signals are widely used in many control applications. Below are some common uses of PWM signals.

- Motor speed control

- Power converters

- Brightness control for LEDs

- Encoded data transmission

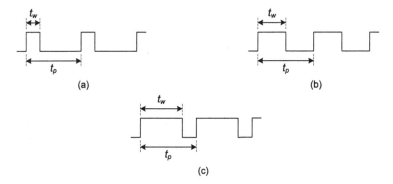

Figure 10.13: Same frequency PWM signal with different duty cycle values, (a) 25%, (b) 50%, and (c) 75%.

The block diagram implementing the PWM functionality is shown in Figure 10.14. The system clock frequency can be scaled, using the optional frequency prescaler, before feeding to the counter. For illustration purpose, we assume that the timer is configured as down counter. When the timer initially starts with a count value of 0, it is immediately loaded with the reload value, which is proportional to the time period (frequency) of the PWM signal. The timer reloading is triggered by a reload signal, which also sets the flip-flop output to logic level high as shown in Figure 10.14. As soon as the counter is reloaded, the reset signal becomes logic low. Now the counter value decrements on each clock cycle and in addition, the counter current value is continuously compared against the duty cycle value. At the beginning, the counter value is higher than that of duty cycle and the comparator output is logic low. Once the counter counts below the duty cycle value, the comparator output goes high and resets the flip-flop. The counter continues to countdown till it reaches 0, which results in an overflow condition (precisely speaking an underflow occurs in this case) and generates the reload signal, which also completes one cycle of PWM signal. The overflow condition reloads the counter and starts next PWM cycle.

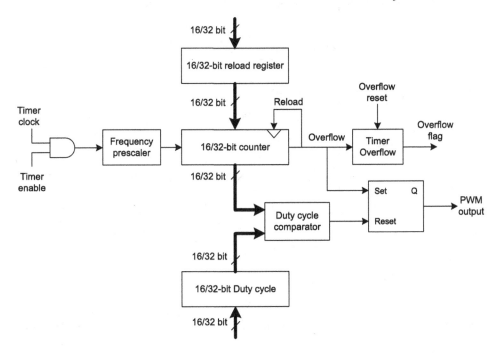

Figure 10.14: Timer configuration as an output device for PWM generation.

The generation of PWM signal as discussed above, is illustrated in Figure 10.15 for 16-bit timer. From Figure 10.15, we observe that the timer reload value $0xzzzz$ determines the frequency (or alternatively time period) of PWM signal, while the pulse width is determined by subtracting the value in duty cycle register from $0xzzzz$.

The PWM waveform is generated by comparing the duty cycle value against the timer current count. If we update the duty cycle register value at the beginning of each PWM cycle, the resulting PWM signal can have an arbitrary pattern embedded in the duty cycle variations. For instance, if we update the duty cycle register values following a sinusoidal pattern then the resulting PWM signal will have pulse widths varying sinusoidally. This is precisely done in PWM based DC to AC inverters.

If the counter values, operating in countdown mode, are plotted as a function of time, the resulting signal appears as a sawtooth waveform (with negative slope), as can be verified from Figure 10.15. If the timer is configured to operate in count up mode, the counter values appear as a sawtooth waveform (with positive slope). The PWM generated using sawtooth signal is called edge aligned PWM. On the other hand if the PWM signal is generated using a triangular waveform, the resulting signal is called center aligned PWM. The triangular wave shape can be constructed by configuring the timer as up-down counter. In this mode, the counter first counts up to a specified value and then counts down to zero, to complete one cycle of PWM signal.

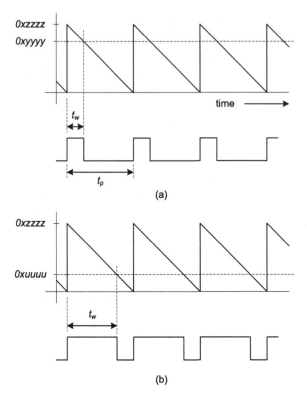

Figure 10.15: PWM signal generation illustration for two different duty cycle values. For a given timer clock input frequency, the PWM frequency is determined by the reload value 0x$zzzz$. The value 0x$yyyy$ in (a) corresponds to 25% duty cycle, while 0x$uuuu$ in (b) results in 75% duty cycle.

PWM Frequency and Resolution Relationship

We already know that PWM frequency (f_{PWM}) is configured by using an appropriate timer reload value for a given timer clock frequency (f_{clk}). Specifically, for given PWM and timer clock frequencies, the reload value (t_{rl}) is obtained as

$$t_{rl} = \frac{f_{clk}}{f_{PWM}}. \tag{10.7}$$

The PWM resolution determines the minimum change in the duty cycle and is quantified in bits as units. For instance, a PWM signal with 8-bit resolution will have the smallest pulse of $\frac{100}{256}$% duty cycle, which is also the smallest possible change in the duty cycle value. Using 10-bit PWM will provide four times better resolution than 8-bit PWM. Since the parameter t_{rl} corresponds to 100% duty cycle of the PWM signal, we can relate it to the PWM resolution (r_{PWM} in bits) as

$$r_{PWM} = \log_2(t_{rl}). \tag{10.8}$$

Combining the expression in (10.9) and (10.7) the relationship between PWM frequency and PWM resolution (in bits) is given by

$$r_{PWM} = \log_2 \left(\frac{f_{clk}}{f_{PWM}} \right).$$

(10.9)

10.8.2 Generating Variable Frequency Signals

The PWM generation using timer, is one specific use of timer as an output compare device. As observed from PWM generation, there are two parameters, t_p and t_w that can be configured, when the timer is used as an output compare device. In case of PWM, the parameter t_w is varied, while t_p is kept constant.

Another possibility is to vary parameters t_p and t_w simultaneously, but the ratio $\frac{t_w}{t_p}$ is kept constant. By keeping this ratio constant, the resulting signal has a fixed duty cycle. Since the parameter t_p determines the periodicity and hence signal frequency, as a result we have a variable frequency signal with constant duty cycle. Using this approach, a variable frequency signal can be generated, where the frequency can be varied either gradually or abruptly.

Finally there are two other possibilities in the choice of parameters t_p and t_w. If parameter t_p is varied while t_w is kept constant, then the resulting signal will be variable duty cycle as well as variable frequency. However, for this case an increase (decrease) in frequency will also result in a corresponding increase (decrease) in duty cycle. This joint variation of signal frequency and duty cycle can be decoupled by making both t_p and t_w variable. The possible choices for parameters t_p, t_w and the signals generated as a result are tabulated in Table 10.5. The condition $t_w < t_p$ should be satisfied for different signals listed in Table 10.5.

Table 10.5: Different signals generated based on parameters t_p and t_w.

Parameter t_p	t_w	$\frac{t_w}{t_p}$	Description
Fixed	Fixed	Fixed	Constant duty cycle as well as fixed frequency signal.
Fixed	Variable	Variable	Variable duty cycle but fixed frequency signal. This is PWM signal.
Variable	Variable	Fixed	Variable frequency but constant duty cycle signal.
Variable	Fixed	Variable	Both frequency as well as duty cycle are variable. These variations are coupled i.e. when frequency increases the duty cycle also increases and vice versa.
Variable	Variable	Variable	Both frequency as well as duty cycle are variable. These variations are decoupled i.e., frequency and duty cycle can be varied arbitrarily.

10.9 General Purpose Timer Modules in TM4C123

In contrast to Systick timer, all other timer modules in Cortex-M core based microcontrollers are integrated as peripheral devices. These timer modules can be configured using associated timer registers in the peripheral address space. The TM4C123 microcontroller includes the following different types of timer modules.

- Basic 16/32-bit timer

- Wide 32/64-bit timer

- Watchdog timer

There are six basic 32-bit general purpose timer modules, which are labeled as Timer0 to Timer5. In addition, there are six wide 64-bit timer modules. Each basic 32-bit timer module can be used as two independent 16-bit timers or it can be configured as one 32-bit timer. In other words, the six basic timer modules provide twelve 16-bit timers. Same is true in case of wide timer modules, where each wide timer module can be configured either as one 64-bit timer or two 32-bit timers. There are two 32-bit watchdog timer modules.

When basic 32-bit timer, for example, Timer0 is used as two independent 16-bit timers, they are labeled as Timer0A and Timer0B. Each 16-bit timer has an associated 8-bit prescaler for basic 32-bit timer modules, while in case of wide 64-bit timer module, each 32-bit timer has a 16-bit prescaler. The use of prescaler depends on the mode of timer operation. In some modes the prescaler acts as true prescaler and holds the least significant 8 bits of the count, while in others it acts as timer extension to effectively extend the 16-bit timer to have a 24-bit counter. When prescaler acts as timer extension, it holds the most significant 8 bits of counter. It is important to mention that use of prescalers is only permitted in split timer configuration with individual two 16-bit timers (for 32-bit basic timer) or two 32-bit timers (for 64-bit wide timers).

This section discusses basic timer configurations and their use for different applications. Most of the basic timer configuration steps are equally applicable for wide timers.

10.9.1   Timer Operating Modes

Before discussing the steps involved in timer configuration, we first briefly introduce an important set of timer operating modes. The selected operating modes of the basic timer modules in TM4C123 microcontroller are listed below. Some other timer operating modes that are not discussed here are, real time clock (RTC) mode and periodic snapshot mode. In addition, wait for trigger mode allows daisy chaining of timers, where TimerN+1 is triggered by TimerN on its time out event. In case of 32-bit timer, Timer0 can trigger Timer1, when wait for trigger mode is configured for Timer1. Similarly for 16-bit timers, Timer0A can trigger Timer0B, which in turn can trigger Timer1A and so on. For further details on these operating modes, the reader is referred to see the TM4C123 microcontroller data sheet. Now we briefly introduce selected operating modes of TM4C123 timers.

- One shot timer mode: In this mode, once the timer is enabled, it starts counting until the reload condition is reached. At this instant, the timer stops counting and is disabled automatically. Timer one shot mode can be used to generate a desired time interval, when an event occurs.

- Periodic timer mode: In periodic mode, when the timer reaches reload condition, it starts counting again after reload and is not disabled automatically.

- Input event count mode: Timer is used to count external events (either periodic or aperiodic) in this mode. An external signal is connected to a GPIO pin that is configured for alternate functionality as a timer input capture/compare pin.

- Input event-time capture mode: This mode is also termed as input capture mode. When an external event occurs on the timer input capture/compare pin, the counter value at that time instant is captured and stored in a register dedicated for this purpose.

- PWM mode: In addition to PWM timer modules, the basic and wide timer modules can also be used for PWM signal generation. A PWM signal is generated on a GPIO pin that is configured for alternate functionality as a timer output capture/compare pin.

A timer can be configured to count up or down in case of one shot, periodic, edge-count or edge-time capture modes of operation, while it is always configured as down counter in case of PWM mode. In one shot and periodic timer modes of operation, the prescaler acts as true prescaler when counting down and as timer extension when counting up. In case of input edge count, edge time capture and PWM modes the prescaler acts as timer extension.

10.9.2   Basic Timer Module Configuration Steps

Both basic and wide timer modules have similar configuration steps. In this section we will limit our discussion to basic timer modules only, which can be easily extended to configure wide timer modules as well. The key steps required to configure both basic as well as wide timer modules are listed below.

- Step 1: Clock gating control for timer module

- Step 2: GPIO configuration as timer capture/compare pin

- Step 3: Timer mode and control configuration

- Step 4: Timer interrupt configuration

Based on the available options in the above mentioned key steps, a timer can be configured and operated in many different ways, as depicted pictorially in Figure 10.16. From Figure 10.16 we can observe that some of the timer configurations are mandatory, while others are optional. For instance, configuration to enable clock gating is mandatory for proper operation of the timer module. On the other hand, alternate function configuration of a GPIO pin, as timer capture compare pin is optional. Similarly the timer interrupt configuration is also optional. In addition, the sequence of these configuration steps can be altered. However, the clock and interrupt enabling related steps should be performed at appropriate point in the configuration sequence.

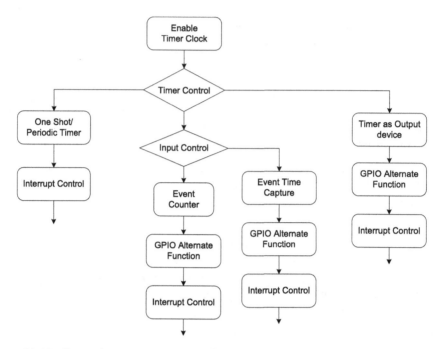

Figure 10.16: General purpose timer configuration tree for different operating modes.

10.9.3   Timer Clock Gating Control Configuration

Each timer module has an associated set of registers to perform the above mentioned configurations. A set of registers used for timer run mode clock gating control are listed in Table 10.6, which are used to control clock gating for different timer modules. The bit 0 to bit 5 of registers RCGC_TIMER_R and RCGC_WTIMER_R are used to enable or disable the clock individually for six basic or six wide timer modules, respectively. Similarly bit 0 and bit 1 of register RCGC_WD_R are used for clock gating of the two watchdog timer modules.

Table 10.6: Timer clock gating control register labels, addresses, reset value, and brief description.

Register label	Address	Reset value	Brief description
RCGC_WD_R	0x400FE600	0x00000000	Watchdog timer clock gating control.
RCGC_TIMER_R	0x400FE604	0x00000000	16/32 bit timer clock gating control.
RCGC_WTIMER_R	0x400FE65C	0x00000000	32/64 bit timer clock gating control.

10.9.4   GPIO Configuration as Timer Capture/Compare Pin

For proper use of a timer module as an input or an output device, the corresponding GPIO pin(s) should be configured for alternate functionality as timer capture/compare pin(s). To use a GPIO pin as timer capture/compare pin, the corresponding GPIO registers should be configured to appropriate values. Specifically four GPIO registers, corresponding to pin direction (GPIO_DIR_R), digital function enabling

(GPIO_DEN_R), alternate function selection (GPIO_AFSEL_R) and port control (GPIO_PCTL_R) for multiplexer selection are configured. To configure a GPIO port pin as timer capture/compare pin, the corresponding bit in the (GPIO_DEN_R) and (GPIO_AFSEL_R) should be set to 1. In addition, the appropriate 4-bit value in GPIO_PCTL_R should be set to 7 to configure that pin as timer pin. The possible choices, for alternate functionality, for different port pins are summarized in Tables 8.2 and 8.3. To configure GPIO pin as timer input capture pin, corresponding bit in GPIO_DIR_R should be cleared, while it should be set for timer PWM mode. The GPIO register descriptions and their addresses are provided in Section 8.4.

10.9.5   Timer Mode and Control Configuration

A set of selected timer registers are provided in Table 10.7, which are used to perform the timer mode and control related configurations. The timer mode registers allow to configure different timer operating modes. The register addresses provided in Table 10.7 are the address offsets from the timer module base address. The actual register address is determined by adding the offset to the corresponding timer module base address. The base addresses for different timer modules are provided in Table 8.4 and are reproduced in Table 10.8 for quick reference.

Table 10.7: General purpose timer mode and control configuration register labels, addresses, reset value and brief description.

Register label	Offset	Reset value	Brief description
GPTM_CONFIG_R	0x000	0x00000000	Timer module configuration register.
GPTM_TA_MODE_R	0x004	0x00000000	Timer A mode control register.
GPTM_TB_MODE_R	0x008	0x00000000	Timer B mode control register.
GPTM_CONTROL_R	0x00C	0x00000000	Timer control register.
GPTM_TA_IL_R	0x028	0xFFFFFFFF	Timer A interval load register.
GPTM_TB_IL_R	0x02C	0x00000000	Timer B interval load register.
GPTM_TA_MATCH_R	0x030	0xFFFFFFFF	Timer A match register.
GPTM_TB_MATCH_R	0x034	-	Timer B match register.
GPTM_TA_PRESCALE_R	0x038	0x00000000	Timer A prescale register.
GPTM_TB_PRESCALE_R	0x03C	0x00000000	Timer B prescale register.
GPTM_TA_PM_R	0x040	0x00000000	Timer A prescale match register.
GPTM_TB_PM_R	0x044	0x00000000	Timer B prescale match register.
GPTM_TA_COUNT_R	0x048	0xFFFFFFFF	Timer A counter register.
GPTM_TB_COUNT_R	0x04C	-	Timer B counter register.
GPTM_TA_VALUE_R	0x050	0xFFFFFFFF	Timer A count value shadow register.
GPTM_TB_VALUE_R	0x054	-	Timer B count value shadow register.

Before discussing the timer mode and control configurations, it is important to first provide the bit field descriptions for different timer configuration registers. We start with the timer configuration (GPTM_CONFIG_R), timer mode (GPTM_TA_MODE_R and GPTM_TB_MODE_R) and timer control (GPTM_CONTROL_R) registers.

Timer Configuration Register

The timer configuration register is responsible to configure global operation of the timer module. The timer module operation as two split (individual) timers or one

Table 10.8: Base addresses for different timer modules.

Start Address	End Address	Description
0x40000000	0x40000FFF	Watchdog timer 0
0x40001000	0x40001FFF	Watchdog timer 1
0x40030000	0x40030FFF	16/32-bit Timer 0
0x40031000	0x40031FFF	16/32-bit Timer 1
⋮	⋮	⋮
0x40035000	0x40035FFF	16/32-bit Timer 5
0x40036000	0x40036FFF	32/64-bit Timer 0
0x40037000	0x40037FFF	32/64-bit Timer 1
0x4004C000	0x4004CFFF	32/64-bit Timer 2
0x4004D000	0x4004DFFF	32/64-bit Timer 3
0x4004E000	0x4004EFFF	32/64-bit Timer 4
0x4004F000	0x4004FFFF	32/64-bit Timer 5

concatenated timer is determined by the configuration of this register. The bit field allocations for the timer configuration register as well as its possible choices are given in Table 10.9. For proper operation, the configuration of this register should not be performed, while the timer is running.

Table 10.9: Timer configuration register, GPTM_CONFIG_R, bit field descriptions.

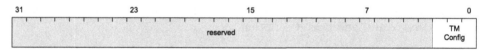

Bit field	Description
TM CONFIG	Timer configuration bit field has the following possible choices. • A value of 0 selects 32-bit for basic timer and 64-bit for wide timer configuration. • A value of 1 selects 32-bit for basic timer and 64-bit for wide timer operating in real time clock (RTC) configuration. • A value of 4 splits 32-bit basic timer into two 16-bit timers and 64-bit wide timer into two 32-bit timers. • All other choices are reserved.

Timer Mode Register

There are two timer mode registers (GPTM_TA_MODE_R and GPTM_TB_MODE_R), one for timer A and second for timer B, for the case when the 32-bit timer is split into two 16-bit timers. The two mode registers follow exactly same bit field allocations. When timers A and B are concatenated to form one 32-bit timer, then only timer A mode register, GPTM_TA_MODE_R is used, while GPTM_TB_MODE_R is ignored. A number of different timer modes can be configured using timer mode register. However, selected timer modes are discussed in this chapter. Accordingly, the relevant bit fields of the timer mode register are described in Table 10.10, which also depicts the bit field allocations. We refer the reader to TM4C123 data sheet for information related to other bit field descriptions of the timer mode register, which are not discussed here.

Table 10.10: Selected bit field allocations and descriptions for timer mode registers. These are applicable for both GPTM_TA_MODE_R and GPTM_TB_MODE_R registers.

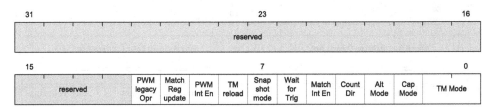

Bit field	Description
PWM Int En	PWM interrupt is enabled by setting this bit to 1. This bit field is only valid in PWM mode of operation.
Count Dir	This bit field configures the timer count direction. Timer counts up when this bit is set to 1 and counts down when cleared.
Alt Mode	Timer alternate mode select bit configures PWM mode when set to 1 and capture/compare mode when cleared to 0.
Cap Mode	Timer capture mode bit selects between external event count and external event time modes of operation. When this bit is set to 1 the timer value is captured in the timer register (GPTM_TA_R or GPTM_TB_R), while clearing this bit configures the timer to count external events.
TM Mode	Timer mode bit field configures one of the three possible timer modes of operation listed below: • Setting to 1 configures one shot timer mode. • Setting to 2 configures periodic timer mode. • Setting to 3 configures capture mode.

Timer Control Register

This register is used to control certain timer features. The two bytes in the lower half word of this register correspond to the two split timers. The bit field allocations as well as their descriptions provided in Table 10.11 are applicable for both timers A and B. This is valid, since bit field allocations in timer control register are the same for both timers A and B.

Timer Interval Load, Prescale, Match, Counter, and Value Registers

We discuss the use of timer interval load, prescale, match, counter, and value registers for different modes of timer operation.

- *One shot and periodic modes*: In these modes, the timer can be configured as up or down counter. In count down mode, when the timer reaches 0x0 (i.e., timeout event), it is reloaded with its start value from the interval load (GPTM_TA_IL_R

Table 10.11: Bit field descriptions for timer control register, GPTM_CONTROL_R.

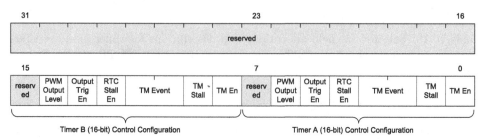

Bit field	Description
PWM Output Level	Timer PWM output is inverted when this bit set to 1 and remains unaffected otherwise.
Output Trig En	Timer output trigger enable bit, when set to 1, enables the triggering of ADC using timer output. Timer mode should be either one shot or periodic for ADC triggering.
RTC Stall En	When RTC stall enable bit is cleared to 0, timer in RTC mode, freezes counting when the processor is halted due to debugger. The timer continues counting in RTC mode when this bit is set to 1.
TM Event	Timer event mode bit field configures one of the three possible timer events listed below: • Setting to 0 configures +ve edge as timer event. • Setting to 1 configures -ve edge as timer event. • Setting to 3 configures both edges as timer events.
TM Stall	When timer stall bit is cleared to 0, the timer continues counting when the processor is halted due to debugger. The timer freezes counting when this bit is set to 1.
TM En	Timer enable bit is used to enable or disable the timer. When this bit is set to 1, timer starts counting.

and GPTM_TB_IL_R) and prescale registers (GPTM_TA_PRESCALE_R and GPTM_TB_PRESCALE_R). For count up mode when the timer count increments to the value in interval load and prescale registers, the timer is reloaded with 0x0. When the timer is configured as one-shot timer and it reaches the time out event (either using count up or down mode), the timer stops counting and clears the timer enable bit in the timer control register. In this mode, an interrupt condition can be generated on each time out event. Based on the interrupt configuration, an additional interrupt condition can also be generated when the timer value equals the value of timer match (GPTM_TA_MATCH_R and GPTM_TB_MATCH_R) and prescale match registers (GPTM_TA_PM_R and GPTM_TB_PM_R).

• *Input event count mode*: In this mode, the timer has a 24-bit counter with upper 8-bits assigned to timer prescale register, while lower 16-bits to timer counter register. In count down mode, the timer match and prescale match registers are initialized in such a way that the difference between the value of timer interval load and prescale registers and the timer match and prescale match registers is equal to the number of events that should be counted. In count up configuration, the timer starts counting from 0 until it reaches the value in timer match and prescale match registers. In count down mode, once the match value is reached, the timer is reloaded with the value in interval load and prescale registers and it stops counting. After this time instant, any further events are ignored till the timer is enabled again. In contrast, the timer in count up mode, continues counting after reloaded with 0.

- *Input event time capture mode*: In this mode, the timer has a 24-bit counter with upper 8-bits assigned to timer prescale register, while lower 16-bits to timer interval load register. The timer is initialized with the value in timer interval load and prescale registers for count down and with 0 when counting up. When an input event is detected on the timer capture/compare pin, the current count from the timer value register (GPTM_TA_VALUE_R or GPTM_TB_VALUE_R) is captured in the timer counter (GPTM_TA_COUNT_R or GPTM_TB_COUNT_R) and prescale snapshot registers. Once an event is captured, the timer continues to count, till the timer reaches the time out. On time out the timer is reloaded from timer interval load and prescale registers for count down mode and with 0 in case of count up mode. An interrupt condition, if configured, is also generated when the event is captured.

- *PWM mode*: In PWM mode timer only runs as down counter and its start value is obtained from the timer interval load and prescale registers. When the timer count reaches the value in timer match and prescale match registers, the PWM output signal is deasserted. The timer continues counting down until it reaches the time out. On time out the timer counter is reloaded with the value in timer interval load and prescale registers and the PWM output signal is also asserted.

10.9.6   Timer Interrupt Configuration

Each timer module has an associated set of registers that are used for interrupt related configuration. Specifically, there is one interrupt mask register, two interrupt status registers and one interrupt clear register as summarized in Table 10.12. To configure different types of timer interrupts, interrupt mask register, GPTM_INT_MASK_R, is used. In addition, the PWM interrupt and RTC match interrupt are configured using timer mode register. The bit field allocations and their descriptions for interrupt mask register, GPTM_INT_MASK_R, are given in Table 10.13.

Table 10.12: Timer interrupt configuration register labels, addresses, reset value, and brief description.

Register label	Offset	Reset value	Brief description
GPTM_INT_MASK_R	0x018	0x00000000	Timer interrupt mask register.
GPTM_RIS_R	0x01C	0x00000000	Timer raw interrupt status register.
GPTM_MIS_R	0x020	0x00000000	Timer masked interrupt status register.
GPTM_INT_CLEAR_R	0x024	0x00000000	Timer interrupt clear register.

The timer interrupt status registers (GPTM_RIS_R, GPTM_MIS_R) and interrupt clear register (GPTM_INT_CLEAR_R) also have the same bit field allocations as that of timer interrupt mask register. As an example, for the bit field, capture A

Table 10.13: Interrupt mask register, GPTM_INT_MASK_R, bit field descriptions.

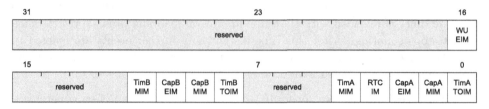

Bit field	Description
WU EIM	64-Bit wide timer write update error interrupt mask. Set this bit to 1 to enable the interrupt.
TimA MIM, TimB MIM	Timer A (B) match interrupt mask bit. Setting this bit to 1 enables the timer match interrupt.
CapA EIM, CapB EIM	This bit is used to mask or unmask the interrupt corresponding to external event time capture. Configuring this bit to 1 enables the event time capture interrupt.
CapA MIM, CapB MIM	This mask bit corresponds to external event count match interrupt. Configuring it to 1 enables the external event count match interrupt.
TimA TOIM, TimB TOIM	This interrupt mask bit corresponds to time out event of the timer. Configuring it to 1 enables the external event count match interrupt.
RTC IM	Setting this bit enables the real time clock (RTC) interrupt.

event interrupt mask (CapA EIM) in mask register, there is a corresponding bit field, capture A event raw interrupt status (CapA ERIS) in the raw interrupt status register. Similarly, masked interrupt status and interrupt clear registers also have the corresponding bit field allocations. The timer interrupt clear register is write 1 to clear, i.e., writing 1 to an arbitrary bit field in timer interrupt clear register, clears corresponding bits in timer raw and masked interrupt status registers.

Now let us illustrate the timer configuration explained above using an example. Assuming it is required to generate 10 ms time intervals and a timer time out interrupt should be generated after each 10 ms interval as well. This can be accomplished using any of the general purpose timers. If we select Timer1 A for this purpose then the corresponding timer configuration is given in Example 10.4.

The implementation in Example 10.4 also illustrates the use of prescaler as discussed below. Since the counter is incremented/decremented by the default system clock frequency of 16 MHz, the timer reload value for 10 ms interval turns out to be $160,000$, which is obtained by using the expression in (10.3). Now this reload value is out of 16-bit timer count limit (of $2^{16} - 1$). To address this issue we can extend the timer count limit by using the associated 8-bit prescaler, which effectively extends the timer to 24-bit counter. Now using a prescaler value of 16 and interval load register value of 10,000 will result in timer interval of 10 ms.

Example 10.4 (General purpose Timer1 A configuration for generating 10ms timer interrupts.). In this example we list the basic steps to configure Timer1 A to generate 10 ms time intervals and the timer interrupt service routine is called after each time interval.

```c
void Timer1_Init(void);
void Timer1A_Handler(void);

// Timer 1 base address and run mode clock control
#define TM_BASE    0x40031000
#define RCGC_TIMER_R            *(volatile unsigned long *)0x400FE604

// General purpose timer register definitions
#define GPTM_CONFIG_R           *(volatile long *)(TM_BASE + 0x000)
#define GPTM_TA_MODE_R          *(volatile long *)(TM_BASE + 0x004)
#define GPTM_CONTROL_R          *(volatile long *)(TM_BASE + 0x00C)
#define GPTM_INT_MASK_R         *(volatile long *)(TM_BASE + 0x018)
#define GPTM_INT_CLEAR_R        *(volatile long *)(TM_BASE + 0x024)
#define GPTM_TA_IL_R            *(volatile long *)(TM_BASE + 0x028)
#define GPTM_TA_PRESCALE_R      *(volatile long *)(TM_BASE + 0x038)

// IRQ 0 to 31 enable and disable registers
#define NVIC_EN0_R              *((volatile long *)0xE000E100)
#define NVIC_DIS0_R             *((volatile long *)0xE000E180)

// Timer1 A interrupt is assigned to NVIC IRQ21
#define NVIC_EN0_INT21     0x00200000

// Timer1 A bit field definitions for mode configuration
#define TIM_16_BIT_CONFIG   0x00000004  // 16-bit timer
#define TIM_PERIODIC_MODE   0x00000002  // Timer mode
#define TIM_A_ENABLE        0x00000001  // Timer A enable control

// Timer1 A time out event interrupt mask and clear
#define TIM_A_TIMEOUT_IM    0x00000001
#define TIM_A_TIMEOUT_IC    0x00000001


// Timer1 A reload & prescale values for 10 ms interval
#define TIM_A_INTERVAL      0x2710    // Timer reload value for
                                      // 10ms interval when system
                                      // clock frequency is 16MHz
                                      // (1/(16*100))*16,000,000
                                      // = 10,000 = 0x2710
#define TIM_A_PRESCALE      0x0F      // 1/16 prescaler

// Configure Timer1 A periodic timer with timeout event interrupt
void Timer1A_Init(void){

  RCGC_TIMER_R |= 0x02;               // Enable clock for timer 1

  GPTM_CONFIG_R = TIM_16_BIT_CONFIG; // Configure 16-bit timer

  // Periodic timer mode with count down
  GPTM_TA_MODE_R |= TIM_PERIODIC_MODE;

  // Configure the timer reload value
  GPTM_TA_IL_R = TIM_A_INTERVAL;
  GPTM_TA_PRESCALE_R = TIM_A_PRESCALE;

  // Timer1 A interrupt configuration
  GPTM_INT_MASK_R |= TIM_A_TIMEOUT_IM;
  NVIC_EN0_R = NVIC_EN0_INT21;        // Enable INT 21 in NVIC
```

```
   // Enable the timer
   GPTM_CONTROL_R |= TIM_A_ENABLE;

}

void Timer1A_Handler(void)
{
   // Clear interrupt flag
   GPTM_INT_CLEAR_R |= TIM_A_TIMEOUT_IC;

   // Perform other necessary tasks
}

int main(void){

 // GPIO_Init();
   Timer1A_Init();                      // Initialize the timer

   while(1){
   }
}
```

The timer interrupt configuration in Example 10.4 enables the Timer1 A interrupt using NVIC interrupt enable register. In addition, the timer time out event interrupt is enabled using the timer interrupt mask register.

10.10   TM4C123 Timer as Input/Output Device

In this section, we will illustrate the use of timers both as input as well as output device. For timer as an input device we use timer capture mode to measure the frequency of an external signal using time period measurement. For timer as an output device, we configure timer to generate PWM signal of 1 kHz frequency.

10.10.1   Measuring Frequency Using Timer as Input Device

To measure the external signal frequency, we configure Timer2 A in event time capture mode to measure the signal time period. In this mode, the timer continuously runs as periodic timer. The timer periodicity is controlled by the reload value stored in timer interval load (GPTM_TA_IL_R) and prescale (GPTM_TA_PRESCALE_R) registers. Configuring GPTM_TA_IL_R with a value of 0xFFFF and GPTM_TA_PRESCALE_R with 0xFF, makes the timer period equal to 1 sec for 16 MHz timer clock frequency. With this configuration, the frequency evaluation of an external signal with a frequency higher than 1 Hz is simplified, since it does not exceed the timer periodicity interval of 1 sec. What will be the maximum frequency that can be measured with this timer configuration, while measurement error does not exceed 1% (due to one count error). Recall that one count error becomes more significant at higher frequencies during time period based frequency measurement. This is left as an exercise at the end of this chapter.

Two consecutive rising or falling edges (or events) correspond to one time period of the external signal. The external signal is connected to GPIO pin PF4, which is configured for its alternate functionality as timer capture/compare pin. On each rising (or correspondingly falling) edge of external signal, the timer current counter value is captured and is stored in the timer count register. In addition, the timer interrupt is also generated on each external event capture. On the capture of second edge, the signal time period is calculated by evaluating the difference between two consecutive captured values as implemented in Example 10.5.

Example 10.5 (Timer configuration for measuring frequency of external signal.). In this example we list the basic steps to configure Timer 2A to measure the frequency of an external signal connected to pin PF4, which is configured as timer capture input pin.

```
// Timer 2 base address
#define TM_BASE              0x40032000

// Peripheral clock enabling for timer and GPIO
#define RCGC_TIMER_R         *(volatile unsigned long *)0x400FE604
#define RCGC2_GPIO_R         *(volatile unsigned long *)0x400FE108
#define CLOCK_GPIOF          0x00000020    //Port F clock control
#define SYS_CLOCK_FREQUENCY  16000000

// General purpose timer register definitions
#define GPTM_CONFIG_R        *(volatile long *)(TM_BASE + 0x000)
#define GPTM_TA_MODE_R       *(volatile long *)(TM_BASE + 0x004)
#define GPTM_CONTROL_R       *(volatile long *)(TM_BASE + 0x00C)
#define GPTM_INT_MASK_R      *(volatile long *)(TM_BASE + 0x018)
#define GPTM_INT_CLEAR_R     *(volatile long *)(TM_BASE + 0x024)
#define GPTM_TA_IL_R         *(volatile long *)(TM_BASE + 0x028)
#define GPTM_TA_MATCH_R      *(volatile long *)(TM_BASE + 0x030)
#define GPTM_TA_PRESCALE_R   *(volatile long *)(TM_BASE + 0x038)
#define GPTM_TA_COUNT_R      *(volatile long *)(TM_BASE + 0x048)

// IRQ 0 to 31 enable and disable registers
#define NVIC_EN0_R           *((volatile unsigned long *)0xE000E100)
#define NVIC_DIS0_R          *((volatile unsigned long *)0xE000E180)

// GPIO alternate function configuration
#define GPIO_PORTF_AFSEL_R *((volatile unsigned long *)0x40025420)
#define GPIO_PORTF_PCTL_R  *((volatile unsigned long *)0x4002552C)
#define GPIO_PORTF_DEN_R   *((volatile unsigned long *)0x4002551C)

// Timer2 A interrupt is assigned to NVIC IRQ23
#define NVIC_EN0_INT23       0x00800000

// Timer2 A bit field definitions for mode configuration
#define TIM_16_BIT_CONFIG    0x00000004 // 16-bit timer
#define TIM_EDGE_TIME_MODE   0x00000004 // Time capture on edge
#define TIM_CAPTURE_MODE     0x00000003 // Timer capture mode

// Timer event type bit filed definitions
#define TIM_A_EVENT_POS_EDGE   0x00000000 // Event is +ve edge
#define TIM_A_EVENT_NEG_EDGE   0x00000004 // Event is -ve edge
#define TIM_A_EVENT_BOTH_EDGES 0x0000000C // Event on both edges
#define TIM_A_ENABLE           0x00000001 // Enable timer A
```

```
// Timer A capture mode interrupt mask/clear bit field definition
#define TIM_A_CAP_EVENT_IM        0x00000004
#define TIM_A_CAP_EVENT_IC        0x00000004

// Reload values for Timer A with prescale
#define TIM_A_INTERVAL            0x0000FFFF
#define TIM_A_PRESCALE            0x000000FF

// Global variables
unsigned int Frequency = 0;


// Initialize SysTick with busy wait running at bus clock.
void Timer2A_Init(void){

  RCGC_TIMER_R |= 0x04;          // Enable clock for timer 2
  RCGC2_GPIO_R |= CLOCK_GPIOF;

  // GPIO port F pin 4 configuration
  GPIO_PORTF_DEN_R |= 0x00000010;
  GPIO_PORTF_AFSEL_R |= 0x00000010; // Alternate function selected
  GPIO_PORTF_PCTL_R |= 0x00070000;  // PF4 as timer capture
                                    // compare input

  GPTM_CONFIG_R = TIM_16_BIT_CONFIG; // Configure 16-bit timer

  //  Timer mode is input time capture with count down
  GPTM_TA_MODE_R |= TIM_EDGE_TIME_MODE + TIM_CAPTURE_MODE;

  GPTM_CONTROL_R |= TIM_A_EVENT_NEG_EDGE;

  // Configure the timer reload value
  GPTM_TA_IL_R = TIM_A_INTERVAL;
  GPTM_TA_PRESCALE_R = TIM_A_PRESCALE;

  // Timer0 A interrupt configuration
  GPTM_INT_MASK_R |= TIM_A_CAP_EVENT_IM;
  NVIC_EN0_R = NVIC_EN0_INT23;            // Enable INT 23 in NVIC

  // Enable the timer
  GPTM_CONTROL_R |= TIM_A_ENABLE;
}

void Timer2A_Handler(void)
{
unsigned int time_period = 0;
static unsigned int time_capture=0;
static char  flag=0;

  GPTM_INT_CLEAR_R |= TIM_A_CAP_EVENT_IC;  // Clear interrupt flag

  if(flag == 0)  // Capture time at the start of signal cycle
  {
    time_capture = GPTM_TA_COUNT_R;
    flag = 1;
  }

  else           // Capture time again at the end of signal cycle
```

```
{
  // Calculate the time period
  if(GPTM_TA_COUNT_R > time_capture)
    time_period =  (0x00FFFFFF - GPTM_TA_COUNT_R) + time_capture;
  else
    time_period =  time_capture - GPTM_TA_COUNT_R;

  // Calculate the signal frequency
  Frequency = SYS_CLOCK_FREQUENCY/time_period;

  flag = 0;              // Clear flag to start new measurement
  }
}

int main(void)
{
  Timer2A_Init();                    // Initialize the timer 2

  while(1)
    {
      // Display the measured frequency
    }
}
```

10.10.2   PWM Generation Using Timer as Output Device

For PWM generation, the Timer1 A is configured in PWM mode, where it runs continuously as periodic timer. The timer periodicity is controlled by the reload value stored in timer interval load (GPTM_TA_IL_R) and prescale (GPTM_TA_PRESCALE_R) registers. To generate a PWM signal of frequency 1 kHz, we require a reload value of 16000, when the timer is operating from 16 MHz clock. For this reload value, we configure the GPTM_TA_IL_R with 16000 and GPTM_TA_PRESCALE_R with 0x00. Recall that the timer runs only as down counter in PWM mode. The % duty cycle of the PWM signal is controlled by configuring the timer match register, GPTM_TA_MATCH_R, with an appropriate value. For instance, to obtain a duty cycle of 75% the GPTM_TA_MATCH_R should be configured with a value of 4000, since the timer is running as down counter.

The implementation of PWM generation with 1 kHz frequency and 75% initial duty cycle, is illustrated in Example 10.6. Since the timer reload value is 16000, the achievable PWM resolution in this case is 13-bit.

Example 10.6 (Timer configuration for genearting PWM signal.). In this example Timer1 A is configured for generating PWM signal on GPIO pin PF2, which is configured as timer output pin. The frequency of the PWM signal is 1 kHz and initial value of the duty cycle is 75%.

```
// Timer 1 base address
#define TM_BASE    0x40031000

// Peripheral clock enabling for timer and GPIO
#define RCGC_TIMER_R          *(volatile unsigned long *)0x400FE604
```

```
#define RCGC2_GPIO_R              *(volatile unsigned long *)0x400FE108
#define CLOCK_GPIOF               0x00000020    //Port F clock control
#define SYS_CLOCK_FREQUENCY       16000000

// General purpose timer register definitions
#define GPTM_CONFIG_R             *(volatile long *)(TM_BASE + 0x000)
#define GPTM_TA_MODE_R            *(volatile long *)(TM_BASE + 0x004)
#define GPTM_CONTROL_R            *(volatile long *)(TM_BASE + 0x00C)
#define GPTM_INT_MASK_R           *(volatile long *)(TM_BASE + 0x018)
#define GPTM_INT_CLEAR_R          *(volatile long *)(TM_BASE + 0x024)
#define GPTM_TA_IL_R              *(volatile long *)(TM_BASE + 0x028)
#define GPTM_TA_MATCH_R           *(volatile long *)(TM_BASE + 0x030)

// GPIO PF2 alternate function configuration
#define GPIO_PORTF_AFSEL_R *((volatile unsigned long *)0x40025420)
#define GPIO_PORTF_PCTL_R  *((volatile unsigned long *)0x4002552C)
#define GPIO_PORTF_DEN_R   *((volatile unsigned long *)0x4002551C)

// Timer config and mode bit field definitions
#define TIM_16_BIT_CONFIG    0x00000004  // 16-bit timer
#define TIM_PERIODIC_MODE    0x00000002  // Periodic timer mode
#define TIM_A_ENABLE         0x00000001  // Timer A enable control

#define TIM_PWM_MODE         0x0000000A  // Timer in PWM mode
#define TIM_CAPTURE_MODE     0x00000004  // Timer capture mode


// Timer1 A reload value for 1 kHz PWM frequency
#define TIM_A_INTERVAL       16000     // Timer reload value for
                                       // 1 kHz PWM frequency at
                                       // clock frequency of 16MHz
                                       // 16,000,000/16000
                                       // = 1 kHz
#define TIM_A_MATCH          4000      // Timer match value for 75%
                                       // duty cycle

// Timer and GPIO intialization and configuration
void Timer1A_Init(void)
{
  // Enable the clock for port F and Timer1
  RCGC2_GPIO_R |= CLOCK_GPIOF;
  RCGC_TIMER_R |= 0x02;

  // Configure PortF pin 2 as Timer1 A output
  GPIO_PORTF_AFSEL_R |= 0x00000004;
  GPIO_PORTF_PCTL_R  |= 0x00000700;  // Configure as timer CCP pin
  GPIO_PORTF_DEN_R   |= 0x00000004;

  // Enable the clock for Timer 1
  GPTM_CONTROL_R &= ~(TIM_A_ENABLE);    // disable timer 1 A

  // Timer1 A configured as 16-bit timer
  GPTM_CONFIG_R |= TIM_16_BIT_CONFIG;

  // Timer1 A in periodic timer, edge count and PWM mode
  GPTM_TA_MODE_R |= TIM_PWM_MODE;
  GPTM_TA_MODE_R &= ~(TIM_CAPTURE_MODE);

  // Make PWM frequency 1 kHz using reload value of 16000
```

```
  GPTM_TA_IL_R = TIM_A_INTERVAL;

  // Configure PWM duty cycle value (should be less than 16000)
  GPTM_TA_MATCH_R = TIM_A_MATCH;

  // Enable timer1 A
  GPTM_CONTROL_R |= TIM_A_ENABLE;
}

// Application main function
int main(void){

  Timer1A_Init();                    // Initialize the timer

  while(1){
     // One of the tasks is responsible to change PWM duty cycle
     // by updating GPTM_TA_MATCH_R register
  }
}
```

10.11   Summary of Key Concepts

This chapter has covered the microcontroller timing interfaces from two different perspectives. One aspect covers the clocking of the microcontroller, while the other discusses the timer modules. Specifically, the following key concepts have been introduced in this chapter.

- Different options available to clock to a microcontroller are discussed, which include both on-chip as well as off-chip possibilities.

- A microcontroller can be operated at different frequencies, despite the fact that oscillator (clock source) generates a fixed frequency signal. This is achieved by using phase-lock loop (PLL) as a frequency multiplier.

- Timer modules can be used to measure timing parameters of unknown signals as well as to generate accurate timing intervals/signals.

- A timer module is a counter driven by a clock derived from the system clock. The timer clock can be same as system clock.

- Timer interrupts can be used as events, generated at the boundaries of intended timing intervals. A timer can have multiple interrupts associated with it.

- Systick timer in TM4C123 microcontroller is a system timer module that can be used to generate timing intervals for the operating system.

- All timer modules in TM4C123, other than Systick, are peripheral timers. The peripheral timers have associated GPIO pins.

- When a peripheral timer is used as an input device, it can measure timing parameters of an external signal connected to the GPIO pin associated with the timer. The use of peripheral timer as input device is illustrated for frequency measurement using period measurement and cycle counting.

- When a peripheral timer is used as an output device, it can generate a digital signal with accurate timing at the GPIO pin associated with the timer. The use of peripheral timer as output device is illustrated for pulse width modulation (PWM).

Review Questions

Question 10.1. List the key application areas where timers are used as basic building blocks.

Question 10.2. What are different methods to generate a clock signal for a microcontroller?

Question 10.3. How can a microcontroller be operated at different clock frequencies, while using a fixed frequency oscillator?

Question 10.4. Is it possible to clock the microcontroller at a frequency lower than that of an oscillator? If the answer is yes, then how and if it is no, then why.

Question 10.5. What is the main difference between a 16-bit and a 32-bit timer?

Question 10.6. What are possible choices to increase the timer resolution?

Question 10.7. How can the timer range be increased? What is the effect of increasing timer range on its resolution?

Question 10.8. Is it possible to run Systick timer in count up mode? Can it be run at a clock frequency of 48 MHz?

Question 10.9. What are the configuration steps to use a timer as an event counter?

Question 10.10. What is one count error and when does it become significant?

Question 10.11. Under what conditions is period measurement more accurate than cycle counting when measuring the frequency of an external signal?

Question 10.12. What are the key applications of pulse width modulation?

Question 10.13. What type of signal is generated if the duty cycle register value does not change while the timer reload register value is increased linearly?

Question 10.14. What are different operating modes of TM4C123 timer module?

Question 10.15. How can the prescaler register in a timer module be used for increasing the timer range?

Question 10.16. What value should be configured to port control register (GPIO_PCTL_R) to use a GPIO pin as timer input/output pin?

Question 10.17. How is the duty cycle of a PWM signal updated in TM4C123 after each timer underflow?

Exercises

Exercise 10.1. What are the main advantages of using an internal oscillator compared to an external oscillator? In addition, list the disadvantages as well.

Exercise 10.2. Write a C program to reconfigure the TM4C123 clock frequency to 64 MHz.

Exercise 10.3. Assume that a timer is running from a clock frequency of 16 MHz. What should the timer reload value be to generate a timer interrupt every 1 ms, when the timer is running in (a) count down mode and (b) count up mode?

Exercise 10.4. An external source generates events aperiodically. We want to generate an interrupt after every fifth event. What timer configuration can be used to solve this problem?

Exercise 10.5. Write a C program for generating 1 kHz square wave on PA5 (i.e., port A pin 5) using the Systick timer. Assume that the Systick timer is operating at a bus clock frequency of 16 MHz.

Exercise 10.6. A 16-bit down counter based timer is driven by a timer clock frequency of 16 MHz. We are interested in generating 100 ms intervals as timer interrupts. What reload value should be used to generate this interval from the timer? You are free to use timer pre-scaler if required.

Exercise 10.7. *Pedestrian Signal Crossing:* A person uses a pedestrian signal to stop the road traffic and cross the road. As soon as the person presses the available user switch on the road side, the traffic signal immediately turns RED from GREEN, while the pedestrian signal turns GREEN from RED. For simplicity, we have omitted the use of intermediate ORANGE signal when switching between RED and GREEN signals. The pedestrian signal remains ON for 15 seconds. Repressing the button during this time has no effect on the signal's duration. After 15 seconds, the traffic signal switches back to GREEN whereas the pedestrian signal switches to RED. The system is to be implemented using ARM Cortex-M processor running at 1 MHz clock. The switch is connected to GPIO port PB3. The traffic GREEN and RED signals are connected on PB4 and PB5, respectively. The pedestrian GREEN and RED signals are connected to GPIO port PB6 and PB7, respectively. The timing diagram for pedestrian switch press behavior is illustrated by Figure 10.17. Assume that port pins PB3 to PB7 are initialized for the desired functionality. Write a C program to implement the above functionality. Clearly specify the required interrupt and timer configurations.

Exercise 10.8. Consider the time period based frequency measurement discussed in Example 10.5. What is the maximum frequency that can be measured reliably using this method, while measurement error does not exceed 1%?

Exercise 10.9. The interrupt entry latency (IEL) is defined as the number of processor clock cycles from the generation of an interrupt signal until the start of execution of first instruction in the corresponding interrupt service routine. It is assumed that no other higher priority interrupt occurs during this time. It is required to measure

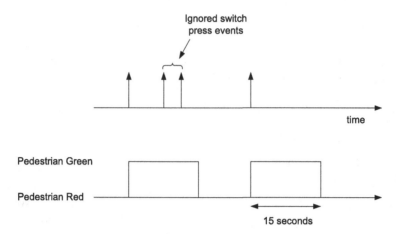

Figure 10.17: Illustration of the pedestrian signal timing diagram.

the IEL of any arbitrary interrupt source. Propose a solution and discuss the steps required for measuring IEL using Systick timer.

Exercise 10.10. How can you configure the priority of Systick timer to 5 on TM4C123? What is the default priority of Systick timer?

Exercise 10.11. In addition to interrupt entry latency, we are also interested in measuring interrupt exit latency. Devise a method, implement it, and then measure the interrupt exit latency.

Exercise 10.12. Consider a scenario where data is transmitted by sending a high frequency signal corresponding to bit 1 and a low frequency signal corresponding to bit 0. This type of communication is termed *frequency shift keying* (FSK). We are interested in implementing this communication using timer module(s) of TivaC microcontroller. Assume that a square wave signal of 100 kHz, 50% duty cycle, is transmitted corresponding to logic 1 (high level) and 50 kHz square wave, 50% duty cycle, is transmitted for logic 0. The data bits are transmitted at a rate of 1 kbps.

- How many timers will be required to implement the above-mentioned data transmission mechanism? Select the timer(s) operating modes.

- List all the configuration steps required for the selected timer(s). In addition, configure the appropriate GPIOs associated with the selected timer for generation and transmission of signals.

- Write a timer interrupt service routine, which implements the above-mentioned data transmission mechanism. Clearly indicate which timer interrupt is configured for this purpose. Assume that data to be transmitted is stored in variable tx_data of type CHAR and is transmitted again, once its transmission is completed.

Exercise 10.13. Consider a 32-bit timer used to count synchronous external events occurring at a rate of 1 MHz. What is the maximum time interval, after which the timer overflow will occur?

Exercise 10.14. Write a C program to generate a variable frequency signal using the timer module. The output signal frequency should vary from 10 kHz to 20 kHz in 1 kHz steps.

Exercise 10.15. Consider a 16-bit timer operating at a clock frequency of 40 MHz. If the desired PWM frequency is 10 kHz, then what is the maximum achievable PWM resolution in this case?

Exercise 10.16. Write a C program to generate a PWM signal with 70% duty cycle with negative half cycle produced first followed by positive half cycle.

Chapter 11

Serial Communication Interfaces

Overview

Different serial communication protocols are widely used for data transfers among digital systems. Serial communication protocols can be either synchronous or asynchronous and both types of these interfaces are integrated in microcontrollers. The popular asynchronous serial interfaces include Universal Asynchronous Receive/Transmit (UART), Universal Serial Bus (USB), Ethernet, and Controller Area Network (CAN), while popular synchronous interfaces include Serial Peripheral Bus (SPI) and Inter-Integrated Circuit (I^{2}C). We will discuss in detail UART, CAN, SPI, and I^{2}C interfaces in this chapter. The coverage of USB and Ethernet interfaces is beyond the scope of this text.

11.1 Fundamentals of Serial Communication

Broadly speaking a microcontroller communicates with other devices using either parallel or serial interface. A *parallel interface* uses multiple parallel lines for data transfer and is primarily used for faster data communications. For instance parallel interfaces are used for connecting memory, graphics display, etc. to the microcontroller. However, parallel interfaces require more wiring effort and are used for scenarios where the distance between two communicating devices is on the order of few meters.

On the other hand *serial interface* uses one line for data transfer simplifying its wiring and can also be used for longer distances, on the order of hundreds of meters. But in case of serial communication, the data transfer rate is relatively slower than parallel interface. In addition, serial communication also requires a parallel to serial converter on the transmitter side and a serial to parallel converter on the receiver side. We have already experimented with simple parallel communication interfacing in Chapter 8, where seven-segment display and LCD were interfaced using 8-bit parallel interface. Some microcontrollers also integrate a memory expansion interface for connecting external memory using parallel communication interface. In this chapter, our focus will be on different serial communication interfaces that are frequently used with microcontrollers in embedded systems.

As mentioned above, the serial communication interfaces transfer data one bit at a time (or bit by bit). *Bit rate* or *baud rate* are the terms used to mention the data transfer rates in serial communication interfaces. There are many different types of serial communication interfaces that are in use today. Below we list a selected set of serial communication interfaces that are available on microcontrollers.

- Universal Asynchronous Receiver/Transmitter (UART)

- Serial Peripheral Interface (SPI) or Synchronous Serial Interface (SSI)

- Inter-Integrated Circuit (IIC or I^{2}C) Bus

- Controller Area Network (CAN) Bus

- Universal Serial Bus (USB)

- Ethernet

The UART interface is one of the simple and widely used serial communication protocols. Specifically UART based standards including RS232, RS422 and RS485 are quite common in industry for long distance communication, which can be on the order of kilometers for some of these standards. The SPI and I^{2}C serial communication is widely used for short distance communications. Both processor to processor as well as processor to device communication can be performed using SPI and I^{2}C protocols, which in general is also true for other serial communication protocols as well. Different devices that can be connected to a microcontroller using SPI and I^{2}C interfaces include, but are not limited to, memories (e.g., EEPROMs, Flash memories, SD cards, etc.), displays (e.g., LCD displays), sensors (accelerometers, acoustic, motion, magnetic, etc.), analog to digital and digital to analog converters, RF communication devices (e.g., IEEE 802.11, IEEE 802.15.4, etc.), and real-time clock chips. The CAN bus serial interface originally developed for automobiles, is also being used widely in the industry. The USB and Ethernet are so prevalent and commonly used that it is even not necessary to mention their uses.

Every serial communication protocol can be classified using different attributes. Two important attributes that are widely used for serial interface classification are synchronization and duplexity. Based on synchronization, the serial communication interfaces can be categorized in the following two broad classes.

1. *Asynchronous Serial Interface*: In case of *asynchronous serial interface*, only data is transmitted from the transmitter to the receiver, at a mutually agreed bit rate. Since the clock is not transmitted by the sender, the receiver generates a local clock signal, of same frequency as that of transmitter, to reliably receive the transmitted data bits. If the transmitter has to switch to a different baud rate, it has to inform this new baud rate to the receiver, before starting data

transmission at new rate. One of the widely used asynchronous serial communication interfaces is Universal Asynchronous Receiver/Transmitter (UART).

2. *Synchronous Serial Interface*: Both data and clock signals are transmitted simultaneously, in case of *synchronous serial interface*. For synchronous serial communication sender can change baud rate without informing the receiver. Two common synchronous serial interfaces available on microcontrollers are Inter-Integrated Circuit (I^{2}C) and Serial Peripheral Interface (SPI).

Interface duplexing is another attribute that is used to classify serial communication interfaces. A *duplex* interface implies that data can be transmitted in either direction between the communicating devices. On the other hand if data can only be transmitted in one direction, then this interface is termed as *simplex* interface. Based on duplex attribute a serial interface can be classified in the following two categories.

1. *Half Duplex*: In case of *half duplex* serial interface, data can only be transmitted in one direction at a time. I^{2}C and Controller Area Network (CAN) bus interfaces are half duplex.

2. *Full Duplex*: When using a *full duplex* serial interface, data can be transmitted in both directions simultaneously. UART and SPI interfaces are full duplex.

The classifications of different serial communication interfaces, based on the above mentioned two attributes, are summarized in Table 11.1.

Table 11.1: Classification of different serial communication interfaces.

Serial interface	Asynchronous	Synchronous	Half duplex	Full duplex
UART	✓			✓
SPI		✓		✓
I^{2}C		✓	✓	
CAN	✓		✓	
USB	✓		✓	
Ethernet	✓			✓

In this chapter we will focus on the UART, I^{2}C, SPI and CAN serial communication interfaces. The other two important interfaces are USB and Ethernet. But due to the breadth of the material and associated complexities, a comprehensive discussion regarding USB and Ethernet interfaces is beyond the scope of this text.

11.2   UART Interface

Universal Asynchronous Receiver/Transmitter (UART) is one of most widely used serial communication interfaces. The UART interface uses asynchronous communication and provides full duplex connectivity. Each data byte is transmitted sequentially bit by bit, as single standalone UART frame. On the receiver side, the received bits are assembled to reconstruct the data byte.

11.2.1   UART Connection

In case of asynchronous data transmission, the sender does not send a clock signal to the receiver. As a result, for full duplex connection, only two data lines (along with third line for common ground) are sufficient for minimum connectivity. This is illustrated in Figure 11.1. In Figure 11.1, the communicating UART devices can be either a microcontroller or any other device equipped with UART interface.

Figure 11.1: Minimum connection requirements for UART interface.

For serial communication, *hand shaking* or *flow control* is essential for reliable data transfers. Consider the scenario, where during data transmission, the receiving device runs out of buffer space and it becomes essential to inform the transmitter to stop further data transmission. This additional control signaling between transmitter and receiver is called flow control. Flow control can be implemented using hardware or software. In *hardware flow control* dedicated hardware connections are used for flow control purpose, while in case of software flow control, no extra hardware is required. Rather a communication protocol implemented as part of the software is used for flow control purpose.

From Figure 11.1, we observe that a minimum of two data lines are required for full duplex data transmission using UART interface. UART has also been used extensively for modem connections. In that context, it is possible to employ hardware flow control. For that purpose, dedicated lines labeled as request to send (RTS) and clear to send (CTS) are used in addition to UART receive (U_RX) and transmit (U_TX) data lines.

11.2.2   UART Communication Protocol Details

A data byte is transmitted by the UART module, by encapsulating it in a UART frame. The UART frame comprise of four distinct bit fields, namely, start bit, data bits to be transmitted, parity bit and stop bit(s). Size of each of these bit fields and their brief description are provided in Table 11.2.

UART Baud Rate

The transmission of UART frame is performed at certain predefined baud rate. Both transmitter and receiver should be configured with same baud rate for reliable data transmission. The UART baud rate is selected by configuring an appropriate value to the baud rate divisor field (baud_divisor) in one of the UART configuration registers as will be explained in the next section. Assuming the frequency of the clock fed

Table 11.2: UART frame bit field descriptions.

Bit field	Size (bits)	Description
Start bit	1	It is a mandatory bit field and marks the start of UART frame.
Data field	5 to 8	This field contains the data to be transmitted. Its size is configurable from 5 to 8 bits. Some UART interfaces support 9 bit data field size as well.
Parity bit	1	It is an optional bit field and is used to transmit the data parity.
Stop bit(s)	1, 1.5 or 2	This is a mandatory bit field and is configurable for different sizes. The minimum size is 1 stop bit, which should be transmitted to mark the end of UART frame.

to the UART module is f_{uart}, then the UART baud rate can be obtained using the following expression.

$$\text{UART Baud} = \frac{f_{\text{uart}}}{k \times \text{baud\_divisor}} \tag{11.1}$$

In (11.1), k is the clock divisor constant and can take different values depending on selected hardware platform. Based on the available configuration parameters, an arbitrary baud rate can be used for data transfers. However, there are some standard baud rates that are used by different applications. Further details regarding baud rate will be provided in the following section.

UART Frame Format

When data of appropriate size (5 to 8 bits) is passed to the UART module for transmission, a Start bit is placed at beginning of the data. The Start bit is the first bit that is transmitted by the transmitter and is marked by the transmission of logic low level for one bit duration. If the UART module is idle before the beginning of transmission, its transmit pin (labeled as U_TX in Figure 11.1) is kept at logic level high. The start of UART frame transmission is marked by a logic high to low transition on U_TX pin generated by the transmitter and is sensed by the receiver. At the completion of Start bit transmission, the first data bit (corresponding to LSB of data) is transmitted followed by bit by bit transmission of the entire data field. After the transmission of last data bit, the optional parity bit, if configured, is transmitted. Finally, one or two stop bits are transmitted to mark the end of UART frame transmission. The transmission of next UART frame can start immediately after the transmission of stop bit(s) corresponding to previous UART frame. A typical UART frame comprising of 1 Start bit, 8 data bits, 1 parity bit and 1 stop bit is shown in Figure 11.2. In Figure 11.2(a) S is the Start bit, D0-D7 are the data bits and D0 is the LSB, P is the parity bit and E is the stop bit.

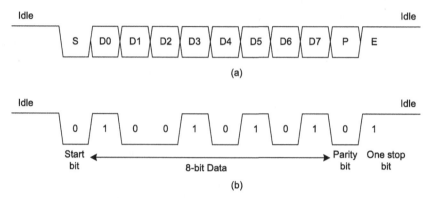

Figure 11.2: (a) Composition of UART frame to transmit 8 bit data, parity and one stop bit. (b) UART frame when transmitting data of value 0xA9, using even parity and one stop bit.

UART Receiver Synchronization

Since no clock signal is transmitted by the UART transmitter, the receiver needs to synchronize itself to the incoming data locally. For that purpose, the UART receiver uses a clock signal that is 16 times the UART baud rate. When the transmission of a new UART frame is initiated by Start bit (i.e., a high to low transition), the UART receiver resets a local counter on the reception of this falling edge. When the counter count equals 8, it corresponds to the mid point of the Start bit. The midpoints of subsequent bits in the UART frame are expected to occur every 16 cycles thereafter. The higher clock signal used by the receiver increases correct bit detection capability by improving the receiver synchronization.

UART Error Types

There are four different types of errors that can occur in case of UART communication. Below we provide brief description regarding each of these errors.

- *Parity Error*: When the received parity bit does not match with the parity calculated from the received data, a *parity error* is generated.

- *Framing Error*: When a Start bit is received but the receiver does not get the corresponding stop bit then a *framing error* is generated.

- *Overrun Error*: When the UART receive buffer is full and new data is received, then the newly received data overwrites the previous data. This situation leads to *overrun error*.

- *Break Error*: When the receiver detects a break condition, it results in *break error*. A break condition is detected when UART RX pin is held low for more than one UART frame transmission time.

11.2.3 RS232 Standard Serial Interface

The RS232 is a standard for asynchronous serial communication following UART protocol. The RS232 uses bipolar negative logic for data transmission, where logic low is mapped to a positive voltage signal while logic high is mapped to negative voltage. When transmitting data using RS232, logic low on UART Tx pin is translated to a positive voltage signal between +3 V to +15 V and is termed as SPACE state on RS232 line. On the other hand logic high is mapped to a negative voltage signal between -3 V to -15 V and is called MARK state. The translation of a UART frame to its corresponding RS232 frame is shown in Figure 11.3.

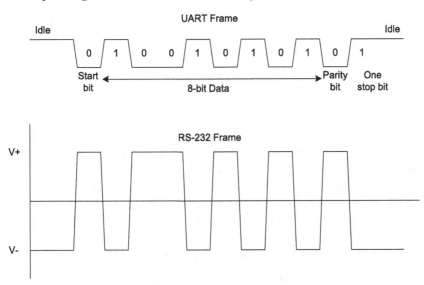

Figure 11.3: Mapping of UART signal to its equivalent RS232 signal. The signal level V+ (V-) can be between +3 (-3) V to +15 (-15) V and corresponds to SPACE (MARK) signal state.

To convert a UART signal to an equivalent RS232 signal, a special RS232 interface chip is used. A simplified circuit diagram to perform this conversion for full duplex connectivity, using RS232 chip, is shown in Figure 11.4. The RS232 standard specifications allow a maximum voltage of 25 V. However, due to the available wide range, many different signal levels including ±5 V, ±10 V, ±12 V, ±15 V, appear in RS232 based serial interfaces. The actual voltage level used by the RS232 interface depends on the line driver that is used for this purpose.

11.3 UART Details on TM4C123 Microcontroller

There are eight UART modules integrated in TM4C123 microcontroller. Of these eight modules only UART 1 module supports hardware flow control. These modules can be used not only for simple UART communications, but also for serial InfraRed (IrDA), ISO smart card as well as RS-485 (9-bit data) interfaces. However, we will confine our discussions to simple UART communications.

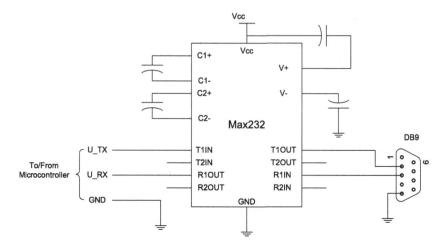

Figure 11.4: The circuit diagram to convert a UART signal to corresponding RS232 signals and vice versa.

The TM4C123 UART interface also provides separate transmit and receive FIFO buffers, allowing to reduce CPU loading due to UART interrupt servicing. The maximum size of both the FIFO buffers is 16. FIFO configuration allows to trigger receive or transmit interrupts corresponding to different trigger levels. For instance 1/4 trigger level for receive FIFO results in Rx interrupt when four bytes have been received. Similarly, a Tx interrupt is generated depending on the emptied level of Tx FIFO. For proper use of UART interface, certain configurations are required. Below we list the basic steps to configure UART module for data exchange.

- Step 1: Clock gating control for UART module

- Step 2: GPIO configuration as UART pin

- Step 3: UART baud rate and frame configuration

- Step 4: UART interrupt configuration

11.3.1 UART Clock Gating Control Configuration

The clock for each of the eight UART modules can be configured separately using RCGC_UART_R register. The register address information along with brief description is provided in Table 11.3. Bit 0 of RCGC_UART_R correspond to UART0 module, while bit 7 correspond to UART7. Writing 1 to an arbitrary bit (from bit 0 to bit 7) enables the clock to corresponding UART module. After enabling the clock to a UART module, there must be a 3 clock cycle delay before accessing the UART module registers.

Table 11.3: UART clock gating control register address and usage.

Register label	Address	Reset value	Brief description
RCGC_UART_R	0x400FE618	0x00000000	UART clock gating control.

11.3.2 GPIO Configuration for UART Alternate Function

Data transfers using UART modules are performed by using GPIO port pins for UART alternate functionality. For this purpose, GPIO registers, corresponding to digital function enabling (GPIO_DEN_R), alternate function selection (GPIO_AFSEL_R) and port control (GPIO_PCTL_R) should be configured with appropriate values. Appropriate bit fields in registers GPIO_DEN_R and GPIO_AFSEL_R should be set to 1. In addition, the corresponding bit field (of size 4-bit) in GPIO_PCTL_R register is set to 1 to configure the port pin as UART Tx or Rx pin. The possible choices, for alternate functionality, for different port pins are summarized in Tables 8.2 and 8.3, while GPIO register descriptions and their addresses are provided in Section 8.4.

11.3.3 UART Control Configuration

Different configurations should be performed before the UART interface can be used for data transfer. The baud rate, frame control and UART control are the key configurations. We will first discuss the UART registers used to perform these configurations, which are listed in Table 11.4. The register addresses provided in Table 11.4 are the address offsets from the UART module base address. The actual register address is determined by adding the offset to the corresponding UART module base address. The base addresses for different UART modules are provided in Table 11.5. Now we explain the key configuration steps to achieve 115200 baud rate, 8 data bits, no parity and one stop bit.

Table 11.4: UART control configuration register labels, addresses, reset value and brief description.

Register label	Offset	Reset value	Brief description
UART_DATA_R	0x000	0x00000000	UART Data Rx/Tx register.
UART_RX_STATUS_R	0x004	0x00000000	Receive status (read only) register.
UART_ERR_CLEAR_R	0x004	0x00000000	Error clear (write only) register.
UART_FLAG_R	0x018	0x00000000	Status flag register.
UART_BAUD_INT_R	0x024	0x00000000	Baud rate divisor, integer part register.
UART_BAUD_FRAC_R	0x028	0x00000000	Baud rate divisor, fraction part register.
UART_LINE_CONTROL_R	0x02C	0x00000000	Line control register.
UART_CONTROL_R	0x030	0x00000300	Status flag register.
UART_CLK_CONFIG_R	0xFC8	0x00000000	Clock source configuration register.

Table 11.5: Base addresses for different UART modules.

Start Address	End Address	Description
0x4000C000	0x4000CFFF	UART 0 module
0x4000D000	0x4000DFFF	UART 1 module
:	:	:
0x40013000	0x40013FFF	UART 7 module

Baud Rate Configuration

For the desired baud rate, we need to first evaluate the baud rate divisor, which can be calculated using the following expression for TM4C123 UART module.

$$\text{UART Baud} = \begin{cases} \frac{f_{\text{uart}}}{16 \times \text{baud\_divisor}} & \text{if HS EN} = 0 \\[2ex] \frac{f_{\text{uart}}}{8 \times \text{baud\_divisor}} & \text{if HS EN} = 1 \end{cases} \tag{11.2}$$

In (11.2), 'HS EN' (high speed enable) is a bit field from UART control register and will be described later in this section. To achieve the desired baud rate, the baud divisor registers (UART_BAUD_INT_R, UART_BAUD_FRAC_R) and clock source configuration register (UART_CLK_CONFIG_R) should be configured with appropriate values. For these registers, the bit field allocations and their descriptions are provided in Table 11.6.

Table 11.6: UART registers for baud rate configuration and their bit field descriptions, (a) UART_BAUD_INT_R register, (b) UART_BAUD_FRAC_R and (c) UART_CLK_CONFIG_R.

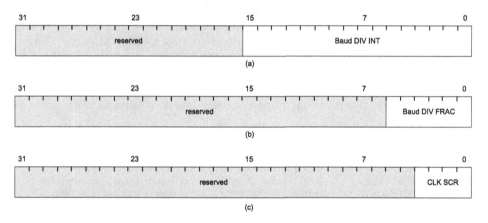

Bit field	Description
Baud DIV INT	UART baud rate divisor configuration register for integer part. Once the baud_divisor is obtained using (11.2), its integer part is configured as 16-bit value to this bit field of UART_BAUD_INT_R register.
Baud DIV FRAC	UART baud rate divisor configuration register for fractional part. This bit field of UART_BAUD_FRAC_R register is configured by multiplying the fractional part of baud_divisor by 64 and adding 0.5 to it.
CLK SCR	Clock source configuration bit field is used to select the UART clock source. Setting this bit field to 0 selects the system clock as UART clock (which is also the default setting), while configuring a value of 5 makes the precision internal oscillator as the UART clock source.

The configuration for the desired UART baud rate of 115200 is illustrated in Example 11.1. We consider the default configurations, for the UART clock source and system clock frequency. By default system clock is the UART clock source and the default system clock frequency is 16 MHz.

Example 11.1 (UART baud rate configuration illustration.).

This C program configures UART_BAUD_INT_R and UART_BAUD_FRAC_R registers for the desired baud rate. For the default system clock frequency of 16 MHz and the desired baud rate of 115200 the baud_divisor is $\frac{16 \times 10^6}{16 \times 115200} =$ 8.68. For this value of baud_divisor, the integer part is 8 that is configured to UART_BAUD_INT_R. The configuration value for fractional part is obtained as $\lfloor (0.68 \times 64 + 0.5) \rfloor = 44$ and is configured to UART_BAUD_FRAC_R register.

```
// UART baud rate configuration
UART_BAUD_INT_R = 0x08;  // configure integer baud divisor as 8
UART_BAUD_FRAC_R = 0x2C; // configure fractional baud divisor as 44
```

Line (Frame) Control Configuration

UART frame format is controlled using UART line control, UART_LINE_CONTROL_R register. The word length, usage of parity and number of stop bits are configured using this register. In addition, enabling of FIFOs and generation of break condition are controlled using this register. The bit field allocations and their descriptions for UART_LINE_CONTROL_R register are provided in Table 11.7.

Table 11.7: UART line control configuration register and its bit field descriptions.

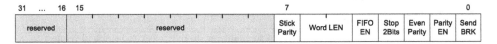

Bit field	Description
Stick Parity	This bit field enables or disables the stick parity. When this bit is set and bits 1 and 2 of this register are also set to 1 then parity bit is transmitted and checked as 0. If bit 1 is set and bit 2 is cleared then parity bit is transmitted and checked as 1.
Word LEN	The word length bit field configures the number of data bits transmitted or received in a UART frame. A value of 0, 1, 2 or 3 configures the data length equal to 5, 6, 7 or 8 bits in the UART frame.
FIFO EN	This bit enables or disables FIFO buffers. When cleared to 0 the FIFOs are disabled.
Stop 2Bits	When this bit is cleared to 0, one stop bit is transmitted at the end of UART frame. Setting it to 1 transmits two stop bits.
Even Parity	Setting to 1 enables even parity, which generates and checks for even number of 1's in the data and parity bits during transmission and reception, respectively.
Parity EN	To enable generation and checking of parity, this bit field should be set to 1.
Send BRK	When set to 1, a logic low signal is generated on the UART Tx line. For proper generation of break condition, the software should set this bit to 1 for at least a duration of two UART frames.

Consider the case, where we are interested in transmitting 8-bit data using UART interface with frame format employing 2 stop bits and no parity. The UART frame/line configuration for this purpose is detailed in Example 11.2.

Example 11.2 (UART frame format configuration using UART line control register.).

For 8-bit data, the Word LEN bit field should be configured with a value of 3 while Stop 2Bits should be set to 1.

```
#define UART_8BIT_DATA        0x0060
#define UART_2STOP_BITS       0x0008

// UART line control configuration
UART_LINE_CONTROL_R = UART_8BIT_DATA + UART_2STOP_BITS;
```

UART Control Configuration

The overall control of the UART module is performed using UART_CONTROL_R register. The bit field allocations for UART_CONTROL_R register are shown in Table 11.8. In addition, bit field descriptions for selected bit fields are also provided in Table 11.8.

Table 11.8: UART control configuration register and selected bit field descriptions.

31 ... 16	15							7							0
reserved	CTS EN	RTS EN	reserved	RTS	reserved	Rx EN	Tx EN	Loop Back EN	reserved	HS EN	End Of TX	Smart CARD	SIR Low PWR	SIR EN	UART EN

Bit field	Description
CTS EN, RTS EN, RTS	These bit fields are used to enable or disable, clear to send and request to send flow control signals.
Rx EN, Tx EN	UART Tx and RX functions can be individually enabled or disabled using these bit fields. To enable UART Tx and/or Rx, these bit fields should be set to 1. In addition, UART EN bit should also be set to 1.
Loop Back EN	Setting this bit to 1 enables the loop back operation by connecting the UART Tx signal to UART Rx signal internally.
HS EN	High speed enable bit is used to select the UART clock scaling factor. Clearing this bit selects a clock scaling factor of 16, while setting it to 1 selects a scaling factor of 8.
End OF Tx	This bit field determines the definition of transmission completion. When set to 1, the transmission of entire data in the UART Tx FIFO indicates end of transmission. On the other hand, clearing it to 0, indicates end of transmission, when the UART FIFO level goes below a threshold value configured by trigger level.
UART EN	Setting this bit to 1 enables the UART interface.

Data Transmission and Reception

The data transmission as well as reception is performed using UART_DATA_R register. A write operation to UART_DATA_R register results in data transmission, while a read operation provides the received data. One can think of UART data register as two different physical registers mapped to the same address, but their access is differentiated based on the read and write operation.

When FIFOs are disabled single character size data is transmitted or received using UART data register. When FIFOs are enabled, their access is also provided through UART data register. For instance, when multiple data is to be transmitted, they can be written to Tx FIFO by performing multiple writes to UART data register and observing the FIFO status on each write to avoid FIFO overflow. The bit field allocations and their descriptions for UART_DATA_R register are provided in Table 11.9. From Table 11.9 we can observe that the UART receive error status is also indicated in this register.

Table 11.9: UART data register bit field allocations and their descriptions.

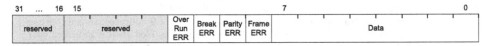

Bit field	Description
Over Run ERR	This is the overrun error status bit. When set to 1, it indicates that new data was received while either FIFO was full or receive buffer was not empty resulting in data loss.
Break ERR	The break error status bit, when set to 1, indicates that a break condition was detected by the receiver.
Parity ERR	The parity error status bit is set to 1 when the received parity does not match with the calculated parity for the received data.
Frame ERR	Framing error bit is set to 1, when a start bit was detected but the corresponding stop bit was not received.
Data	The data to be transmitted is written to this 8-bit data register, while data received by UART is read from this data register.

UART Status and Error Handling

The UART receive error status is available as part of the UART_DATA_R register. The four receive status bits are also accessible using UART_RX_STATUS_R register, which is dedicated for this purpose. Bits 0 to 3 of this register correspond to framing, parity, break and overrun errors, respectively. To clear any of these UART errors, UART_ERR_CLEAR_R register is used. Writing an 8-bit arbitrary value to UART_ERR_CLEAR_R register, clears all the UART receive error flags.

The current operating state of UART module is indicated using UART_FLAG_R register. Bit field assignments and their brief descriptions are provided in Table 11.10. Status bits in UART_FLAG_R register indicate whether transmitter/receiver are empty and ready to accept new data or their respective buffers/FIFOs are full.

11.3.4 UART Interrupt Configuration

Each UART module has an associated set of registers that are used for interrupt related configuration. Specifically, there is one interrupt mask register, two interrupt status registers and one interrupt clear register as summarized in Table 11.11. In addition, for FIFO based transmission and reception, there is an interrupt FIFO level select register. To configure different types of UART interrupts, interrupt mask

Table 11.10: UART flag register bit field allocations and their descriptions.

Bit field	Description
Tx FE	This bit field corresponds to transmit FIFO empty status flag. When this bit is 0 transmitter has data to transmit. If this flag is 1 and FIFO is disabled then it indicates that transmit holding register is empty, otherwise transmit FIFO is empty.
Rx FF	This bit is receive FIFO full status flag. A value of 0 indicates receiver can receive data. If it is set to 1 and FIFO is disabled then it indicates that receive holding register is full otherwise receive FIFO is full.
Tx FF	This bit is transmit FIFO full status flag. A value of 0 indicates transmitter can transmit new data. If it is set to 1 and FIFO is disabled then it indicates that transmitter holding register is full otherwise transmit FIFO is full.
Rx FE	This bit indicates receive FIFO empty status. When this bit is 0, the receiver has data. If this flag is 1 and FIFO is disabled then it indicates that receiver holding register is empty, otherwise receive FIFO is empty.
BUSY	A value of 1 indicates that UART is busy in transmitting data. This bit is set to 1 as soon as UART transmit FIFO becomes nonempty.
CTS	Clear to send flag is used for hardware flow control. When it is set to 1, it indicates that clear to send signal is asserted.

register, UART_INT_MASK_R, is used. The bit field allocations and their descriptions for interrupt mask register, UART_INT_MASK_R, are given in Table 11.12.

Table 11.11: UART interrupt configuration register labels, addresses, reset value and brief description.

Register label	Offset	Reset value	Brief description
UART_INT_MASK_R	0x038	0x00000000	UART interrupt mask register.
UART_RIS_R	0x03C	0x00000000	UART raw interrupt status register.
UART_MIS_R	0x040	0x00000000	UART masked interrupt status register.
UART_INT_CLEAR_R	0x044	0x00000000	UART interrupt clear register.
UART_INT_FIFO_LS_R	0x034	0x00000022	UART interrupt FIFO level select register.

The UART interrupt status registers (UART_RIS_R, UART_MIS_R) and interrupt clear register (UART_INT_CLEAR_R) also have the same bit field allocations as that of UART interrupt mask register. As an illustration, for the receive interrupt (Rx IM), bit field in the mask register, there is a corresponding bit field, receive raw interrupt status (Rx RIS) in the raw interrupt status register. Similarly, masked interrupt status and interrupt clear registers also have the corresponding bit field allocations. The UART interrupt clear register is write 1 to clear, i.e., writing 1 to an arbitrary bit field in UART interrupt clear register, clears the corresponding bits in the UART raw interrupt status and masked interrupt status registers.

When UART transmit and receive FIFOs are enabled, then a transmit or receive interrupt can correspond to different situations depending on the current configuration of UART interrupt FIFO level select register, UART_INT_FIFO_LS_R. The bit field allocations and their descriptions for this register are given in Table 11.13.

Based on the above-mentioned configuration steps, we configure the UART 2 module

Table 11.12: Interrupt mask register, UART_INT_MASK_R, bit field descriptions.

31 ... 16	15										7			0
reserved	reserved	9Bit IM	reserved	OE IM	BE IM	PE IM	FE IM	RT IM	Tx IM	Rx IM	reserved		CTS IM	reserved

Bit field	Description
9Bit IM	This bit enables or disables 9 bit mode interrupt. Setting it to 1 enable the interrupt.
OE IM, BE IM, PE IM, FE IM	These bits are used to enable or disable overrun error, break error, parity error and framing error interrupts. When set to 1 the corresponding error interrupt is enabled.
RT IM	Receive timeout interrupt is used for FIFO based data reception. This interrupt is generated when receive FIFO is not empty and no new data is received for a specific time duration. The value of the time duration depends on the configuration of HSE bit field in UART control register. This interrupt is cleared either by writing 1 to the corresponding bit in UART_INT_CLEAR_R or when the FIFO becomes empty by reading all of the data.
Tx IM, Rx IM	Tx and Rx interrupt mask bits are used to enable/disable UART transmit completion or new data reception interrupts. Configuring it to 1 enables the corresponding UART interrupt.
CTS IM	UART clear to send interrupt mask bit is used to implement interrupt driven flow control. This hardware based hand shaking is only supported for UART 1 module.

for data transmission and reception. To keep the implementation complexity to minimum, FIFOs have not been used. Example 11.3 provides the C code implementation, which can be considered as simple UART device driver.

> **Example 11.3** (UART device driver for data transmission and reception.).
> This example uses UART 2 module for illustration. The configuration parameters used are: 115200 data baud rate, 8 bit data size, no parity and one stop bit.
> The UART 2 module Tx pin is mapped to GPIO pin PD7, which is locked by default. This example also illustrates the procedure to unlock a locked GPIO port pin. The UART Rx synchronization is implemented as interrupt driven, while Tx is busy waiting based.

```
// Clock gating control related register definitions
#define SYSCTL_RCGC_UART_R *((volatile unsigned long *)0x400FE618)
#define SYSCTL_RCGC2_R     *((volatile unsigned long *)0x400FE108)
#define CLOCK_GPIOD        0x00000008   //Port D clock control
#define RCGC_UART2         0x00000004   // Uart2 clock control

// Uart base address and register definitions
#define UART_BASE          0x4000E000

#define UART_DATA_R        *(volatile long *)(UART_BASE + 0x000)
#define UART_RX_STATUS_R   *(volatile long *)(UART_BASE + 0x004)
#define UART_ERR_CLEAR_R   *(volatile long *)(UART_BASE + 0x004)
#define UART_FLAG_R        *(volatile long *)(UART_BASE + 0x018)
#define UART_BAUD_INT_R    *(volatile long *)(UART_BASE + 0x024)
#define UART_BAUD_FRAC_R   *(volatile long *)(UART_BASE + 0x028)
```

Table 11.13: Interrupt FIFO level select register, UART_INT_FIFO_LS_R, bit field descriptions.

Bit field	Description
Rx INT FIFO LS	This bit field corresponds to receive interrupt FIFO level selection. When the UART Rx FIFO is filled to a specific level, a receive interrupt is generated. Below is the list of possible configuration values corresponding to different triggering points for receive interrupt. • A value of 0 triggers interrupt when Rx FIFO is 1/8 full. • A value of 1 triggers interrupt when Rx FIFO is 1/4 full. • A value of 2 triggers interrupt when Rx FIFO is 1/2 full. • A value of 3 triggers interrupt when Rx FIFO is 3/4 full. • A value of 4 triggers interrupt when Rx FIFO is 7/8 full.
Tx INT FIFO LS	This bit field corresponds to transmit interrupt FIFO level selection. When the UART Tx FIFO is emptied to a specific level, a transmit interrupt is generated. Below is the list of possible configuration values corresponding to different triggering points for transmit interrupt. • A value of 0 triggers interrupt when Tx FIFO is 7/8 empty. • A value of 1 triggers interrupt when Tx FIFO is 3/4 empty. • A value of 2 triggers interrupt when Tx FIFO is 1/2 empty. • A value of 3 triggers interrupt when Tx FIFO is 1/4 empty. • A value of 4 triggers interrupt when Tx FIFO is 1/8 empty.

```
#define UART_LINE_CONTROL_R   *(volatile long *)(UART_BASE + 0x02C)
#define UART_CONTROL_R        *(volatile long *)(UART_BASE + 0x030)
#define UART_CLK_CONFIG_R     *(volatile long *)(UART_BASE + 0xFC8)

// Uart interrupt configuration register definitions
#define UART_INT_MASK_R       *(volatile long *)(UART_BASE + 0x038)
#define UART_RIS_R            *(volatile long *)(UART_BASE + 0x03C)
#define UART_MIS_R            *(volatile long *)(UART_BASE + 0x040)
#define UART_INT_CLEAR_R      *(volatile long *)(UART_BASE + 0x044)
#define UART_INT_FIFO_LS_R    *(volatile long *)(UART_BASE + 0x034)

// GPIO Port D alternate function configuration
#define GPIO_PORTD_AFSEL_R *((volatile unsigned long *)0x40007420)
#define GPIO_PORTD_PCTL_R  *((volatile unsigned long *)0x4000752C)
#define GPIO_PORTD_LOCK_R  *((volatile unsigned long *)0x40007520)
#define GPIO_PORTD_CR_R    *((volatile unsigned long *)0x40007524)
#define GPIO_PORTD_DEN_R   *((volatile unsigned long *)0x4000751C)

// IRQ 32 to 63 Set Enable Register
#define  NVIC_EN1_R          *((volatile unsigned long *)0xE000E104)
#define  NVIC_EN1_INT33      0x00000002   // Interrupt 33 enable

// Uart bit field definitions
#define UART_ENABLE          0x001
```

```
#define UART_TX_ENABLE      0x100
#define UART_RX_ENABLE      0x200

// Function prototypes
void Delay(unsigned long counter);
void UART_Init(void);
void UART_Transmit(unsigned char tx_data);

// UART intialization and GPIO alternate function configuration
void UART_Init(void)
{
  // Enable the clock for port D
  SYSCTL_RCGC2_R |= CLOCK_GPIOD;

  // Unlock Port D pin 7 since default setting is NMI
  GPIO_PORTD_LOCK_R = 0x4C4F434B;
  GPIO_PORTD_CR_R |= 0x00000080;

  GPIO_PORTD_DEN_R |= 0x0C0;        // Assert DEN for port D

  // Configure PortD pins 6 and 7 as UART 2
  GPIO_PORTD_AFSEL_R |= 0x000000C0;
  GPIO_PORTD_PCTL_R |= 0x11000000;

  // Enable the clock for Uart 2
  SYSCTL_RCGC_UART_R |= RCGC_UART2;
  Delay(10);

  // Configure UART2 baudrate, data size, parity, stop bits
  /* BRD = 16,000,000 / (16 * 115,200) = 8.68
  integer_part = 8
  fraction_part = floor(0.68 * 64 + 0.5) = 44 */
  UART_BAUD_INT_R = 0x08;
  UART_BAUD_FRAC_R = 44;

  // word size is 8-bit, no parity and one stop bit
  UART_LINE_CONTROL_R = 0x0060;

  // clear the UART Rx interrupt flag
  UART_INT_CLEAR_R &= ~(0x0780);

  // Enable UART 2
  UART_CONTROL_R |= UART_RX_ENABLE + UART_TX_ENABLE + UART_ENABLE;

  // Unmask receive interrupt
  UART_INT_MASK_R |= 0x10;

  // Enable UART 2 interrupt in NVIC
  NVIC_EN1_R = NVIC_EN1_INT33;
}

// Uart interrupt handler
void UART2_Handler(void){
unsigned char rx_data = 0;

  UART_INT_CLEAR_R &= ~(0x010); // Clear receive interrupt

  rx_data = UART_DATA_R;            // get the received data byte
```

```
    UART_Transmit(rx_data+1);       // send data that is received
}

// This function implements the UART transmission
void UART_Transmit(unsigned char tx_data)
{
   // Busy waiting being done to check if TX is empty
   while((UART_FLAG_R & 0x80) == 0)
   {}

   // Send back the received data by writing it to TX buffer
   UART_DATA_R = tx_data;
}

// This function implements the delay
void Delay(unsigned long counter)
{
   unsigned long i = 0;
   for(i=0; i< counter; i++);
}

// Application main function
int main(void){

   UART_Init();                     // Initialize the timer

   while(1){
      // One of the tasks is responsible to transmit/receive data
   }
}
```

The UART interface discussed in the previous section is a point to point serial communication protocol. When a device has to communicate with multiple other devices using UART interface, it requires as many UART interfaces as the number of devices. A microcontroller having a limited number of UART interfaces, can only connect to a few devices using UART. The solution to this problem is to migrate from a point to point serial interface to a serial bus interface. In the following we will discuss SPI, I²C and CAN serial communication interfaces. All of these three communication interfaces are serial bus interfaces.

11.4   I²C Interface

The *Inter-Integrated Circuit*, IIC or I²C is a synchronous half-duplex serial communication protocol [Leens(2009)] developed by Philips. Multiple devices can be connected to I²C bus using only two wires. The I²C protocol specifications define multiple bus speeds, including standard mode at 100 kbps, fast mode at 400 kbps, fast-plus mode at 1 Mbps and high-speed mode at 3.33 Mbps. Each communicating device can be in one of the two possible modes, the master mode or slave mode. In addition, a device can switch between master and slave modes, making the I²C bus truly a multi-master serial bus interface. A master initiates communication with a slave device. The data can be either requested from or sent to the slave device.

11.4.1   I²C Bus Protocol

I²C is a multi-master serial bus protocol requiring two signal lines. The two I²C signal lines are labeled as *serial clock* (SCL) and *serial data* (SDA). Each device connected to an I²C bus is expected to have a unique address assigned to it. This address is used by the master device to communicate with a specific slave device. The key attributes associated with I²C bus protocol are listed below.

- Each slave device has a 7-bit unique address assigned to it.

- The two signal lines, SCL and SDA, are bidirectional.

- Data is transmitted as a sequence of 8-bit bytes.

- In addition to data transmission, control signaling including communication start/stop, data send/receive control as well as its acknowledgment and slave device address communication is also performed using the same pair of lines.

I²C Signaling Sequence for Data Transfer

A master device initiates data transfer on I²C bus by generating a START condition. The START condition can be considered as an alert to all the devices connected to the bus. After the START condition, the master device transmits the ADDRESS of intended slave device followed by the Receive/Send (R/$\overline{S}$) control signal. The Send/Receive control signal informs the slave device whether it should expect incoming data or it should respond with data. After receiving the slave address, all the devices will compare the slave address with their own address. The slave device for which the address matches, will respond with an acknowledge (ACK) signal. On the other hand, those devices for which the slave address does not match, they continue waiting till the bus become idle after issuance of STOP condition.

After receiving an ACK signal, the master device starts either data transmission or reception depending on the control signal issued after the slave address. In case of transmission, the master will transmit data one byte at a time and slave device will respond with an ACK signal after receiving each data byte as shown in Figure 11.5. When the master is finished with all data transmission, it will generate a STOP condition. In case of data reception by the master device, the slave device sends first data byte after sending the ACK signal. The master device responds with an ACK signal after receiving each data byte as can be seen from Figure 11.5. Once the master has received the data, it sends NACK signal followed by STOP condition releasing the bus for other nodes.

The sequence of operations to perform data transfer between master and slave devices can be observed from Figure 11.6. It can be observed that each data transfer requires a sequence of nine bits involving 8 bits of data and 1 ACK bit. After the START condition, the first byte is always transmitted by the master device. The most significant 7 bits of this byte are the slave address, while the LSB of this byte is the R/$\overline{S}$ signaling bit.

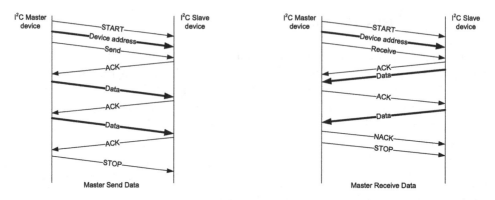

Figure 11.5: Illustration of data transmission and reception, for two bytes of data, by a master device.

Figure 11.6: Sequence of operations to perform n bytes of data transfer.

START STOP Conditions and Data Validity

The START and STOP conditions are generated uniquely for proper differentiation from other possible transitions on the I^2C bus. When the I^2C bus is idle, both SCL and SDA lines are in logic high state. According to I^2C bus specifications, data changes on SDA line can only happen during the low half cycle of SCL signal. Any data changes on the SDA line during the positive half cycle of SCL signal are invalid, which is shown pictorially in Figure 11.7. For proper data transfer it is required that data remains stable on SDA line, while the SCL line is in high state.

To generate a START condition, a high to low transition is generated on SDA line, during the positive half cycle of SCL line. Similarly, a STOP condition is generated by low to high transition on SDA line, during the positive half cycle of SCL line. This is depicted in Figure 11.8.

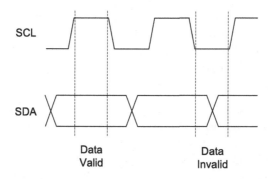

Figure 11.7: Validity of data on I^2C bus.

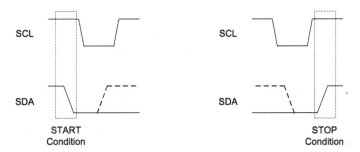

Figure 11.8: Generation of START and STOP conditions for data transfer.

11.4.2 I^2C Bus Connectivity

The device connectivity to I^2C bus is shown in Figure 11.9. The SCL and SDA lines are configured as open drain with pull-up resistors installed as shown in Figure 11.10.

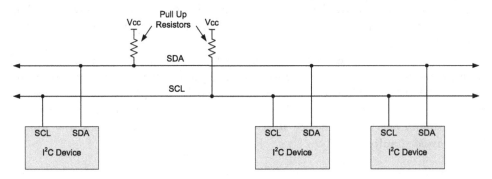

Figure 11.9: Illustration of multiple devices connected to I^2C bus.

The logic low and high levels are generated physically by pulling the line to ground or releasing the line. Effectively, a device connected to an I^2C bus only drives logic low levels. All the devices that are connected to I^2C bus, they listen to it continuously. A master device aiming to initiate transmission on I^2C bus, when detects a START condition, it will wait till a STOP condition is detected, before attempting bus access again. Similarly the slave devices on I^2C bus will match the device address received after the START condition to find out whether they are the intended devices for data transfer. A slave device that is not addressed, waits for the STOP condition before it listens to the bus again.

A nice feature of the I^2C bus is its built in bus arbitration capability [Leens(2009)]. This is implemented by having the master device to listen to the bus, while it is initiating communication on the bus. Now consider the scenario, when two devices attempt to start communication simultaneously to their intended slave devices. A start condition is generated by pulling the SDA line low (while SCL is kept at logic level high) simultaneously, followed by the transmission of slave address. Assume that the two salve addresses used by the master devices are 1110010 and 1111011.

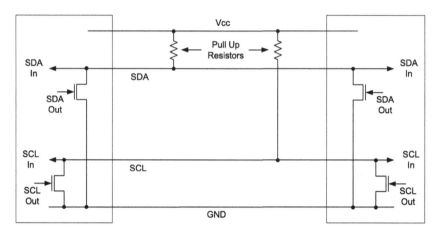

Figure 11.10: Illustration of internal signal driving of I²C bus lines [Leens(2009)].

The transmission of slave addresses (MSB is transmitted first) from the two masters continues bit by bit till the point, where the slave addresses become different. At that point one master node drives logic high, while the other drives logic low (corresponding to fourth bit location of the slave address, when counting from the MSB side). While the two master nodes drive the bus for different logic levels, they receive only logic level low due to the open-drain with pull-up based hardware configuration. As a result, that master device who transmitted logic low will win the bus arbitration process and the other master device will back off by postponing its transmission. Since an I²C device listens to the bus continuously and can recognize it being idle, any possibility of a master device starting transmission, while the bus was already busy is avoided automatically.

11.5 I²C Details on TM4C123 Microcontroller

There are four I²C modules integrated in TM4C123 microcontroller. Each of these modules can be configured as I²C master, slave or even both master and slave simultaneously. All of the I²C modules support four data transmission rates of 100 kbps, 400 kbps, 1 Mbps and 3.33 Mbps. When the module is configured as a master module it supports, multi-master capability, performs bus arbitration and uses 7-bit addressing. For proper use of I²C interface, certain configurations are required. Below we list the basic steps to configure I²C module for data transfer.

- Step 1: Clock gating control for I²C module
- Step 2: GPIO configuration for I²C alternate function
- Step 3: I²C master device configuration
- Step 4: I²C interrupt configuration

The I²C module in TM4C123 can be configured both as a master as well as a slave device. Here we will only discuss the configuration of I²C module as a master device.

In most of the applications, the microcontroller I^{2}C interface is used to communicate with different types of slave devices, e.g., memories, RTC (real time clock) chip, digital sensors, etc. The configuration of I^{2}C module in TM4C123 as a slave device is quite straight forward and the interested reader is suggested to consult the device specific data sheet.

11.5.1 I^{2}C Clock Gating Control Configuration

The clock for each of the four I^{2}C peripheral modules can be configured separately using RCGC_I2C_R register. The register address information along with brief description is provided in Table 11.14. Bit 0 of RCGC_I2C_R register correspond to I2C module 0, while bit 3 correspond to I2C module 3. Writing 1 to a bit (from bit 0 to bit 3) enables the clock for the corresponding I^{2}C module. After enabling the clock to an I^{2}C module, there must be a 3 clock cycle delay before accessing the I^{2}C module registers.

Table 11.14: I^{2}C clock gating control register address and usage.

Register label	Address	Reset value	Brief description
RCGC_I2C_R	0x400FE620	0x00000000	I^{2}C clock gating control.

11.5.2 GPIO Configuration for I^{2}C Alternate Function

Data transfer using I^{2}C bus interface is performed by configuring GPIO port pins for I^{2}C alternate functionality. For this purpose, GPIO registers, corresponding to digital enabling (GPIO_DEN_R), alternate function selection (GPIO_AFSEL_R) and port control (GPIO_PCTL_R) should be configured with appropriate values. The bit fields for registers GPIO_DEN_R and GPIO_AFSEL_R should be set to 1 to configure GPIO pins for I^{2}C functionality. In addition, the corresponding bit field in GPIO_PCTL_R register is set to 3 to configure the port pin as SCL or SDA line.

Conventionally, both SCL and SDA lines are configured as open drain with external pull-up resistors. However, in TM4C123 microcontroller, only the GPIO pin for SDA line should be configured as open drain. This can be accomplished using GPIO open drain configuration register, GPIO_OD_R. The SCL line on the other hand, has an active pull-up installed internally and should *not* be configured as open drain. The possible choices, for alternate functionality, for different port pins are summarized in Tables 8.2 and 8.3, while GPIO register descriptions and their address offsets are provided in Section 8.4. Example 11.4 illustrates the steps to configure GPIO pins for I^{2}C alternate functionality.

Example 11.4 (GPIO configurations for I^{2}C alternate functionality.).

I^{2}C module 3 uses GPIO pins PD0 (for SCL) and PD1 (for SDA). This program illustrates the GPIO pin configurations for I^{2}C functionality.

```
#define RCGC_GPIO_R        *((volatile unsigned long *)0x400FE608)
```

```
#define CLOCK_GPIOD            0x00000008    //Port D clock control

// GPIO Port D alternate function configuration
#define GPIO_PORTD_AFSEL_R *((volatile unsigned long *)0x40007420)
#define GPIO_PORTD_PCTL_R  *((volatile unsigned long *)0x4000752C)
#define GPIO_PORTD_DEN_R   *((volatile unsigned long *)0x4000751C)
#define GPIO_PORTD_OD_R    *((volatile unsigned long *)0x4000750C)

  // Enable the clock for port D
  RCGC_GPIO_R |= CLOCK_GPIOD;

  // Assert DEN for port D
  GPIO_PORTD_DEN_R |= 0x03;

  // Configure Port D pins 0 and 1 as I2C 3
  GPIO_PORTD_AFSEL_R |= 0x00000003;
  GPIO_PORTD_PCTL_R  |= 0x00000033;

  // Configure SDA pin (PD1) as open darin
  GPIO_PORTD_OD_R |= 0x00000002;
```

11.5.3   I^{2}C Master Device Configuration

There are two basic configurations, which need to be performed for data transfers using I^{2}C module as master device and are listed below.

- The basic step is to enable the module as master device, along with the desired clock frequency configuration.

- When master module is used to send data to the slave device, it is required to configure (a) slave address, (b) send command and (c) data to be transmitted. On the other hand when it is used to receive data from the slave device, it is required to configure (a) slave address and (b) receive command.

To perform the above mentioned configuration steps a set of I^{2}C master module registers are used, which are listed in Table 11.15. The register addresses provided in Table 11.15 are the address offsets from the I^{2}C module base address. The actual register address is determined by adding the offset to the corresponding I^{2}C module base address. The base addresses for different I^{2}C modules are provided in Table 11.16. Next we discuss the above-mentioned configuration steps.

Table 11.15: I^{2}C configuration register labels, addresses, reset value and brief description.

Register label	Offset	Reset value	Brief description
I2CM_SLAVE_ADDR_R	0x000	0x00000000	I^{2}C slave address register.
I2CM_CONTROL_R	0x004	0x00000000	Master control (write only) register.
I2CM_STATUS_R	0x004	0x00000020	Master status (read only) register.
I2CM_DATA_R	0x008	0x00000000	Master data register.
I2CM_TIME_PERIOD_R	0x00C	0x00000001	Master SCL clock time period register.
I2CM_CONFIG_R	0x020	0x00000000	Master configuration register.

Table 11.16: Base addresses for different I^2C modules.

Start address	End address	Description
0x40020000	0x40020FFF	I^2C 0 module address range
0x40021000	0x40021FFF	I^2C 1 module address range
0x40022000	0x40022FFF	I^2C 2 module address range
0x40023000	0x40023FFF	I^2C 3 module address range

Master Module and Clock Configuration

We use register I2CM_CONFIG_R to configure I^2C module as master, while the clock frequency is configured using I2CM_TIME_PERIOD_R register. The bit field allocation and its description for configuration register is provided in Table 11.17 and for that of time period register is given in Table 11.18.

Table 11.17: I^2C configuration register and its bit field descriptions.

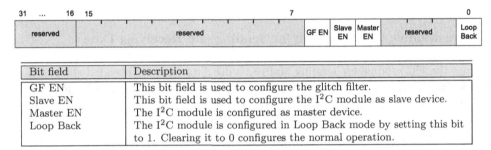

Bit field	Description
GF EN	This bit field is used to configure the glitch filter.
Slave EN	This bit field is used to configure the I^2C module as slave device.
Master EN	The I^2C module is configured as master device.
Loop Back	The I^2C module is configured in Loop Back mode by setting this bit to 1. Clearing it to 0 configures the normal operation.

Table 11.18: I^2C clock time period configuration register I2CM_TIME_PERIOD_R, and its bit field descriptions.

Bit field	Description
HS	This is high speed (HS) enable bit. When cleared to 0, the bit field SCL Period (in this register) is used to define the SCL clock time period for 100 kbps, 400 kbps and 1 Mbps. When set to 1, the SCL Period bit field is used to configure SCL clock time period for 3.33 Mbps.
SCL Period	This bit field is used to configure the I^2C clock time period.

The SCL Period bit field in register I2CM_TIME_PERIOD_R is used to configure for the desired clock frequency using the following expression.

$$1 + \text{SCL Period} = \frac{f_{I^2C}}{2(\text{SCL\_LP} + \text{SCL\_HP})\text{I2C\_CLK\_Freq}} \qquad (11.3)$$

In (11.3), SCL_LP and SCL_HP are, respectively the low and high (half cycle) periods and are fixed at 6 and 4. The I2C_CLK_Freq in (11.3) is the desired clock frequency, while f_{I2C} is by default the system clock frequency. For a system clock frequency of 16 MHz and desired I2C_CLK_Freq of 100 kbps, SCL Period is evaluated as 7.

Example 11.5 (I^2C module configuration for master device and clock configuration for 100 kbps.).

This C program configures I2CM_CONFIG_R and I2CM_TIME_PERIOD_R registers for master device and clock frequency of 100 kbps.

```
// Enable I2C 3 master function
I2CM_CONFIG_R = 0x0010;

/* Configure I2C 3 clock frequency
(1 + TIME_PERIOD) = SYS_CLK/(2*(SCL_LP + SCL_HP) * I2C_CLK_Freq)
TIME_PERIOD = 16,000,000/(2(6+4)*100000) - 1 = 7
*/
I2CM_TIME_PERIOD_R = 0x07;    // High speed is disabled
```

Master Module Configuration for Sending and Receiving Data

To use the I^2C module for sending and receiving data with an intended slave device, the address of the slave device is configured to the I2CM_SLAVE_ADDR_R register. The bit field descriptions for the I2CM_SLAVE_ADDR_R are provided in Table 11.19. The bit 1 to bit 7 of this register are used to configure the slave address, while LSB (bit 0) of this register is the $R/\overline{S}$, which specifies whether the operation is send or receive.

Table 11.19: I^2C slave address register bit field allocations and their descriptions.

Bit field	Description
Slave Address	This bit filed specifies the 7-bit salve address.
$R/\overline{S}$	This bit specifies whether the master operation is send (logic low) or receive (logic high).

The data register I2CM_DATA_R, is used to send as well as receive data from the slave device. A data byte written to this register is used to send data to the slave, while data received from the slave device is obtained by reading this register. The bit field description for the I2CM_DATA_R register is provided in Table 11.20.

The I^2C master module controls the bus operations by using I2CM_CONTROL_R register. The bit filed allocations and their descriptions for I2CM_CONTROL_R register are provided in Table 11.21. When START bit is set to 1 the I^2C master generates a START or repeated START condition. The STOP bit determines, whether the data transfer cycle is terminated (in case of STOP bit set to 1) or continues with more data transfers (when STOP bit is cleared). When the Run bit is set to 1, it enables the I^2C master module to send and receive data.

Table 11.20: I^{2}C data register bit field descriptions.

Bit field	Description
Data	The data sent or received is written to or read from this bit field.

Table 11.21: I^{2}C control register bit field allocations and their descriptions.

31 ... 16	15	7				0
reserved	reserved	HS	ACK	STOP	START	Run

Bit field	Description
HS	If cleard to 0, the master operates in standard (100 kbps), fast mode (400 ksps) or fast-plus mode (1 Mbps). When set to 1, high speed mode (3.33 Mbps) is enabled.
ACK	When set to 1, the received data byte is acknowledged automatically by the master device, otherwise it is not acknowledged automatically.
STOP	When set to 1 the I^{2}C master module generates a STOP condition.
START	When set to 1 the I^{2}C master module generates a START condition.
Run	A value of 1 indicates that the master can transmit or receive data.

Now consider the case, when master needs to send one byte data. In this case START, Run and STOP bits should be set to 1, while R/$\overline{\text{S}}$ bit in slave address register should be cleared. Since slave sends ACK when master is sending data, the ACK bit from master is don't care. Similarly to send multiple data bytes START and Run bits should be set to 1. To receive single data byte from the slave, R/$\overline{\text{S}}$ bit should be set to 1 along with START, Run and STOP bits. Since only one byte data is being requested from the slave, the master device clears its ACK bit to send NACK signal. However, in case of receiving multiple data bytes, the ACK bit should also be set by the master module. A set of important commands that can be configured to I2CM_CONTROL_R register of the master module are listed in Table 11.22, where the label 'X' denotes don't care condition.

Table 11.22: Selected send and receive I^{2}C master module commands.

R/$\overline{\text{S}}$	ACK	STOP	START	Run	Command description
Send data commands					
0	X	1	1	1	Single-send command
0	X	0	1	1	Burst-send start command
X	X	0	0	1	Burst-send continue command
X	X	1	0	1	Burst-send stop command
X	X	1	0	0	Stop command
Receive data commands					
1	0	1	1	1	Single receive command
1	1	0	1	1	Burst-receive start command
X	1	0	0	1	Burst-receive continue command
X	0	1	0	1	Burst-receive stop command

The status of I^{2}C master module can be obtained by reading the I2CM_STATUS_R register. It is important to realize that I^{2}C master module control and status registers are, respectively, write only and read only registers, which are accessible using the same address. The bit filed allocations and their descriptions for I2CM_STATUS_R register are provided in Table 11.23. Before starting a new operation, the availability of the master module is determined by checking the state of Busy bit in the I2CM_STATUS_R register.

Table 11.23: I^{2}C status register bit field allocations and their descriptions.

31 ... 16	15		7							0
reserved	reserved		CLK TimOut ERR	Bus Busy	Idle	ARB Lost	Data ACK	ADDR ACK	Error	Busy

Bit field	Description
CLK TimOut ERR	When set to 1 it indicates that a clock time out error has occurred. This error is cleared if the master send a STOP condition or it is reset.
Bus Busy	When set to 1 the bus is busy, otherwise it is idle.
Idle	A value of 1 indicates that I^{2}C module is idle.
ARB lost	A value of 1 indicates that I^{2}C master lost the bus arbitration.
Data ACK	If set to 1 then the data was not acknowledged, otherwise it was acknowledged.
ADDR ACK	When cleared to 0, it indicates that transmitted address was acknowledged. If set to 1 then the address was not acknowledged.
Error	A value of 1 specifies that an error occurred during the last operation.
Busy	When set to 1, this bit indicates that I^{2}C master module is busy. A 0 specifies idle condition.

Using Example 11.6, we illustrate the use of I^{2}C master module to receive one byte data from intended slave.

Example 11.6 (I^{2}C master module data reception from slave device.).

This C function illustrates the use of I^{2}C master module to receive one data byte from the receiver using data, slave address, status and control registers. To receive one data byte, the master module first informs the slave device about the address of data (to be read) by performing a write operation.

```
#define CMD_SINGLE_RECEIVE          0x00000007
#define BUSY_STATUS_FLAG            0x01

// Receive one byte of data from I2C slave device
unsigned char I2C_Receive_Data(unsigned char Slave_addr, unsigned
    char Data_addr)
{
  unsigned char value = 0;

  // Wait for the master to become idle
  while(I2CM_STATUS_R & BUSY_STATUS_FLAG)  { };

  // Configure slave address for write operation
  I2CM_SLAVE_ADDR_R = (Slave_addr << 1);
  I2CM_DATA_R = Data_addr;              // Data to be written
  I2CM_CONTROL_R = CMD_SINGLE_SEND;     // Command to be performed

  // Wait until master is done
```

```
    while(I2CM_STATUS_R & 0x01)  { };

    // Configure the slave address for read operation
    I2CM_SLAVE_ADDR_R = ((Slave_addr << 1) | 0x01);
    I2CM_CONTROL_R = CMD_SINGLE_RECEIVE;  // Command to be performed

    // Wait until master is done
    while(I2CM_STATUS_R & BUSY_STATUS_FLAG)  { };
    value = (unsigned char) I2CM_DATA_R;  // Read the data received

    return value;
}
```

11.5.4 I²C interrupt configuration

Each I²C module has an associated set of interrupt configuration registers. Specifically, there is an interrupt mask register, two interrupt status registers and one interrupt clear register as summarized in Table 11.24. There are two types of I²C master device interrupts, which can be configured using interrupt mask register, I2C_INT_MASK_R. The bit field descriptions for interrupt mask register are given in Table 11.25.

Table 11.24: I²C interrupt configuration register labels, addresses, reset value and brief description.

Register label	Offset	Reset value	Brief description
I2CM_INT_MASK_R	0x010	0x00000000	I2C interrupt mask register.
I2CM_RIS_R	0x014	0x00000000	I2C raw interrupt status register.
I2CM_MIS_R	0x018	0x00000000	I2C masked interrupt status register.
I2CM_INT_CLEAR_R	0x01C	0x00000000	I2C interrupt clear register.

Table 11.25: I²C interrupt mask register bit field descriptions.

Bit field	Description
CLK TimOut IM	This is bit 1 of register I2C_INT_MASK_R. When set to 1, a clock timeout error interrupt is generated from the I²C master module.
Master IM	This is bit 0 of register I2C_INT_MASK_R. When set to 1, this bit allows interrupts from the I²C master device to NVIC.

The I²C interrupt status registers (I2C_RIS_R, I2C_MIS_R) and interrupt clear register (I2C_INT_CLEAR_R) also have the same bit field allocations as that of I²C interrupt mask register. As an illustration, for the master interrupt mask (Master IM) bit field in the mask register, there is a corresponding bit field, master raw interrupt status (Master RIS) in the raw interrupt status register. Similarly, masked interrupt status and interrupt clear registers also have the corresponding bit field allocations. The I²C interrupt clear register is write 1 to clear, i.e., writing 1 to the specified bit field in I²C interrupt clear register, clears the corresponding bits in the I²C raw interrupt status and masked interrupt status registers.

Finally we illustrate the use of I²C master module to interface with a real time clock (RTC) chip, DS1307 from Maxim Integrated. The microcontroller can read timing data from RTC chip using I²C bus interface. The RTC data is accessed using the RTC chip internal registers. The DS1307 internal register at address 0x00 contains seconds, while minutes and hours information is available from registers at addresses 0x01 and 0x02 respectively. The timing data in these registers is in BCD format. Example 11.7 provides the device driver for I²C master module, to read data from RTC chip.

Example 11.7 (I²C device driver for RTC (DS1307) chip.). This example illustrates the use of I²C master module to read timing data from RTC chip. When powered, the RTC chip clock is disabled internally. To enable the clock, bit 7 of register 0 of RTC chip should be cleared. This is achieved by writing 0 to this register. The write operation to a register of RTC chip requires sending of two data bytes. The RTC chip treats the first data byte sent as the register address, while the next byte is treated as the data value to be written to that register. The timing data is read using single byte read function. However, it is possible to read the timing data using burst read mode as well.

```
// Clock gating control related register definitions
#define RCGC_I2C_R            *((volatile unsigned long *)0x400FE620)
#define RCGC_GPIO_R           *((volatile unsigned long *)0x400FE608)
#define CLOCK_GPIOD           0x00000008   //Port D clock control
#define CLOCK_I2C3            0x00000008   // I2C3 clock control

// I2C base address and register definitions
#define I2C_BASE              0x40023000

#define I2CM_SLAVE_ADDR_R     *(volatile long *)(I2C_BASE + 0x000)
#define I2CM_CONTROL_R        *(volatile long *)(I2C_BASE + 0x004)
#define I2CM_STATUS_R         *(volatile long *)(I2C_BASE + 0x004)
#define I2CM_DATA_R           *(volatile long *)(I2C_BASE + 0x008)
#define I2CM_TIME_PERIOD_R    *(volatile long *)(I2C_BASE + 0x00C)
#define I2CM_CONFIG_R         *(volatile long *)(I2C_BASE + 0x020)
#define I2CM_BUS_MONITOR_R    *(volatile long *)(I2C_BASE + 0x02C)
#define I2CM_CONFIG2_R        *(volatile long *)(I2C_BASE + 0x038)

// I2C interrupt configuration register definitions
#define I2C_INT_MASK_R        *(volatile long *)(I2C_BASE + 0x038)
#define I2C_RIS_R             *(volatile long *)(I2C_BASE + 0x03C)
#define I2C_MIS_R             *(volatile long *)(I2C_BASE + 0x040)
#define I2C_INT_CLEAR_R       *(volatile long *)(I2C_BASE + 0x044)

// GPIO Port D alternate function configuration
#define GPIO_PORTD_AFSEL_R *((volatile unsigned long *)0x40007420)
#define GPIO_PORTD_PCTL_R  *((volatile unsigned long *)0x4000752C)
#define GPIO_PORTD_DEN_R   *((volatile unsigned long *)0x4000751C)
#define GPIO_PORTD_OD_R    *((volatile unsigned long *)0x4000750C)

// I2C bit field definitions
#define I2C_ENABLE                 0x01
#define BUSY_STATUS_FLAG           0x01

// I2C command definitions for control register
#define CMD_SINGLE_SEND            0x00000007
#define CMD_SINGLE_RECEIVE         0x00000007
#define CMD_BURST_SEND_START       0x00000003
```

```
#define CMD_BURST_SEND_CONT              0x00000001
#define CMD_BURST_SEND_FINISH            0x00000005

// Function prototypes
void I2C_Init(void);
unsigned char I2C_Receive_Data(unsigned char Slave_addr,
                               unsigned char Data_addr);
void I2C_Send_Data(unsigned char Slave_addr,
                   unsigned char Data_addr, unsigned char Data);

// I2C intialization and GPIO alternate function configuration
void I2C_Init(void)
{
  RCGC_GPIO_R |= CLOCK_GPIOD;   // Enable the clock for port D
  RCGC_I2C_R |= CLOCK_I2C3;     // Enable the clock for I2C 3
  GPIO_PORTD_DEN_R |= 0x03;     // Assert DEN for port D

  // Configure Port D pins 0 and 1 as I2C 3
  GPIO_PORTD_AFSEL_R |= 0x00000003;
  GPIO_PORTD_PCTL_R |= 0x00000033;

  GPIO_PORTD_OD_R |= 0x00000002; // SDA (PD1) pin as open darin

  I2CM_CONFIG_R = 0x0010;        // Enable I2C 3 master function

  /* Configure I2C 3 clock frequency
  (1 + TIME_PERIOD) = SYS_CLK/(2*(SCL_LP + SCL_HP) * I2C_CLK_Freq)
  TIME_PERIOD = 16,000,000/(2(6+4)*100000) - 1 = 7  */
  I2CM_TIME_PERIOD_R = 0x07;
}


// Receive one byte of data from I2C slave device
unsigned char I2C_Receive_Data(unsigned char Slave_addr, unsigned
   char Data_addr)
{
  unsigned char value = 0;

   // Wait for the master to become idle
  while(I2CM_STATUS_R & BUSY_STATUS_FLAG)  { };

  // Configure slave address for write operation, data and control
  I2CM_SLAVE_ADDR_R = (Slave_addr << 1);
  I2CM_DATA_R = Data_addr;           // Data to be written
  I2CM_CONTROL_R = CMD_SINGLE_SEND;   // Command to be performed

  // Wait until master is done
  while(I2CM_STATUS_R & 0x01)  { };

  // Configure the slave address for read operation
  I2CM_SLAVE_ADDR_R = ((Slave_addr << 1) | 0x01);
  I2CM_CONTROL_R = CMD_SINGLE_RECEIVE;  // Command to be performed

  // Wait until master is done
  while(I2CM_STATUS_R & BUSY_STATUS_FLAG)  { };
  value = (unsigned char) I2CM_DATA_R;  // Read the data received

  return value;
}
```

```
// This function writes one data byte to Data_addr of I2C slave
void I2C_Send_Data(unsigned char Slave_addr,
                   unsigned char Data_addr, unsigned char Data)
{
    // Wait for the master to become idle
    while(I2CM_STATUS_R & BUSY_STATUS_FLAG) { };

    // Configure slave address for write operation, data and control
    I2CM_SLAVE_ADDR_R = (Slave_addr << 1);
    I2CM_DATA_R = Data_addr;                      // Data to be written
    I2CM_CONTROL_R = CMD_BURST_SEND_START; // Command to be performed

    // Wait until master is done
    while(I2CM_STATUS_R & BUSY_STATUS_FLAG) { };

    // Configure data to be written and send command
    I2CM_DATA_R = Data;
    I2CM_CONTROL_R = CMD_BURST_SEND_CONT;

    // Wait until master is done
    while(I2CM_STATUS_R & BUSY_STATUS_FLAG)  { };

    // Send the finsh command
    I2CM_CONTROL_R = CMD_BURST_SEND_FINISH;

    // Wait until master is done
    while(I2CM_STATUS_R & BUSY_STATUS_FLAG)
    { };
}
```

Example 11.8 illustrates the use of I^2C device driver to read seconds, minutes, and hours data.

Example 11.8 (Reading timing data from RTC chip.).

```
#define DEV_ADDRESS                  0x68

// Global variable definitions
unsigned char reg_addr = 0x00;
unsigned char seconds, minutes, hours;

int main(void)
{

  // Initialize I2C moude 3
  I2C_Init();

  // Configure the RTC chip to turn on the clock
  I2C_Send_Data(DEV_ADDRESS,  reg_addr, 0);

  // infinite loop
  while(1) {

    seconds = I2C_Receive_Data(DEV_ADDRESS, reg_addr++);
    minutes = I2C_Receive_Data(DEV_ADDRESS, reg_addr++);
```

```
    hours = I2C_Receive_Data(DEV_ADDRESS, reg_addr);
    reg_addr = 0;
  }
}
```

11.6   Serial Peripheral Interface (SPI)

Serial Peripheral Interface (SPI), also termed synchronous serial interface, is a serial bus protocol originally developed by Motorola. The SPI serial bus implements full duplex communication using master-slave paradigm. The SPI bus, unlike I²C bus, is a single master bus protocol and requires four wires to communicate between a pair of devices. There are no standard specifications for SPI protocol and as a result different variants of this protocol exist [Leens(2009)]. For instance, there is Freescale SPI, TI SPI and another closely related communication protocol, named as Microwire.

Simple SPI protocol based communication between a master and a slave device uses four signal lines, which are briefly described below.

- SCLK: This is a serial clock signal that is transmitted by the bus master to the slave device(s).

- SS: This signal is used to select or enable a slave device for SPI communication. This signal is active low i.e., a logic low on this line enables the device for SPI communication.

- MOSI: The data from a master to a slave device, is transmitted using Master-Out-Slave-In (MOSI) data line.

- MISO: The data to a master device from a slave, is transmitted using Master-In-Slave-Out (MISO) data line.

The SPI bus connections, for point to point communication between master and single slave device are shown in Figure 11.11. In case of multiple devices, a separate slave select (SS) signal is used by the bus master, to enable the desired slave device. If n number of slave devices are connected to the bus master, then the number of master device lines required for SPI bus connectivity are $n + 3$. This scenario is shown in Figure 11.12

Figure 11.11: SPI master connected to single SPI slave. This is called point-to-point topology.

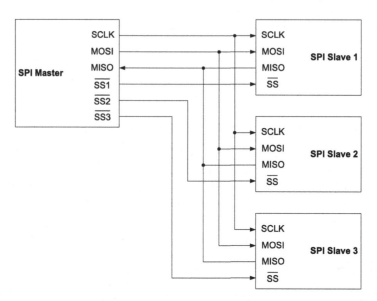

Figure 11.12: SPI master connected to multiple slave devices. This is called bus topology.

11.6.1 SPI Modes of Operation

SPI protocol supports four different modes of communication, namely MODE 0, 1, 2, 3. Each mode is defined using two parameters related to SCLK signal. The two parameters are *serial clock-phase* (SPHA) and *serial clock-polarity* (SPOL). The SPOL parameter is used to define the polarity or state of SCLK signal when data is not being transferred. The SCLK signal can be either at logic low or at logic high level, depending on the SPOL parameter configuration. The SPHA parameter defines the SCLK signal edge that is used to toggle data on MOSI line as well as data sampling of the MISO line by the master. Four possible combinations of these two parameters provide four different operating modes, which are shown in Figure 11.13.

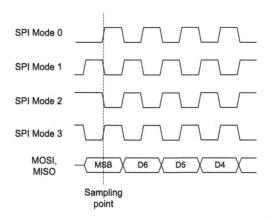

Figure 11.13: Modes of operation for SPI communication protocol.

11.6.2 SPI Signal Timing

Due to single bus master protocol followed by the SPI interface, all the bus communications with the slave devices are initiated by the master device. When the SPI master intends to send/receive data to/from a slave device, it pulls the corresponding SS line low. This is followed by the activation of the clock signal at the desired frequency. The master transmits data using MOSI line, while the incoming data from the selected slave is received by sampling MISO line. The timing diagram for SPI communication, following Freescale mode 0 format, is shown in Figure 11.14.

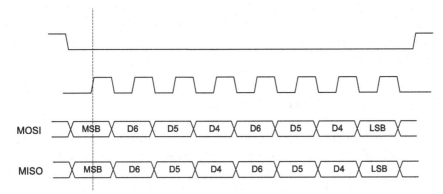

Figure 11.14: SPI signal timing diagram when mode 0 is used for communication.

For proper communication between a master and a slave device, same parameter values (for SCLK frequency, SPHA and SPOL) should be configured by both the devices. In case of SPI bus, where multiple slave devices are connected to the master, it is quite possible that different SPI parameters supported by the slave devices, are not same. In that case, the master device may have to reconfigure its parameters before starting communication with a different slave device.

Since SPI protocol does not have standard specifications, which leads to certain limitations. For instance, there is neither an acknowledgment mechanism for data reception confirmation, nor any flow control. To integrate these features, a higher layer protocol is required. In addition, the maximum data rate for SPI interface is not defined and it mostly depends on the I/O capabilities of the microcontroller used for SPI communication.

11.7 SPI Details on TM4C123 Microcontroller

There are four SPI modules integrated in TM4C123 microcontroller. Each of these modules can be configured as SPI master or slave and supports three different types of SPI frame formats based on Freescale SPI, MICROWIRE or Texas Instruments synchronous serial interface. The discussion in this section will be confined to Freescale frame format. The interested reader is referred to the device data sheet [TI(2014)], for further details related to other frame formats. For proper use of SPI interface, certain configurations need to be performed. Below we list the basic steps to configure an SPI module for data transfer.

- Step 1: SPI clock gating control

- Step 2: GPIO configuration for SPI alternate function

- Step 3: SPI master device configuration

- Step 4: SPI interrupt configuration

The SPI module in TM4C123 can be configured both as a master as well as a slave device. Here we will only discuss the configuration of SPI module as a master device. The configuration of SPI module in TM4C123 as a slave device is quite straight forward and the interested reader is suggested to consult the device specific data sheet for that purpose.

11.7.1 SPI Clock Gating Control

The clock for each of the four SPI modules can be configured separately using register RCGC_SPI_R. The register address information along with brief description is provided in Table 11.26. Bit 0 of RCGC_SPI_R register correspond to SPI module 0, while bit 3 correspond to SPI module 3. Writing 1 to a bit (from bit 0 to bit 3) enables the clock for the corresponding SPI module. After enabling the clock to an SPI module, there must be a 3 clock cycle delay before accessing the SPI module registers.

Table 11.26: SPI clock gating control register address and brief description.

Register label	Address	Reset value	Brief description
RCGC_SPI_R	0x400FE61C	0x00000000	SPI clock gating control register.

11.7.2 GPIO Configuration for SPI Alternate Function

Any data transfers using SPI bus are performed by configuring GPIO port pins for SPI alternate functionality. For that purpose, GPIO registers, corresponding to digital enable (GPIO_DEN_R), alternate function selection (GPIO_AFSEL_R) and port control (GPIO_PCTL_R) should be configured with appropriate values. The bit fields for registers GPIO_DEN_R and GPIO_AFSEL_R should be set to 1 to configure GPIO pins for SPI functionality. In addition, the corresponding bit field in GPIO_PCTL_R register is set to 2 to configure the port pin as SPI SCLK, SS, MOSI and MISO line. There is one exception to GPIO_PCTL_R register configuration for SPI functionality. To configure SPI module 3, the GPIO_PCTL_R register should be configured to a value of 1 rather 2 for SPI function.

The possible choices, for alternate functionality, for different port pins are summarized in Tables 8.2 and 8.3, while GPIO register descriptions and their address offsets are provided in Section 8.4.

11.7.3 SPI Master Device Configuration

The basic configurations that need to be performed for data transfers, using SPI module as master device, are listed below.

- One key configuration involves the enabling of the module as master device, along with the selection among the different frame formats supported by the SPI module.

- The next step is to configure the desired clock frequency (or baud rate), clock phase and its polarity.

- Finally the size of data that is transferred in each frame is configured.

To perform the above-mentioned configuration steps, certain SPI module registers are used, which are listed in Table 11.27. The register addresses provided in Table 11.27 are the address offsets from the SPI module base address. The actual register address is determined by adding the offset to the corresponding SPI module base address. The base addresses for different SPI modules are provided in Table 11.28. Next we discuss the above mentioned configuration steps.

Table 11.27: SPI configuration register labels, addresses, reset value and brief description.

Register label	Offset	Reset value	Brief description
SPI_CONTROL0_R	0x000	0x00000000	SPI protocol control register.
SPI_CONTROL1_R	0x004	0x00000000	SPI function control register.
SPI_DATA_R	0x008	0x00000000	Master data register.
SPI_STATUS_R	0x00C	0x00000003	SPI status (read only) register.
SPI_CLK_PERSCALE_R	0x010	0x00000000	SPI clock prescale register.
SPI_CLK_SCR_R	0xFC8	0x00000000	SPI clock source configuration register.

Table 11.28: Base addresses for different SPI modules.

Start address	End address	Description
0x40008000	0x40008FFF	SPI 0 module address range
0x40009000	0x40009FFF	SPI 1 module address range
0x4000A000	0x4000AFFF	SPI 2 module address range
0x4000B000	0x4000BFFF	SPI 3 module address range

Master Mode and Clock Configuration

We use SPI_CONTROL0_R and SPI_CONTROL1_R registers to perform SPI function and protocol related configurations. The bit field allocations and their descriptions for SPI_CONTROL0_R register are provided in Table 11.29 and for that of SPI_CONTROL1_R register are given in Table 11.30. The SPI_CONTROL0_R register is specifically used to configure the SPI protocol mode, frame format and data size. In addition, SPI_CONTROL0_R register will also be used for clock frequency configuration as explained later in this section. The SPI_CONTROL1_R register is used to configure different SPI modes, including master, slave or loop back modes.

Table 11.29: SPI control register for protocol configurations and its bit field descriptions.

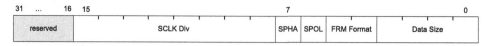

Bit field	Description
Data Size	This bit field configures the size of data used by SPI frames. The data size can be configured from 4-bit to 16-bit. This bit field is configured with a value of 0x3 for 4-bit data and 0xF for 16 bit data size.
FRM Format	This bit field is used to select one of the three supported frame formats. A value of 0 selects Freescale frame format, while values of 1 and 2, respectively, select Texas Instruments and MICROWIRE frame formats.
SPOL	This bit controls the clock polarity for Freescale SPI frame format.
SPHA	This bit controls the clock phase for Freescale SPI frame format. Using this bit along with SPOL bit, four different modes for SPI communication, when using Freescale frame format, are defined.
SCLK Div	This bit filed along with clock prescale register is used to configure the SPI clock frequency. It can be assigned values from 0 to 255. See expression in (11.4) for further details.

The SPI clock frequency is configured using the registers SPI_CLK_PERSCALE_R, SPI_CONTROL0_R and SPI_CLK_SCR_R. Bit 0 to bit 7 of SPI_CLK_PERSCALE_R register are used as clock prescaler. This register can only accept even values from 2 to 254. The SPI_CLK_SCR_R configuration register is simply used to select between system clock and precision internal oscillator as reference clock source for SPI clock frequency. System clock is selected as the clock source for SPI module by configuring SPI_CLK_SCR_R register to a value of 0.

Now SCLK Div bit field of SPI_CONTROL0_R along with SPI_CLK_PERSCALE_R register are used to configure the desired SPI clock frequency using the following expression.

$$\text{SPI Clock} = \frac{f_{\text{sys\_clock}}}{(\text{SPI\_CLK\_PERSCALE\_R})(1 + \text{SCLK Div})} \qquad (11.4)$$

In (11.4), SPI Clock represent the desired SPI clock frequency, $f_{\text{sys\_clock}}$ is the system clock frequency, while the prescale factor denoted by the SPI_CLK_PERSCALE_R register can take even values from 2 to 254. For a system clock frequency of 16 MHz and desired SPI clock of 1 MHz, one possibility is to configure SPI_CLK_PERSCALE_R with a value of 16, while assigning 0 to SCLK Div bit field. Example 11.9 illustrates the SPI module configuration.

Example 11.9 (SPI module configuration for master device and clock frequency.).

This SPI module initialization routine configures SPI module as master operating at 1 MHz clock frequency. In addition, 8-bit data size, mode 0 of Freescale frame format are also configured.

```
// SPI bit field definitions
```

Table 11.30: SPI control register for function configuration and its bit field descriptions.

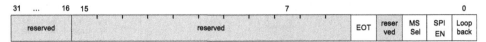

Bit field	Description
EOT	This bit field indicates end-of-transmission (EOT) in two different ways using Tx Raw Interrupt Status flag bit (TxRIS). When this bit is configured to 0, the TxRIS flag indicates that Tx FIFO is at least half empty. When set to 1, the TxRIS flag indicates end to transmission, which is equivalent to Tx FIFO being completely empty. This bit field is valid for master mode only.
MS Sel	This bit is used to select between master and slave modes of operation. When set to 1, slave mode is selected.
SPI EN	This bit field is used to enable or disable the SPI module. It is set to 1 to enable the SPI module.
Loop back	Setting this bit field to 1, enables the loop back mode of operation. For normal operation this bit field should be cleared.

```
#define SPI_ENABLE              0x02

// Disable SPI and perform control configuration
SPI_CONTROL1_R &= ~(SPI_ENABLE);

// Freescale frame format, mode 0, 8 bit data and 0 SCLK divisor
SPI_CONTROL0_R |= 0x0007;

SPI_CLK_PRESCALE_R = 0x0010;    // Prescale system clock by 16
SPI_CLK_CONFIG_R = 0x00;        // Clock source is system clock

// Enable SPI master module
SPI_CONTROL1_R |= SPI_ENABLE;
```

Data Transfer and Status

The data transmission and reception is performed through separate FIFOs. Each FIFO is of 16-bit size and can store 8 data elements. The SPI_DATA_R register is used to access both transmit (Tx) and receive (Rx) FIFOs. Reading SPI_DATA_R register reads the Rx FIFO, while a write operation to the SPI_DATA_R register, writes the data to Tx FIFO. When transmitting data of size less than 16 bits, data should be right justified before writing it to the SPI_DATA_R register.

The current state of Tx and Rx FIFOs can be observed using SPI_STATUS_R register. The bit field allocations and their descriptions for SPI_STATUS_R register are provided in Table 11.31. Separate FIFO status flags are available for both Tx and Rx FIFOs to indicate their empty or full state.

Table 11.31: SPI status register to observe the current state of the SPI module and its Tx and Rx FIFOs.

Bit field	Description
Busy	When reading this bit returns 1, it indicates that SPI module is busy in transmission, reception or both. A value of 0 indicates that SPI module is currently idle.
Rx FF	This bit indicates that Rx FIFO is full. This bit is cleared automatically by reading data.
Rx FNE	When read as 1, this bit indicates that Rx FIFO is not empty.
Tx FNF	When read as 1 it indicates that Tx FIFO is not full and is full otherwise.
Tx FE	This bit when read as 1, it indicates that Tx FIFO is empty.

11.7.4 SPI Interrupt Configuration

Each SPI module has an associated set of interrupt configuration registers. Specifically, there is an interrupt mask register, two interrupt status registers and one interrupt clear register as summarized in Table 11.32. There are four types of SPI device interrupts, which can be configured using interrupt mask register, SPI_INT_MASK_R. The bit field descriptions for raw interrupt status register are given in Table 11.33.

Table 11.32: SPI interrupt configuration register labels, addresses, reset value, and brief description.

Register label	Offset	Reset value	Brief description
SPI_INT_MASK_R	0x014	0x00000000	SPI interrupt mask register.
SPI_RIS_R	0x018	0x00000000	SPI raw interrupt status register.
SPI_MIS_R	0x01C	0x00000000	SPI masked interrupt status register.
SPI_INT_CLEAR_R	0x020	0x00000000	SPI interrupt clear register.

The SPI interrupt mask register, SPI_INT_MASK_R and masked interrupt status register, SPI_MIS_R have the same bit field allocations as that of raw interrupt status register. The SPI_INT_MASK_R register is used to enable or disable the interrupts, while SPI_MIS_R register shows only the status of unmasked interrupts. The interrupt clear register (SPI_INT_CLEAR_R) has only two bit fields that are used to clear the Rx time out and Rx overflow interrupts.

For the Tx FIFO raw interrupt status (TxRIS) bit field in the SPI_RIS_R register, there is a corresponding bit field, Tx FIFO interrupt mask (TxIM), in the interrupt mask register. The SPI interrupt clear register is write 1 to clear, i.e., writing 1 to the specified bit field in SPI interrupt clear register, clears the corresponding bits in the SPI raw interrupt status and masked interrupt status registers. Example 11.10 illustrate the configuration for SPI master module for data transfer.

Example 11.10 (SPI device driver for data transmission and reception.). This example illustrates the use of SPI master to transmit and receive data. The SPI

Table 11.33: SPI raw interrupt status register bit field allocations and their descriptions.

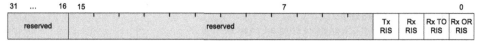

Bit field	Description
Tx RIS	This bit indicates the Tx FIFO raw interrupt status. When Tx RIS is read as 1 and EOT bit in register SPI_CONTROL1_R is zero then it indicates that Tx FIFO is half empty or less. When EOT bit is 1 and Tx RIS is also 1 then it indicates that Tx FIFO is empty. This interrupt is cleared automatically when Tx FIFO is more than half full (for EOT bit cleared to 0) or Tx FIFO has some data (when EOT bit is set to 1).
Rx RIS	This bit is set to 1 when Rx FIFO is at least half full. This interrupt is cleared when Rx FIFO is less than half full after reading data.
Rx TO RIS	This bit is set to 1 when a time duration equal to 32 cycles of SPI clock have lapsed after Rx FIFO became non-empty. This interrupt is cleared by writing 1 to the corresponding bit field of SPI_INT_CLEAR_R register.
Rx OR RIS	This bit is set to 1 when Rx FIFO over flows. This interrupt is cleared by writing 1 to the corresponding bit field of SPI_INT_CLEAR_R register.

module is configured for Freescale frame format, mode 0, 8-bit data size and 1 MHz SPI clock frequency.

```
// Clock gating control related register definitions
#define RCGC_SPI_R          *((volatile unsigned long *)0x400FE61C)
#define RCGC_GPIO_R         *((volatile unsigned long *)0x400FE608)
#define CLOCK_GPIOD         0x00000008   //Port D clock control
#define CLOCK_SPI1          0x00000002   // SPI 1 clock control

// SPI base address and register definitions
#define SPI_BASE            0x40009000

#define SPI_CONTROL0_R       *(volatile long *)(SPI_BASE + 0x000)
#define SPI_CONTROL1_R       *(volatile long *)(SPI_BASE + 0x004)
#define SPI_DATA_R           *(volatile long *)(SPI_BASE + 0x008)
#define SPI_STATUS_R         *(volatile long *)(SPI_BASE + 0x00C)
#define SPI_CLK_PRESCALE_R   *(volatile long *)(SPI_BASE + 0x010)
#define SPI_CLK_CONFIG_R     *(volatile long *)(SPI_BASE + 0xFC8)

// SPI interrupt configuration register definitions
#define SPI_INT_MASK_R       *(volatile long *)(SPI_BASE + 0x014)
#define SPI_RIS_R            *(volatile long *)(SPI_BASE + 0x018)
#define SPI_MIS_R            *(volatile long *)(SPI_BASE + 0x01C)
#define SPI_INT_CLEAR_R      *(volatile long *)(SPI_BASE + 0x020)

// GPIO Port D alternate function configuration
#define GPIO_PORTD_AFSEL_R  *((volatile unsigned long *)0x40007420)
#define GPIO_PORTD_PCTL_R   *((volatile unsigned long *)0x4000752C)
#define GPIO_PORTD_DEN_R    *((volatile unsigned long *)0x4000751C)

// SPI bit field definitions
#define SPI_ENABLE                   0x02
#define SPI_BUSY_FLAG                0x10
#define SPI_TX_FIFO_EMPTY_FLAG       0x01
#define SPI_TX_FIFO_NOT_FULL_FLAG    0x02
#define SPI_RX_FIFO_NOT_EMPTY_FLAG   0x04
```

```c
#define SPI_DONE                        0x08

// Function prototypes
void SPI_Init(void);
unsigned char SPI_Receive_Data(void);
void SPI_Send_Data(unsigned char Data);

// SPI intialization and GPIO alternate function configuration
void SPI_Init(void)
{
  RCGC_GPIO_R |=  CLOCK_GPIOD;   // Enable the clock for port D
  RCGC_SPI_R |= CLOCK_SPI1;      // Enable the clock for SPI 1
  GPIO_PORTD_DEN_R |= 0x0F;      // DEN for port D pins 0-3

  // Configure Port D pins 0-3 for SPI 1
  GPIO_PORTD_AFSEL_R |= 0x0000000F;
  GPIO_PORTD_PCTL_R |= 0x0000222;

  // Disable SPI and perform control configuration
  SPI_CONTROL1_R &= ~(SPI_ENABLE);

  // Freescale frame format, mode 0, 8 bit data & 1 MHz SPI clock
  SPI_CONTROL0_R |= 0x0007;
  SPI_CLK_PRESCALE_R = 0x0010;   // Prescale system clock by 16
  SPI_CLK_CONFIG_R = 0x00;       // Clock source is system clock

  SPI_CONTROL1_R |= SPI_ENABLE;  // Enable SPI master module
}

// This function returns one data element if Rx FIFO is non-empty
unsigned char SPI_Receive_Data(void)
{
  unsigned char value = 0;

  // Check if data is received
  if(SPI_STATUS_R & SPI_RX_FIFO_NOT_EMPTY_FLAG)
    value = (SPI_DATA_R & 0x00FF);
  return value;
}

// This function transmits one data element if Tx FIFO has space
void SPI_Send_Data(unsigned char Data)
{
  // Check if Tx FIFO has space
  if(SPI_STATUS_R & SPI_TX_FIFO_NOT_FULL_FLAG)
    SPI_DATA_R = (Data & 0x00FF);
}

// Application main function
int main(void){

  SPI_Init();                    // Initialize the timer

  while(1){
    // One of the tasks is responsible to exchange data
  }
}
```

11.8 Controller Area Network (CAN) Interface

The controller area network (CAN) is a multi-cast serial bus interface, originally developed by Bosch GmbH in the mid eighties and later standardized as ISO 11898 in early nineties. CAN bus was originally developed for automotive applications. However, it has found applications in plant automation, robotics, agriculture, construction, medical, military, etc. due to its highly reliable communication capability. Unlike other serial bus interfaces, CAN bus is a message based rather than a node based communication interface. In other words, CAN messages are not transmitted from one node to other node using node ID. Rather an arbitrary node on the bus can broadcast a message using a specific message ID. On receiving a CAN message, one or more nodes can process the received data and respond accordingly. The message ID based communication also allows to add new as well as remove existing nodes in CAN network, without requiring to update the existing nodes. Next we summarize the key features of the CAN bus interface.

- The maximum data rate supported by the CAN bus is 1 Mbps. However, the actual data rate used by the CAN bus is dependent on the physical length of the bus. For instance, the data rate of 1 Mbps requires the bus length shorter than 40 m.

- CAN uses carrier sense multiple access (CSMA) protocol with collision resolution (CR) capability due to bitwise bus arbitration, as will be explained later in this section.

- CAN bus is multi-master, supports message prioritization and provides latency guarantees.

- It is possible to configure CAN network either as peer-to-peer or master-slave.

- CAN uses non-return to zero (NRZ) signaling with bit stuffing.

- The CAN bus supports many different types of physical layers. Among them the most widely used is the one described by the CAN ISO 11898-2 standard, which uses two wire bus with balanced signaling and is termed as high-speed CAN.

- Another physical layer is based on CAN ISO 11898-3 standard, which supports lower bus speeds and is fault tolerant to sustain communication even if one of the bus wires is damaged physically. This physical layer also uses balanced signaling and is sometimes termed as low-speed CAN.

11.8.1 CAN Network

The nodes in a CAN network are connected to the bus using CAN *transceiver*, which is responsible for translating the logic levels to balanced signaling on the bus and vice versa. In addition, a CAN transceiver is also responsible for performing bitwise arbitration. An example CAN network, with multiple nodes connected to the CAN bus, using CAN transceivers is illustrated in Figure 11.15.

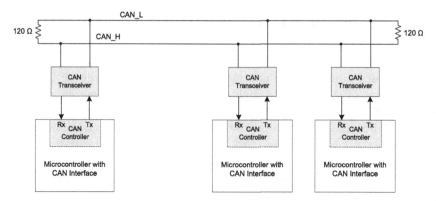

Figure 11.15: Example CAN network illustrating node connectivity and use of CAN transceivers.

The logic level signals from the CAN controller are converted to balanced differential signals by the CAN transceiver before transmitting to the CAN bus. The CAN bus uses two wires for data transmission, which are labeled as CAN high (CAN_H) and CAN low (CAN_L). A logic high (i.e., bit '1') from the CAN controller is translated to a *recessive state* on the CAN bus, while a logic low is converted to the *dominant state* as shown in Figure 11.16. The CAN bus is expected to have a nominal impedance of 120Ω and should be terminated by the same value at both ends.

The CAN receiver, during frame transmission, requires enough transitions to maintain synchronization. For this purpose, when five consecutive bits of same polarity appear in a CAN frame, an opposite polarity bit is inserted after the five bits. This scheme is termed as *bit stuffing*. Bit stuffing is required, since CAN uses non-return to zero (NRZ) line coding.

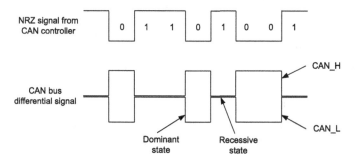

Figure 11.16: Illustration of CAN bus recessive and dominant states using balanced differential signaling.

11.8.2   CAN Frames

The data transfers in CAN network are performed by using short messages, which are called *frames*. The maximum data payload, which can be transmitted using a single CAN frame is eight bytes. There are four types of CAN frames listed below.

- CAN data frame

- CAN remote frame

- CAN error frame

- CAN overload frame

The data and remote frames can be either standard CAN frames or extended CAN frames. A standard CAN frame uses an 11-bit message identifier, while an extended frame uses a 29-bit message identifier apart from other differences as will be discussed later. A CAN controller complying CAN 2.0B active standard is capable of transmitting and receiving both extended as well as standard frames. A CAN 2.0B passive controller can transmit and receive standard frames. However, it cannot transmit extended frames and can only receive extended frames and then discards them after acknowledging them. Finally, a CAN 2.0A controller can only transmit and receive standard CAN frames. In the following, we introduce the above-mentioned different types of CAN frames and provide details related to their frame formats.

CAN Data Frame

The CAN *data frames* are the most common type of frames found in a CAN network. Each data frame can be of either standard or extended type as shown in Figure 11.17. Different fields of a CAN data frame are defined in Table 11.34.

The transmission of each data frame is initiated by start of frame (SOF) field, which consists of one dominant bit. The SOF is followed by arbitration field, which comprises of the message identifier (ID) and some other special bits. In case of standard CAN message, the 11-bit message ID is followed by RTR (remote transmission request) bit, while in case of an extended frame an SRR (substitute remote request) bit is transmitted after 11-bit base ID. Both RTR bit, in case of standard data frame, and SRR bit in case of extended frame, are transmitted as recessive bits. This follows with the transmission of IDE (identifier extension) bit in case of both standard and extended frames. The IDE bit is transmitted as dominant in case of standard frame, while it is transmitted as recessive in case of extended frame. As a result, if there is a collision of standard and extended frames, the standard frame takes priority over the extended frame due to its transmission as dominant IDE bit. This happens due to the fact that wired AND logic is implemented by the CAN bus. In wired AND logic if a single node drives the bus to dominant state (logical 0), then the entire bus is in dominant state irrespective of the number of CAN nodes driving the bus to recessive state (logical 1).

The IDE bit in standard data frame belongs to the control field, while it is part of the arbitration field in case of extended frame, as can be observed from Figure 11.17. The remaining 18 ID bits of an extended frame are transmitted after the IDE bit followed by the transmission of RTR bit. This completes the transmission of arbitration field for extended data frames.

The control field is transmitted after the arbitration field, which mainly consists of data length code (DLC). Four bits are allocated for DLC, which indicates the number of data bytes in the following data field. The data field contains any number of data bytes from 0 to 8. The CRC field follows the data field, which comprises of 15-bit CRC and a CRC delimiter. After the CRC field an acknowledgment (ACK) field follows that is made up of an ACK bit and an ACK delimiter. The transmitter sends a recessive ACK bit, which is overdriven by the receiver with a dominant ACK bit, if the receiver has received the message correctly with a matching CRC. The ACK field is followed by end of frame (EOF), which consists of seven consecutive recessive bits.

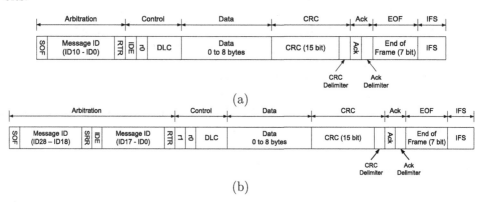

Figure 11.17: Two different CAN data frame formats, (a) standard CAN data frame and (b) extended data frame.

Table 11.34: Field definitions for CAN data frame (both standard and extended).

Field	Description
Start of Frame (SOF)	This one bit field marks the start of frame by generating a high to low transition on the Tx pin of the CAN controller. Equivalently, a recessive to dominant state transition is generated on the CAN bus.
Arbitration field	The arbitration field in standard CAN frames consists of 11 message ID bits followed by remote transmission request (RTR) bit. In extended frames the arbitration field consists of 29-bit message ID, SRR-bit, IDE-bit and RTR-bit.
Control field	The control field is a six bit field. In standard CAN frame, it includes IDE bit, a reserved bit and four bit DLC. In case of extended frame, control field has two reserved bits and four bit DLC.
Data field	The data field contains between 0 to 8 data bytes in both standard as well as extended CAN frames.
CRC field	For both standard and extended CAN frames, there is a 15-bit CRC followed by one bit CRC delimiter.
Ack field	This two bit field comprises of an Ack bit followed by Ack delimiter bit.
End of Frame field	The end of frame is a collection of seven consecutive recessive bits, which marks the completion of data frame.
Inter Frame Space	The inter frame space (IFS) field is not part of the CAN data frame. The CAN standard requires an IFS of a minimum of 3 recessive bits, before initiating the transmission of a new CAN frame.

CAN Remote Frame

Consider a scenario, where a CAN node requires some specific data from another CAN node. In this case, the intended recipient can generate a data request using CAN *remote frame*. The message ID used by the remote frame is same as the message ID of intended data frame. On receiving a CAN remote frame, each CAN node on the bus will look for a corresponding data frame with matching message ID. The node having a data frame with matching message ID will respond. A remote frame is quite similar to data frame, with the following important differences.

- A remote frame is explicitly differentiated from a data remote by setting the RTR bit in the arbitration field to recessive state. Recall that RTR bit was set to dominant state in case of data frames.

- A remote frame does not have a data field.

- The DLC field in a remote frame is configured to the intended data length of the corresponding data frame that will be transmitted in response.

The frame format for a CAN remote frame is shown in Figure 11.18. It can be observed that most of the fields in a remote frame are same as the data frame except the few differences mentioned above.

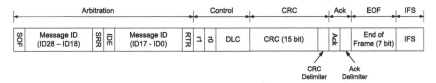

Figure 11.18: Frame format for CAN remote frame.

CAN Error Frame

The CAN *error frame* consists of two fields. The first field comprises of superposition of *error flags* from different CAN nodes, while the second field is error delimiter. The CAN error flags can be of the following two types.

- Active error flag: The active error flag consists of six consecutive dominant bits.

- Passive error flag: The passive error flag consists of six consecutive recessive bits.

Both active as well as passive error flags transmit six consecutive bits of same polarity, thus violating the bit stuffing rule. When other CAN nodes on the bus detect the transmission of CAN error frame, they also transmit error frames on their part. These error frames from other nodes on the bus partly overlap the error delimiter field. The error delimiter field consists of 8 consecutive recessive bits. It is also important to mention that when two or more error frames are transmitted simultaneously and at least one of them is error active then it is the error active frame that will be transmitted on the CAN bus. The message transmitter when detects an error frame, it retransmits the same message once the bus becomes idle.

CAN Overload Frame

The frame format for CAN *overload frame* is similar to CAN error frame. It has an overload flag and an overload delimiter. An overload frame is transmitted due to an overload condition. The overload condition occurs when a CAN receiver on the bus is not ready to receive a new data or remote frame. A node can transmit at most two consecutive overload frames to delay the transmission of data or remote frames from other nodes on the CAN bus. The overload flag consists of six consecutive dominant bits, while overload delimiter has eight recessive bits.

11.8.3   CAN Bus Arbitration

The transmission of a data or remote frame can be initiated by an arbitrary node, when the node detects the bus in idle state, i.e., recessive state. It is possible that two or more nodes start transmitting at the same time. In that case, the conflict is resolved using the so called bitwise *bus arbitration*. The bus arbitration is performed using the arbitration field of a CAN message frame. In bitwise arbitration, each node involved in transmission, compares its transmitted bit level with the one received from the bus. If the transmitted and received bit levels are same, the node continues its transmission. However, when a node transmits a recessive level and receives a dominant level, it recognizes that another node of higher priority is also transmitting and loses bus arbitration to that node. During bus arbitration, it is possible to transmit a recessive level and receive a dominant level due to the *wired AND* nature of the CAN bus. The above mentioned scenario is equivalent to the fact that a node transmitting a CAN frame with a higher message ID loses arbitration to a node transmitting a lower message ID frame. The node that loses bus arbitration, postpones its transmission. Finally, the node transmitting a CAN frame with lowest message ID wins the bus arbitration. The winner of bus arbitration is completely unaware of the fact that it was involved in a bus conflict. An example scenario, illustrating the bitwise arbitration among three competing nodes is shown in Figure 11.19. From Figure 11.19 we can observe that node 3 is the winner of bus arbitration. For simplicity, we have used logical signal levels to illustrate the bus arbitration, however, actual transmission on the CAN bus is based on balanced differential signaling.

It is quite possible that the transmission of a data frame and a remote frame with same message ID is initiated by two different nodes on the CAN bus. In this scenario, the node transmitting a data frame wins the bus arbitration. This is based on the fact that the RTR bit is dominant in case of data frame and recessive in case of remote frame.

11.8.4   CAN Bit Timing

The CAN bit timing is based on the bit (baud) rate used for data transmission on the CAN bus. The nominal bit time is equal to the inverse of the CAN bit rate. Each nominal bit time is decomposed into four different segments as shown in Figure 11.20. Each of these four segments are defined as integer multiples of a base time unit

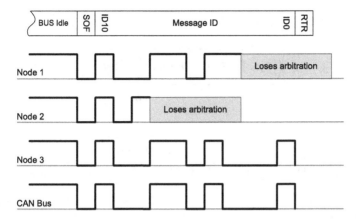

Figure 11.19: Illustration of bitwise bus arbitration using standard CAN frames with three contending nodes.

termed as *time quantum*, t_Q. The duration of one time quantum is nominally equal to one cycle of CAN controller clock, which in turn is derived from system clock. The four segments are described below.

- Sync: The first segment of each CAN bit is the synchronization segment, which is equal to one t_Q.

- Prop: The propagation segment accounts for the propagation delay and ensures that the earliest sampling by any node is delayed enough that different bit values transmitted from all the nodes have reached all the nodes on the bus. The propagation segment can be configured to any value from $1t_Q$ to $8t_Q$.

- Phase 1 and 2: The two phase segments are used to adjust the sampling instant and compensate for any phase errors. The phase 1 segment can take values from $1t_Q$ to $8t_Q$. On the other hand the phase 2 segment takes the maximum of the phase 1 and information processing time. The information processing time is required to be equal to or less than $2t_Q$.

Another important parameter of interest related to CAN bit timing is the *synchronization jump width* (SJW). A CAN controller expects the occurrence of an edge (due to bit transition) during the synchronization segment of bit interval. When an expected edge is not detected during synchronization segment, then CAN controller may prolong or shorten the duration of bit time by an integer multiple of t_Q. The maximum value that can be used for this bit duration adjustment is termed as synchronization jump width. The SJW can be configured to a value between 1 and $\min\{4, Phase1\}$. Due to the flexibility in configuring different segments, one bit duration can vary from $4t_Q$ to $25t_Q$.

11.8.5 CAN Error Handling and Fault Confinement

CAN message errors and bus fault confinement are used to manage CAN network as well as the integrity of any data transmissions. There are five different types of errors

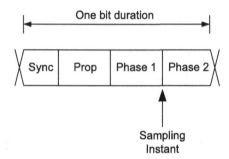

Figure 11.20: Decomposition of a CAN bit time into different segments.

as discussed below.

- *Bit Error*: Recall that each CAN node after transmitting a bit also monitors the bus for the received bit polarity. A *bit error* occurs when the bit transmitted by a node is different from the one monitored on the bus. However, one key exception occurs during the transmission of remote or data frames, where a node transmits a recessive bit during arbitration or acknowledgment field while a dominant bit is received. Another exception is possible during the transmission of error frames, where a node transmits a passive error flag (recessive state) and receives a dominant bit.

- *Stuff Error*: A *stuff error* is detected if six consecutive bits of same level are detected in that portion of the message, which has been encoded using bit stuffing.

- *CRC Error*: A *CRC error* occurs when the calculated 15-bit CRC value at the receiver does not match with the CRC received from the transmitter.

- *Acknowledgment Error*: An *acknowledgment error* is detected if the transmitter does not receive a dominant state during the acknowledgment slot.

- *Form Error*: There are few regions in a CAN frame which follow a fixed pre-defined format. For instance, the CAN standard defines the bit levels for CRC delimiter, ACK delimiter, end of frame and intermission. Based on these fields, extra rules for error checking have been defined. When an invalid sequence is detected, in one of these fixed fields by a CAN controller, the controller signals a *form error*.

The bit error and stuff errors are bit level errors, while CRC, acknowledgment and form errors are message level errors. In case of CRC error, the transmission of error flag is started after the ACK delimiter, while in case of other errors, the transmission of an error flag starts during the next bit slot. The above-mentioned errors are used to maintain transmit and receive error counters. The state of a CAN node is defined based on the current values of these error counters. The rules to increment/decrement these error counters are discussed later in this section. Based on these error counters, each CAN node can be in one of the following three states.

- Error active: A CAN node is in error active state, if both transmit and receive error counters are no more than 127. A CAN node in error active state, can transmit as well as receive message frames and whenever it detects an error it transmits an error active flag. Each node on the CAN bus starts in error active state with its error counters reset to 0.

- Error passive: If any of the two error counters increases above 127, then the state of the corresponding CAN node becomes error passive. A CAN node in error passive state, transmits a passive error flag on the detection of an error condition. In addition, a CAN node in error passive state has to wait for a certain delay after each transmission on the bus, before it can initiate its own transmission.

- Bus off: If the node transmit error counter becomes 256 or higher then the node state is switched to bus off state. In bus off state a node cannot transmit any message on the CAN bus.

Different rules are followed to increment and decrement transmit and receive error counters. For instance, a node experiencing an error during its transmission will increment its transmit error counter by 8, while an error condition during reception results in incrementing the receive error counter by 1. On transmitting and receiving a message correctly, the respective error counters are decremented by 1. When CAN node in error active state experiences eight consecutive transmission attempts resulting in error conditions, then its state will change from error active to error passive, provided no successful transmissions have occurred on the bus during its transmission attempts. Further transmission attempts may force this node to switch to bus off state. A node in error passive state, can switch back to error active state on the successful transmission or reception of message frames, provided both transmit and receive error counters are less than 128. Similarly, a CAN node in buss off state can switch to error active state, with both transmit and receive error counters reset to 0 if it observes 128 occurrences of 11 consecutive recessive bit states.

11.9 CAN Details on TM4C123 Microcontroller

There are two CAN modules integrated in TM4C123 microcontroller. Each of these modules supports CAN protocol version 2.0 part A and B. Both 11-bit standard and 29-bit extended message identifiers are supported by the TM4C123 CAN modules. The two CAN modules have the following key features.

- Maximum data rate of 1 Mbps

- Thirty two message objects for transmission and reception

- Each message object has an individual identifier mask

- Configurable automatic retransmission to support event triggered or time triggered retransmissions

The working of CAN module can be explained in the context of its architecture. Broadly, each CAN module in TM4C123 has the following three main building blocks.

- CAN protocol controller

- Message memory or buffers

- CAN register interface for the module

The CAN protocol controller lies at the heart of CAN module. It is responsible for implementing all the protocol details while transmitting and receiving message frames. On one side the CAN protocol controller is connected to the CAN transceiver chip through CAN_H and CAN_L pins, while on the other side it is interfaced with the message object memory (message RAM). The message RAM in the CAN module comprise of a set of 32 message memory blocks. Each of the memory blocks holds the current message configuration, message filtering, its status as well as message data. The message RAM is not directly accessible in TM4C123. Rather it is accessed using one of the two message object interfaces available in the CAN module, as can be seen from Figure 11.21.

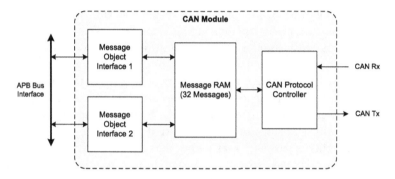

Figure 11.21: The block diagram illustrating the architecture of CAN module.

The configuration of CAN module has four basic steps listed below. Among these the third step is the most involved, which covers all the CAN protocol as well as message handing related configurations and will be discussed in detail. The other configuration steps are quite similar to the configuration steps discussed for other communication interfaces in this chapter.

- Step 1: CAN clock gating control

- Step 2: GPIO configuration for CAN alternate function

- Step 3: CAN protocol and message configuration

- Step 4: CAN interrupt configuration

11.9.1   CAN Clock Gating Control

The clock for each of the CAN modules can be configured separately using register RCGC_CAN_R. The register address information along with brief description is provided in Table 11.35. Bit 0 of RCGC_CAN_R register correspond to CAN module 0, while bit 1 correspond to CAN module 1. Writing 1 to either of the bits enables the clock for the corresponding CAN module. After enabling the clock to a CAN module, there must be a 3 clock cycle delay before accessing the CAN module registers.

Table 11.35: CAN clock gating control register address and brief description.

Register label	Address	Reset value	Brief description
RCGC_CAN_R	0x400FE634	0x00000000	CAN clock gating control register.

11.9.2 GPIO Configuration for CAN Alternate Function

Data transfers through CAN bus are performed by configuring GPIO port pins for CAN alternate functionality. For that purpose, GPIO registers, corresponding to digital enabling (GPIO_DEN_R), alternate function selection (GPIO_AFSEL_R) and port control (GPIO_PCTL_R) should be configured with appropriate values. The bit fields for registers GPIO_DEN_R and GPIO_AFSEL_R should be set to 1 to configure GPIO pins for CAN functionality. In addition, the corresponding bit field in GPIO_PCTL_R register is set to 8 to configure the port pin as CAN TX and RX line.

The possible choices, for alternate functionality, for different port pins are summarized in Tables 8.2 and 8.3, while GPIO register descriptions and their address offsets are provided in Section 8.4.

11.9.3 CAN Protocol and Message Configuration

The CAN module configuration involves multiple steps, which can be grouped into two sets. One set of configurations are performed during initialization, while the other configurations are required during normal operation. During CAN module initialization, we need to configure the CAN baud rate as well as each of the message objects in the message RAM.

To perform the above mentioned configurations, a set of CAN module registers are used, which are listed in Table 11.36. The register addresses provided in Table 11.36 are the address offsets from the CAN module base address. The actual register address is determined by adding the offset to the corresponding CAN module base address. The base addresses for the two CAN modules are provided in Table 11.37. Next we discuss the above mentioned configuration steps.

There are two separate (but identical) message object interfaces in each CAN module. Each of these two message interfaces can be used to configure a message object in the message RAM. In addition these message interfaces are used to transmit or receive a CAN message frame between the application and message RAM. However, in Table 11.36, only the register addresses for message object interface 1 are provided.

CAN Baud Rate Configuration

CAN module can be configured to operate either in normal mode or in initialization mode. For baud rate configuration, the CAN module should be in initialization mode. However, message frame transmission and reception is not permitted during initialization mode.

Table 11.36: CAN configuration register labels, addresses, reset value, and brief description.

Register label	Offset	Reset value	Brief description
CAN_CONTROL_R	0x000	0x00000001	CAN control register.
CAN_STATUS_R	0x004	0x00000003	CAN status (read only) register.
CAN_ERR_COUNTER_R	0x008	0x00000000	CAN error counter register.
CAN_BIT_TIMING_R	0x00C	0x00002301	CAN baud rate configuration register.
CAN_INTERRUPT_R	0x010	0x00000000	CAN interrupt status register.
CAN_BAUD_PRE_EXT_R	0x018	0x00000000	CAN baud prescaler extension register.
CAN message object interface registers			
CAN_IF1_CMD_REQ_R	0x020	0x00000001	CAN interface command req. register.
CAN_IF1_CMD_MASK_R	0x024	0x00000000	CAN interface command mask register.
CAN_IF1_MASK1_R	0x028	0x0000FFFF	CAN interface message mask 1 register.
CAN_IF1_MASK2_R	0x02C	0x0000FFFF	CAN interface message mask 2 register.
CAN_IF1_ARBIT1_R	0x030	0x00000000	CAN interface arbitration 1 register.
CAN_IF1_ARBIT2_R	0x034	0x00000000	CAN interface arbitration 2 register.
CAN_IF1_MSG_CONT_R	0x038	0x00000000	CAN interface message control register.
CAN_IF1_DATA1_R	0x03C	0x00000000	CAN interface message data 1 register.
CAN_IF1_DATA2_R	0x040	0x00000000	CAN interface message data 2 register.
CAN_IF1_DATA3_R	0x044	0x00000000	CAN interface message data 3 register.
CAN_IF1_DATA4_R	0x048	0x00000000	CAN interface message data 4 register.

Table 11.37: Base addresses for two CAN modules.

Start address	End address	Description
0x40040000	0x40040FFF	CAN 0 module address range
0x40041000	0x40041FFF	CAN 1 module address range

To enter the initialization mode the 'Init Mode' bit field in CAN_CONTROL_R register should be set to 1. In addition, the 'CC EN' bit field in CAN_CONTROL_R should also be set to 1. Setting 'CC EN' to 1 allows to configure the CAN_BIT_TIMING_R and CAN_BAUD_PRE_EXT_R registers for desired baud rate. The bit field allocations and their descriptions for CAN_CONTROL_R register are provided in Table 11.38, while that of CAN_BIT_TIMING_R and CAN_BAUD_PRE_EXT_R registers are given in Table 11.39.

Key steps involved in configuring CAN baud rate and bit timing are listed below.

- Select the desired baud rate. The bit time (1/baud rate) is preferred to be an integer multiple of CAN module input clock, f_{CAN} (which can be system clock as well).

- The basic time unit for bit time is the time quantum, t_Q, which is defined as $t_Q = \frac{[\text{BRPE}:\text{BRP}]}{f_{CAN}}$, where [BRPE: BRP] is the combined bit field used as baud rate prescaler.

- Using the functional parameters introduced in Section 11.8.4, the bit time is defined as
$$\text{bit time} = (\text{Sync} + \text{Prop} + \text{Phase1} + \text{Phase2})t_Q.$$

Using the programmable parameters defined in CAN_BIT_TIMING_R, the bit

Table 11.38: CAN control register and its bit field descriptions.

31 ... 16	15	7	Test	CC EN	AR DIS	reserved	Err IE	Status IE	CAN IE	Init Mode	0
reserved	reserved										

Bit field	Description
Test	This bit field is used to enable the CAN module test mode. CAN module operates normally when this bit is cleared.
CC EN	This bit is used to control the configuration of CAN bit timing control registers. When set to 1, the configuration of CAN_BIT_TIMING_R and CAN_BAUD_PRE_EXT_R registers is permitted.
AR DIS	This bit field is used to enable or disable automatic retransmission of disturbed messages. It is set to 1 to disable automatic retransmission.
Err IE, Status IE	These bits are used to enable or disable error and status interrupts. Setting any of these bits to 1 enables the corresponding interrupt. A change in bus off and error warning bit fields of CAN_STATUS_R results in error interrupt, while TxOK, RxOK or a change in error code bit fields of the CAN_STATUS_R register result in a status interrupt.
CAN IE	This bit is used to enable or disable the CAN module interrupts.
Init Mode	CAN module enters initialization mode when this bit is set to 1. For normal operation this bit should be cleared.

time is defined as

$$\text{bit time} = (\text{TSEG1} + \text{TSEG2} + 3)t_Q.$$

Next we illustrate the configuration for desired baud rate using an example. Consider the case where the system clock frequency is 16 MHz and the desired baud rate is 500 kbps.

Example 11.11 (CAN baud rate and bit timing configuration illustration.).

This example illustrates the configuration of registers CAN_BIT_TIMING_R and CAN_BAUD_PRE_EXT_R for the desired baud rate. For the default system clock frequency of 16 MHz and the desired baud rate of 500 kbps, we require bit time = $8\,t_Q$.

```
// CAN bit field definitions
#define CAN_CONTROL_INIT            0x0001
#define CAN_CONTROL_CC_EN           0x0040

// CAN baudrate calculations for 500 kbps baudrate and system
// clock of 16 MHz. We require bit_time = 8 * tq, which gives
// tq = 0.25us and baudrate prescaler of 4.
// bit_time = (TSEG1 + TSEG2 + 3) * tq = (3 + 2 + 3) * tq
#define CAN_BAUD_PRESCALER          0x0003 // Prescaler is 4-1=3
#define CAN_BAUD_TSEG1              0x0003 // TSEG1 is 4-1=3
#define CAN_BAUD_TSEG2              0x0002 // TSEG2 is 3-1=2
#define CAN_BAUD_SJW               0x0001 // SJW is 2-1=1

// Turn On CAN Init mode
CAN_CONTROL_R |= CAN_CONTROL_INIT;
```

Table 11.39: Bit field descriptions for CAN bit timing (CAN_BIT_TIMING_R) and baud rate prescale (CAN_BAUD_PRE_EXT_R) registers.

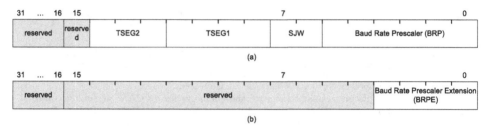

Bit field	Description
TSEG2	This bit field is used to define 'Phase 2' part of the bit time duration, which is the time interval after sampling instant (see Figure 11.20 for details). This bit field is configured using the expression given by TSEG2 = Phase 2 - 1.
TSEG1	This bit field is used to define the sum of 'Phase 1' and 'Prop' parts of the bit time duration before the sampling instant. Their relationship is given by TSEG1 = Phase 1 + Prop - 1
SJW	This synchronization jump width (SJW) bit field is used to apply an adjustment for a phase error, detected by the CAN module during start of frame. The adjustment can be applied to TSEG1 and/or TSEG2 by the value configured to SJW.
Baud Rate Prescaler & Baud Rate Prescaler Extension	These bit fields are used collectively to define the base time unit called time quantum, t_Q. This is implemented by dividing the baud rate prescaler (BRP) appended with baud rate prescaler extension (BRPE) by the clock frequency of the CAN module. The BRPE is implemented using a separate register and is appended as most significant bits to BRP.

```
// Configure CAN bit timing register using parameter values
// defined above for CAN baudrate of 500 kbps
// Enable bit time register configuration using control register
CAN_CONTROL_R |= CAN_CONTROL_CC_EN;
CAN_BIT_TIMING_R = (CAN_BAUD_TSEG2 << 12) | (CAN_BAUD_TSEG1 << 8)
                 | (CAN_BAUD_SJW << 6) | CAN_BAUD_PRESCALER;
CAN_BAUD_PRE_EXT_R = 0x0;
```

CAN Message Object Configuration

There are a total of 32 message objects in the message RAM. Each of these message objects should be configured with appropriate values before it can be used for data transmission or reception. If a message object is not used by the application, then it should be configured as not valid during initialization. To configure a message object, it is not required to have the CAN module in initialization mode, rather message configuration can be done on the fly.

As mentioned earlier, there are two message interfaces that can be used to configure message objects in the message RAM. However, our discussion below, related to message configuration, will only use message interface 1.

To configure a message object as invalid, it is required to just clear the Msg_Valid bit in the CAN_IF1_ARBIT2_R register. On the other hand to configure a message object in message RAM, all its fields should be initialized with appropriate values. In particular, the message ID, mask filter, control and data fields of a message object should be configured with appropriate values. For that purpose, CAN message interface registers, CAN_IF1_ARBIT1_R, CAN_IF1_ARBIT2_R are used to configure message ID, CAN_IF1_MASK1_R, CAN_IF1_MASK2_R registers are used for message filter mask, CAN_IF1_CMD_REQ_R, CAN_IF1_CMD_MASK_R, and CAN_IF1_MSG_CONT_R registers are used to control message configuration as well as their transfer, while registers CAN_IF1_DATA1_R to CAN_IF1_DATA4_R are used for data payload. Next we first introduce each of these message interface registers, while their use for message configuration and transfer will be illustrated later.

The bit field allocations and their descriptions for registers CAN_IF1_CMD_REQ_R and CAN_IF1_CMD_MASK_R are provided in Table 11.40 and Table 11.41 respectively. The CAN_IF1_CMD_REQ_R is used to initiate the message transfer between message interface 1 registers and the selected message object from message RAM. The CAN_IF1_CMD_MASK_R register is used to mention the direction of transfer and to select which message interface registers will be involved in the transfer.

Table 11.40: CAN message interface command request register and its bit field descriptions.

Bit field	Description
Busy	The busy flag is used to indicate that a transfer between message interface registers and message RAM is in progress. Once the transfer is initiated by a write operation to the 'Message NUM' bit field of this register, the busy flag is set. As soon as the transfer is finished, busy flag is cleared automatically.
Message NUM	This message number bit field is used to select one of the 32 message objects in the message RAM for data transfer.

Different messages in CAN network are differentiated based on their message identifiers. The message IDs are also used to determine the message priority. To configure the message IDs, message interface arbitration registers, CAN_IF1_ARBIT1_R and CAN_IF1_ARBIT2_R are used. The bit field allocations and their descriptions for these registers are outlined in Table 11.42. When 11-bit message IDs are used only CAN_IF1_ARBIT2_R register needs to be configured, while for extended message IDs both arbitration registers are configured.

A node in the CAN network can receive selected messages based on its application requirements. For that purpose message acceptance filleting can be applied. To activate message acceptance filtering, message interface mask registers, CAN_IF1_MASK1_R and CAN_IF1_MASK2_R are used. The bit field allocations and their descriptions for these registers are outlined in Table 11.43. When 11-bit message IDs are used only CAN_IF1_MASK2_R register needs to be configured, while for extended message IDs both arbitration registers may require configuration.

Table 11.41: CAN message interface command mask register and its bit field descriptions.

Bit field	Description
WR/$\overline{\text{RD}}$	This bit is used to specify the direction of data transfer between message interface registers and message RAM. A value of 1 transfers data from CAN message interface registers to one of the message objects (specified by the Message NUM bit field) in message RAM.
Mask	This bit is used to update mask bits in the message interface mask registers. When set to 1 the mask registers from the selected message object in message RAM are copied to the message interface mask registers and remain unchanged otherwise.
ARB	This bit is used to update message ID (arbitration) bits in the message interface arbitration registers. When set to 1 the arbitration registers from the selected message object in message RAM are copied to the message interface arbitration registers.
CTRL	The CTRL (control) bit field when configured to 1, copies the message control bits from CAN_IF1_MSG_CONT_R register to the corresponding message interface registers.
Clear INT Pend	When this bit is set to 1 and (a) WR/$\overline{\text{RD}}$ = 1, then 'INT Pend' is cleared in the message object, (b) WR/$\overline{\text{RD}}$ = 0, then 'INT Pend' is first copied to CAN_IF1_MSG_CONT_R register and then it is cleared. When 'Clear INT Pend' bit is 0 and (a) WR/$\overline{\text{RD}}$ = 1, then 'INT Pend' remain unchanged in the message object (in message RAM), (b) WR/$\overline{\text{RD}}$ = 0, then interrupt pending status is transferred to the CAN_IF1_MSG_CONT_R.
New data/Tx RQST	When this bit is set to 1 and (a) WR/$\overline{\text{RD}}$ = 1, then transmission is requested, (b) WR/$\overline{\text{RD}}$ = 0, then 'New Data' status is first copied to CAN_IF1_MSG_CONT_R register and then it is cleared. When 'New data/Tx RQST' bit is 0 and (a) WR/$\overline{\text{RD}}$ = 1, then a transmission is not requested, (b) WR/$\overline{\text{RD}}$ = 0, then new data status is transferred from message buffer (in message RAM) to the CAN_IF1_MSG_CONT_R.
Data A, Data B	When 'Data A' bit is 0 no data field update is performed. When this bit is set to 1 and (a) WR/$\overline{\text{RD}}$ = 1, then message data bytes 0-3 are transferred from message object to message interface registers CAN_IF1_DATA1_R and CAN_IF1_DATA2_R, (b) WR/$\overline{\text{RD}}$ = 0, then message data bytes 0-3 are transferred from message interface registers CAN_IF1_DATA1_R and CAN_IF1_DATA2_R to the selected message object in the message RAM. The above description is valid for 'Data B' bit and correspondingly message interface registers CAN_IF1_DATA3_R and CAN_IF1_DATA4_R.

The CAN_IF1_MSG_CONT_R register controls the message transfer between message interface registers and message RAM. The bit field allocations and their descriptions for CAN_IF1_MSG_CONT_R register are given in Table 11.44.

The message interface data registers are used to exchange data payload with the message RAM. In particular message interface 1 has four data registers, CAN_IF1_DATA1_R to CAN_IF1_DATA4_R. Each of these data registers contains two data bytes of the payload in two least significant bytes. At a given time, it is possible that all these four registers contain data, then the valid number of data bytes are determined from the data length code (DLC) bit field of CAN_IF1_MSG_CONT_R register.

Based on the above mentioned configuration steps, each CAN module can be config-

Table 11.42: CAN message interface arbitration registers and their bit field descriptions.

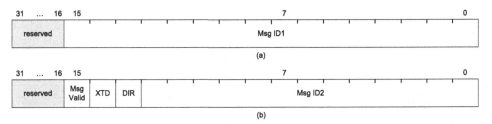

Bit field	Description
Msg Valid	When cleared to 0, the message is not valid and is not considered by the message handler. Setting it to 1 makes the message valid and used by the CAM message handler for data transmission reception purpose. This bit should be set to 1 only for those messages, which have been configured with valid values for different message fields.
XTD	When this bit is cleared to 0 a standard message ID of 11-bit is used. Setting it to 1 requires the message to use an extended 29-bit ID.
DIR	This bit controls the message direction (0 for receive) as well as its type as data or remote frame. When DIR bit is 0 and 'Tx RQST' bit in CAN_IF1_MSG_CONT_R register is set to 1, this message object receives a remote frame with the same message ID. On reception of a data frame of same message ID, the message is stored in the message object. When DIR bit is set to 1 and 'Tx RQST' bit in CAN_IF1_MSG_CONT_R register is also set to 1, then the respective message object is transmitted as data frame. On reception of a remote frame request with matching ID (with DIR bit 1) the 'Tx RQST' bit of this message is set if 'Remote EN' bit field in CAN_IF1_MSG_CONT_R register is also set.
Msg ID1, Msg ID2	For 29-bit extended message identifier, the message ID1 and ID2 bit fields from two arbitration registers are combined as [Msg ID2:Msg ID1]. For 11-bit message ID, bits 2 to 12 of Msg ID2 bit field are used for message identifier and ID bit field of register CAN_IF1_ARBIT1_R is ignored.

ured for data transmission and reception. In addition, it is also required to configure all those message objects (in message RAM) to be invalid that are not being used by the application. For illustration purpose, we provide a simple initialization of message objects, where all the message objects in the message RAM are configured as invalid and it is also ensured that their interrupt and new data flags are cleared during this process. It is important to mention that message object configuration does not mandate to put the CAN module in Init Mode. However, during initialization, it would be appropriate to have the CAM module in Init Mode to avoid any unwanted bus activity. Example 11.12 below illustrates the initialization of all message objects in the message RAM as invalid.

Example 11.12 (CAN message object initialization.).

We illustrates the initialization of message objects in message RAM as invalid and clear any of the message related pending interrupts and new data status flags.

```
// CAN control register bit field definitions
#define CAN_CONTROL_INIT            0x0001
```

Table 11.43: CAN message interface filtering mask registers and their bit field descriptions.

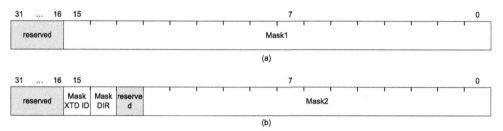

Bit field	Description
Mask XTD ID	When this bit is set to 1, the XTD bit of CAN_IF1_ARBIT2_R is used for message acceptance filtering and is ignored otherwise.
Mask DIR	When this bit is set to 1, the DIR bit of CAN_IF1_ARBIT2_R is used for message acceptance filtering and is ignored otherwise.
Mask1, Mask2	For 29-bit extended message identifier, the Mask1 and Mask2 bit fields from two mask registers are combined as [Mask2:Mask1] for message filtering. For 11-bit message ID, bits 2 to 12 of Mask2 bit field are used for message filtering and Mask1 bit field of register CAN_IF1_MASK1_R is ignored. For each Mask1 and Mask2 bit set to 1, the corresponding ID bit from arbitration register is used in message acceptance filtering.

```
// Bit field definitions for CAN interface registers
#define CAN_IF1_BUSY                   0x8000
#define CAN_IF1_CMD_MASK_WR_NRD        0x0080
#define CAN_IF1_CMD_MASK_ARB          0x0020
#define CAN_IF1_CMD_MASK_CONT         0x0010
#define CAN_IF1_CMD_MASK_CLR_PEND_INT 0x0008
#define CAN_IF1_CMD_MASK_NEW_DATA     0x0004

unsigned long msg_num;

// Turn On CAN Init mode
CAN_CONTROL_R |= CAN_CONTROL_INIT;

// Wait if the CAN controller is busy
while(CAN_IF1_CMD_REQ_R & CAN_IF1_BUSY)
{ }

// Initialize all the messages as not valid. For this interface
// 1 is used to configure all messages as invalid.
CAN_IF1_CMD_MASK_R |= CAN_IF1_CMD_MASK_WR_NRD
                    + CAN_IF1_CMD_MASK_ARB
                    + CAN_IF1_CMD_MASK_CONT;

CAN_IF1_ARBIT2_R = 0;
CAN_IF1_MSG_CONT_R = 0;

for(msg_num = 1; msg_num <= 32; msg_num++)
{
   // Wait for busy bit to clear
   while(CAN_IF1_CMD_REQ_R & CAN_IF1_BUSY)
   { }
```

Table 11.44: CAN message interface control register and its bit field descriptions.

31 ... 16	15									7		0
reserved	New Data	Msg Lost	INT Pend	Use Mask	Tx IE	Rx IE	Remote EN	Tx RQST	EOB	reserved		DLC

Bit field	Description
New Data	If this bit is set to 1, it indicates that new data has been written to the data field of this message by either message handler or user program. This bit is cleared by the processor.
Msg Lost	When this bit is 0, it shows no new message has been lost since this bit was cleared last time by the processor.
INT Pend	When this bit is set to 1 it shows that this message is the source of interrupt. The interrupt identifier in the CAN_INTERRUPT_R register points to this message object as the source of interrupt provided there is no other higher priority interrupt has been generated in the meantime.
Use Mask	The mask is used for message acceptance filtering when this bit is set and ignored otherwise.
Tx IE, Rx IE	The transmit and receive interrupts can be enabled/disabled using these bits. Setting the bit to 1 enables the corresponding interrupt and 'INT Pend' bit of this register is set as a result of successful transmission (if Tx IE is set) or reception (if Rx IE is set).
Remote EN	When this bit is set to 1, then on receiving a remote frame with matching ID, the 'Tx RQST' of this register is set to transmit this message frame.
Tx RQST	When this bit is set to 1, it requests the transmission of this message frame. If 'WR/$\overline{\text{RD}}$' and 'Tx RQST' bits in CAN_IF1_CMD_MASK_R are set the this bit is ignored.
EOB	When set to 1, it indicates that this message frame comprise of either a single message object or is the last message object of a FIFO buffer.
DLC	This data length code (DLC) bit field specifies the number of bytes in the data frame. The valid values are 0 to 8.

```
    // Initiate programming the message object
    CAN_IF1_CMD_REQ_R = msg_num;
}


// Make sure that the interrupt and new data flags are updated
// for the message objects.
CAN_IF1_CMD_MASK_R |= CAN_IF1_CMD_MASK_CLR_PEND_INT
                    + CAN_IF1_CMD_MASK_NEW_DATA;

for(msg_num = 1; msg_num <= 32; msg_num++)
{
    // Wait for busy bit to clear
    while(CAN_IF1_CMD_REQ_R & CAN_IF1_BUSY)
    { }

    // Initiate programming the message object
    CAN_IF1_CMD_REQ_R = msg_num;
}
```

CAN Message and Error Status

The are multiple status indicators that might be of use to the application for data transfer on the CAN bus. Some status flags are related to the bus activity, while others are related to message transmission/reception and node state itself. In the CAN module, different registers are allocated for status purpose. The CAN_STATUS_R and CAN_ERR_COUNTER_R registers are two important status registers, which provide important status information. Specifically, the CAN_STATUS_R provides information that is useful while servicing a CAN module interrupt. The CAN_ERR_COUNTER_R contains information related to transmit and receive error counters, which determines the current state of the CAN node.

The bit field allocations and their descriptions for CAN_STATUS_R register are provided in Table 11.45. When servicing a CAN interrupt the information regarding the interrupt source can be obtained from the CAN_STATUS_R. For instance, a change in the 'Bus OFF' and 'Err Warn' bit fields can result in an error interrupt, while 'Rx OK', 'Tx OK' and 'LEC' bit fields can be the source of status interrupt provided the interrupts are enabled by configuring the appropriate bit fields of CAN_CONTROL_R.

Table 11.45: CAN status register and its bit field descriptions.

31 ... 16	15 7						0
reserved	reserved	Bus OFF	Err Warn	Err Passive	Rx OK	Tx OK	Last Error Code

Bit field	Description
Bus OFF	When this bit is 1, it indicates that the CAN node is in buss off state.
Err Warn	If this bit is equal to 1, then at least one of the error counters (Tx or Rx) has reached the error warning limit of 96.
Err Passive	If this bit is read as 1, it shows that the CAN node is in error passive state.
Rx OK, Tx OK	If these bit are read as 1 since the last time they were cleared, it indicates successful reception (Rx OK) and/or transmission (Tx OK) of a CAN message frame. These bits should be cleared by writing a 0.
Last Error Code (LEC)	This bit field provides the error code for the last error that occurred on the bus. Different possible values for the error codes are: • 0: No error • 1: Stuff error • 2: Format error • 3: ACK error • 4: Bit 1 error, the device sends 1 by monitors a 0 on the bus • 5: Bit 0 error, the device sends 0 by monitors a 1 on the bus • 6: CRC error

The bit field allocations and their descriptions for CAN_ERR_COUNTER_R register are provided in Table 11.46.

11.9.4 CAN Interrupt Configuration

Two types of interrupts, error interrupts and status interrupts, are generated by the CAN module. The interrupt related configurations and its servicing are performed using registers CAN_CONTROL_R, CAN_STATUS_R, and CAN_INTERRUPT_R. Both error as well as status interrupts can be enabled/disabled using CAN_CONTROL_R

Table 11.46: CAN error counter register and its bit field descriptions.

31	...	16	15				7				0
reserved		Rx Err Passive		Rx Error Counter				Tx Error Counter			

Bit field	Description
Rx Err Passive	When this bit becomes 1, it indicates that the receive error counter has reached the error passive level (i.e., 128 or higher).
Rx Error Counter	This bit field is the receive error counter and can take values between 0 and 127.
Tx Error Counter	This bit field is the transmit error counter and can take values between 0 and 255.

register. Specifically, 'Err IE' and 'Status IE' are respectively, used for enabling/disabling of error and status interrupts. The 'CAN IE' bit in the CAN_CONTROL_R register is used to enable/disable different interrupts from the CAN module.

The status interrupt can be generated either due to successful transmission/reception of a message frame or it can be due to one of the errors listed in the 'LEC' bit field of CAN_STATUS_R. Since the status interrupt due to successful transmission or reception of a message frame can be generated by any of the 32 message objects, the CAN_INTERRUPT_R register indicates which message object is the source of interrupt. The CAN_INTERRUPT_R register has a 16 bit Int ID bit field, for which valid values are 0x0001 to 0x0020 (corresponding to any of the message objects) and 0x8000 for status interrupt. If interrupts from multiple message objects are pending, then CAN_INTERRUPT_R register points to the highest priority pending interrupt. It is important to mention here that reading the CAN_STATUS_R register clears the CAN_INTERRUPT_R register.

The transmit/receive interrupts for each message object in the message RAM can be enabled/disabled individually. For that purpose, 'Tx IE' and 'Rx IE' bits in the CAN_IF1_MSG_CONT_R should be set up accordingly at the time of configuring the message object in message RAM.

Next we illustrate, using Example 11.13, all the steps involved in CAN module configuration along with message transmission and reception. For that purpose, we require a baud rate of 500 kbps, an interrupt driven message reception and no message filtering.

Example 11.13 (CAN device driver for data transmission and reception.). This example illustrates the use of CAN module 0 to transmit and receive data. Specifically, CAN module initialization, its configuration for message transmission and reception are illustrated. For message reception, the message object is configured in the message RAM first and on message reception an interrupt is generated.

```
// Clock gating control related register definitions
#define RCGC_CAN_R        *((volatile unsigned long *)0x400FE634)
#define RCGC_GPIO_R       *((volatile unsigned long *)0x400FE608)
#define CLOCK_GPIOE       0x00000010  // Port E clock control
```

```c
#define CLOCK_CAN0              0x00000001   // Can0 clock control

// CAN base address and register definitions
#define CAN_BASE                0x40040000

#define CAN_CONTROL_R       *(volatile long *)(CAN_BASE + 0x000)
#define CAN_STATUS_R        *(volatile long *)(CAN_BASE + 0x004)
#define CAN_BIT_TIMING_R    *(volatile long *)(CAN_BASE + 0x00C)
#define CAN_BAUD_PRE_EXT_R  *(volatile long *)(CAN_BASE + 0x018)

// CAN Interface 1 configuration registers
#define CAN_IF1_CMD_REQ_R   *(volatile long *)(CAN_BASE + 0x020)
#define CAN_IF1_CMD_MASK_R  *(volatile long *)(CAN_BASE + 0x024)
#define CAN_IF1_MASK1_R     *(volatile long *)(CAN_BASE + 0x028)
#define CAN_IF1_MASK2_R     *(volatile long *)(CAN_BASE + 0x02C)
#define CAN_IF1_ARBIT1_R    *(volatile long *)(CAN_BASE + 0x030)
#define CAN_IF1_ARBIT2_R    *(volatile long *)(CAN_BASE + 0x034)
#define CAN_IF1_MSG_CONT_R  *(volatile long *)(CAN_BASE + 0x038)
#define CAN_IF1_DATA1_R     *(volatile long *)(CAN_BASE + 0x03C)
#define CAN_IF1_DATA2_R     *(volatile long *)(CAN_BASE + 0x040)
#define CAN_IF1_DATA3_R     *(volatile long *)(CAN_BASE + 0x044)
#define CAN_IF1_DATA4_R     *(volatile long *)(CAN_BASE + 0x048)

// CAN interrupt status register definition
#define CAN_INTERRUPT_R     *(volatile long *)(CAN_BASE + 0x010)

// GPIO Port E alternate function configuration
#define GPIO_PORTE_AFSEL_R *((volatile unsigned long *)0x40024420)
#define GPIO_PORTE_PCTL_R  *((volatile unsigned long *)0x4002452C)
#define GPIO_PORTE_DEN_R   *((volatile unsigned long *)0x4002451C)

// IRQ 32 to 63 Set Enable Register
#define NVIC_EN1_R          *((volatile unsigned long *)0xE000E104)
#define NVIC_EN1_INT39      0x00000080    // Interrupt 39 enable

// CAN control and status register bit field definitions
#define CAN_CONTROL_INIT            0x0001
#define CAN_CONTROL_CC_EN           0x0040
#define CAN_STATUS_RX_OK            0x0010

// Bit field definitions for CAN interface registers
#define CAN_IF1_BUSY                0x8000
#define CAN_IF1_CMD_MASK_WR_NRD     0x0080
#define CAN_IF1_CMD_MASK_ARB        0x0020
#define CAN_IF1_CMD_MASK_CONT       0x0010
#define CAN_IF1_CMD_MASK_CLR_PEND_INT 0x0008
#define CAN_IF1_CMD_MASK_NEW_DATA   0x0004

// CAN baudrate calculations for 500 kbps baudrate and system
// clock of 16 MHz. We require tq = 0.25us, which gives baudrate
// prescaler of 4 and bit_time = 8 * tq.
// bit_time = (TSEG1 + TSEG2 + 3) * tq = (3 + 2 + 3) * tq
#define CAN_BAUD_PRESCALER          0x0003 // Prescaler is 4-1=3
#define CAN_BAUD_TSEG1              0x0003 // TSEG1 is 4-1=3
#define CAN_BAUD_TSEG2              0x0002 // TSEG2 is 3-1=2
#define CAN_BAUD_SJW               0x0001 // SJW is 2-1=1

typedef struct
{
```

```
    unsigned long MsgID;           // CAN message ID, 11 or 29 bit
    unsigned long MsgIDMask;       // CAN message ID mask
    unsigned long MsgLen;          // CAN message data length field
    unsigned long MsgFlags;        // CAN message control and status
    unsigned char *MsgData;        // CAN message data
} CAN_Msg_Object;

// Function prototypes
void CAN_Init(void);
void CAN_Config_Rx_Message(CAN_Msg_Object can_msg);
void CAN_Config_Tx_Message(CAN_Msg_Object can_msg);
void Delay(unsigned long counter);

unsigned char Can_data_buffer[8] = {0};

// CAN intialization and GPIO alternate function configuration
void CAN_Init(void)
{
  unsigned long msg_num;

  // Configure GPIO pins for CAN module 0
  RCGC_GPIO_R |= CLOCK_GPIOE;          // Enable clock for port E
  GPIO_PORTE_DEN_R |= 0x030;           // DEN for pins PE4 and PE5
  GPIO_PORTE_AFSEL_R |= 0x00000030;    // Configure PE4 and PE5
  GPIO_PORTE_PCTL_R |= 0x00880000;     // as CAN module 0

  // Configure CAN module for baudrate and message initilization
  RCGC_CAN_R |= CLOCK_CAN0;            // Enable clock for CAN 0
  Delay(3);
  CAN_CONTROL_R |= CAN_CONTROL_INIT;   // Turn On CAN Init mode

  // Wait if the CAN controller is busy
  while(CAN_IF1_CMD_REQ_R & CAN_IF1_BUSY) { }

  // Initialize all the messages as not valid. Use CAN message
  // interface 1 to configure all messages as invalid.
  CAN_IF1_CMD_MASK_R |= CAN_IF1_CMD_MASK_WR_NRD
                      + CAN_IF1_CMD_MASK_ARB
                      + CAN_IF1_CMD_MASK_CONT;

  // Clear arbitration and message control registers
  CAN_IF1_ARBIT1_R = 0;
  CAN_IF1_ARBIT2_R = 0;
  CAN_IF1_MSG_CONT_R = 0;

  // Configure all messages objects in message RAM as invalid
  for(msg_num = 1; msg_num <= 32; msg_num++)
  {
     // Wait for busy bit to clear
     while(CAN_IF1_CMD_REQ_R & CAN_IF1_BUSY) { }

     // Initiate programming the message object
     CAN_IF1_CMD_REQ_R = msg_num;
  }

  // Make sure that the interrupt and new data flags are updated
  // for all the message objects in message RAM
  CAN_IF1_CMD_MASK_R |= CAN_IF1_CMD_MASK_CLR_PEND_INT
                      + CAN_IF1_CMD_MASK_NEW_DATA;
```

```
   for(msg_num = 1; msg_num <= 32; msg_num++)
   {
      // Wait for busy bit to clear
      while(CAN_IF1_CMD_REQ_R & CAN_IF1_BUSY) { }

      // Initiate programming the message object
      CAN_IF1_CMD_REQ_R = msg_num;
   }

   // Enable bit time register configuration using control register
   CAN_CONTROL_R |= CAN_CONTROL_CC_EN;

   // Configure CAN bit timing register for 500 kbps baud rate
   CAN_BIT_TIMING_R = (CAN_BAUD_TSEG2 << 12)|(CAN_BAUD_TSEG1 << 8)
                      |(CAN_BAUD_SJW << 6)|CAN_BAUD_PRESCALER;
   CAN_BAUD_PRE_EXT_R = 0x00;

   // Enable CAN 0 interrupts for error, status and CAN controller
   CAN_CONTROL_R |= 0x0E;

   // Enable CAN interrupt in NVIC
   NVIC_EN1_R = NVIC_EN1_INT39;

   // Exit CAN Initialization mode
   CAN_CONTROL_R &= ~(CAN_CONTROL_INIT |CAN_CONTROL_CC_EN);
}

// CAN interrupt handler
void CAN0_Handler(void){
unsigned long status = 0;
unsigned long int_source = 0;

   // Get the source of interrupt information
   int_source = CAN_INTERRUPT_R;
   status = CAN_STATUS_R;                // Clear interrupt register

     if(status & CAN_STATUS_RX_OK)
     {
       CAN_IF1_MSG_CONT_R = (0x080 | 0x08);

       // Initiate programming the message object
       CAN_IF1_CMD_REQ_R = 1;

       // Wait for busy bit to clear
       while(CAN_IF1_CMD_REQ_R & CAN_IF1_BUSY) { }

       // Get the received data from CAN data interface registers
       Can_data_buffer[0] = (char)(CAN_IF1_DATA1_R & 0xFF);
       Can_data_buffer[1] = (char)((CAN_IF1_DATA1_R >> 8) & 0xFF);
       Can_data_buffer[2] = (char)(CAN_IF1_DATA2_R & 0xFF);
       Can_data_buffer[3] = (char)((CAN_IF1_DATA2_R >> 8) & 0xFF);
       Can_data_buffer[4] = (char)(CAN_IF1_DATA3_R & 0xFF);
       Can_data_buffer[5] = (char)((CAN_IF1_DATA3_R >> 8) & 0xFF);
       Can_data_buffer[6] = (char)(CAN_IF1_DATA4_R & 0xFF);
       Can_data_buffer[7] = (char)((CAN_IF1_DATA4_R >> 8) & 0xFF);
     }
}
```

```
// CAN message configuration for reception
void CAN_Config_Rx_Message(CAN_Msg_Object can_msg)
{
  // configure the message
  CAN_IF1_CMD_MASK_R  = (CAN_IF1_CMD_MASK_WR_NRD | 0x70);
  CAN_IF1_MASK1_R = 0;
  CAN_IF1_MASK2_R  = can_msg.MsgIDMask;
  CAN_IF1_ARBIT1_R = 0;
  CAN_IF1_ARBIT2_R  = 0x8000 | (can_msg.MsgID << 2);

  // Configure Rx interrupt and EOB
  CAN_IF1_MSG_CONT_R =  (0x400 | 0x080 | can_msg.MsgLen);

  // Initiate programming message object to first message buffer
  CAN_IF1_CMD_REQ_R = 1;

  // Wait for busy bit to clear
  while(CAN_IF1_CMD_REQ_R & CAN_IF1_BUSY)
  { }
}

// CAN message configuration for transmission
void CAN_Config_Tx_Message(CAN_Msg_Object can_msg)
{
  // configure the message
  CAN_IF1_CMD_MASK_R  = (CAN_IF1_CMD_MASK_WR_NRD | 0x73);
  CAN_IF1_MASK1_R = 0;
  CAN_IF1_MASK2_R  = can_msg.MsgIDMask;;
  CAN_IF1_ARBIT1_R = 0;
  CAN_IF1_ARBIT2_R  = 0xA000 | (can_msg.MsgID << 2);

  // Configure Tx request, EOB and data length
  CAN_IF1_MSG_CONT_R =  (0x100 | 0x080 | can_msg.MsgLen);

  // Configure the data to be transmitted
  CAN_IF1_DATA1_R = can_msg.MsgData[0]+(can_msg.MsgData[1] << 8);
  CAN_IF1_DATA2_R = can_msg.MsgData[2]+(can_msg.MsgData[3] << 8);
  CAN_IF1_DATA3_R = can_msg.MsgData[4]+(can_msg.MsgData[5] << 8);
  CAN_IF1_DATA4_R = can_msg.MsgData[6]+(can_msg.MsgData[7] << 8);

  // Initiate programming message object to second message buffer
  CAN_IF1_CMD_REQ_R = 2;

  // Wait for busy bit to clear
  while(CAN_IF1_CMD_REQ_R & CAN_IF1_BUSY)
  { }
}

// This function implements the delay
void Delay(unsigned long counter)
{
  unsigned long i = 0;

  for(i=0; i< counter; i++);
}
```

Next we illustrate the use of CAN device driver for data transmission. Example 11.14 implements repeated transmission of a fixed length 'Hello' message. For proper functioning of this demo another CAN node, configured (at least) for message reception is required.

Example 11.14 (CAN device driver usage for data transmission.). This example illustrates the use of CAN device driver for data transmission. The transmission is not interrupt driven and each message transmission is treated as stand alone independent transmission.

```
// CAN message object structure
typedef struct
{
    unsigned long MsgID;        // CAN message ID, 11 or 29 bit
    unsigned long MsgIDMask;    // CAN message ID mask
    unsigned long MsgLen;       // CAN message data length field
    unsigned long MsgFlags;     // CAN message control and status
    unsigned char *MsgData;     // CAN message data
} CAN_Msg_Object;

// External functions used by this example
extern void EnableInterrupts(void);
extern void DisableInterrupts(void);
extern void Delay(unsigned long counter);
extern void CAN_Init(void);
extern void CAN_Config_Rx_Message(CAN_Msg_Object can_msg);
extern void CAN_Config_Tx_Message(CAN_Msg_Object can_msg);

// CAN message data used for transmission
unsigned char msgData[8]={'H','e','l','l','o'}; // 8 byte buffer
    for tx message data

int main()
{
CAN_Msg_Object can_msg;

  // Disable global interrupts
  DisableInterrupts();

  // Initialize CAN interface 0
  CAN_Init();

  // Configure the CAN message
  can_msg.MsgID = 0;
  can_msg.MsgIDMask = 0;
  can_msg.MsgLen = 0x5;
  can_msg.MsgData = msgData;

  // Enable global interrupts
  EnableInterrupts();

  // Repeated periodic data transmissions
  while(1)
  {
    CAN_Config_Tx_Message(can_msg);
    Delay(500000);
  }
}
```

11.10 Summary of Key Concepts

Serial communication interfaces can be either synchronous or asynchronous. In addition, these interfaces can be either half- or full-duplex.

UART Interface

- UART is asynchronous, full-duplex point to point serial communication protocol.

- The data transmission rate is also called baud rate. The baud rate is configured on the two communicating devices before the start of data transfer.

- Each data element (in general 8-bit byte) in transmitted separately as UART frame, with a start, stop and optional parity bit.

- The start bit is responsible to synchronize the receiver.

- For long distance data transmission, UART signals are converted to standard RS232 signals using dedicated transceiver chip.

I^{2}C Interface

- The I^{2}C is a synchronous serial bus and supports multi-master architecture.

- Each slave module is assigned a unique address.

- Data transfers are initiated by a master module and the first data byte transferred is the slave address.

- The bus arbitration resolves any competition among different active master modules, which are trying to access the bus.

SPI Interface

- The serial peripheral interface (SPI) is also referred as synchronous serial interface.

- SPI is a single master bus architecture. When multiple slaves are interfaced, each slave device is selected using a dedicated slave select line.

- Data transfer rates are controlled by the master module.

CAN Interface

- The controller area network (CAN) is a broadcast bus, where all the nodes hear all the transmissions on the bus.

- The maximum data rate on the CAN bus is 1 Mbps provided the bus physical length is limited to 40 meters.

- The CAN messages are assigned message IDs that not only tell about the data contained in the message but also define the priority of the message. Using a lower value for message ID increases the priority.

- In case of simultaneous transmission from more than one CAN node, the node transmitting the highest priority message (correspondingly lowest message ID) is successful in its transmission due to bitwise bus arbitration.

- The CAN data frames are used to exchange data among the nodes with a maximum payload of 8 bytes.

- CAN bus also supports remote frames that can be used to request a certain type of data from those node(s) who are source of that data.

Review Questions

Question 11.1. What is the difference between synchronous and asynchronous serial communication interface?

Question 11.2. How does the full-duplex serial interface differ from half-duplex interface?

Question 11.3. Why is serial communication interface preferred over its parallel counterpart?

Question 11.4. What are the key differences between UART and RS232 based serial interfaces?

Question 11.5. If the UART Rx pin is connected to ground by accident, then what type of error condition will be generated by the UART interface?

Question 11.6. Consider a situation where a new UART frame is received, while the previously received frame is not processed. What type of error should be expected in such a scenario?

Question 11.7. While transmitting one data byte using UART frame, is it possible to transmit the MSB as the first transmitted data bit after the start bit?

Question 11.8. What is the significance of the ratio of UART interface clock frequency to UART baud rate?

Question 11.9. When TM4C123 is operating at a system clock frequency of 16 MHz, what is the maximum achievable baud rate?

Question 11.10. Why should UART interrupt flag be cleared before exiting the corresponding UART ISR? This justification is equally valid for other serial communication interfaces as well.

Question 11.11. Is I^2C bus single- or multi-master? What about SPI and CAN buses?

Question 11.12. How are START and STOP conditions generated on an I²C bus interface?

Question 11.13. Is it possible for two devices, connected to the I²C bus, to have the same slave addresses?

Question 11.14. What value should be configured to port control register (GPIO_PCTL_R) to use a GPIO pin as I²C interface pin?

Question 11.15. What are different modes of operation for SPI bus?

Question 11.16. How do we define SPI clock phase and its polarity?

Question 11.17. In case of simultaneous transmissions from two or more nodes, how are collisions resolved on the CAN bus?

Question 11.18. How much is the permissible CAN bus length, when the data transmission rate is 1 Mbps?

Question 11.19. What are different types of CAN frames? Which CAN frames can have non-zero data field?

Question 11.20. What are different time segments that constitute one bit duration?

Question 11.21. What are different error types that can be experienced by a CAN frame?

Question 11.22. Is it possible for a CAN node to transmit a frame, while being in error passive state?

Exercises

Exercise 11.1. What are the advantages/disadvantages of an asynchronous serial interface compared to a synchronous serial interface?

Exercise 11.2. Consider a UART interface operating at 115200 baud, with 8 data and 1 stop bits. How much time will it take to transmit one data byte on this interface?

Exercise 11.3. For the system clock frequency of 48 MHz and desired baud rate of 460800, what values should be configured to registers UART_BAUD_INT_R and UART_BAUD_FRAC_R?

Exercise 11.4. Write a C program to configure UART1 module of TM4C123, for both transmission and reception of data, with baud rate 19200, 8 data bits, even parity, and 2 stop bits. Assume that the processor and UART module are operating at a system clock frequency of 40 MHz.

Exercise 11.5. A data buffer named DATA_BUFF[50] of type CHAR and size 50 bytes is used to store bytes that are transmitted using UART interface. A variable BYTE_COUNT of type INT represents the number of bytes in DATA_BUFF waiting for transmission. The bytes are to be transmitted using UART1 configuration preformed in Exercise 11.4. A function named UART_Send_Data() writes data to DATA_BUFF, updates variable BYTE_COUNT, and enables transmission interrupt. Write the interrupt service routine UART1_Handler() responsible to transmit all the data bytes in DATA_BUFF.

Exercise 11.6. Assume that an interrupt driven UART transmission and reception is being performed. At some arbitrary time, UART interrupt occurs and as a response the UART interrupt service routine is entered. At this time instant, we observe that both transmit and receive interrupt flags are active. Which interrupt (corresponding to either UART transmit or receive) should be processed first and what is the justification for your choice?

Exercise 11.7. Consider the transfer of one data byte from the transmitting node to receiving node. If I^2C interface is used to make this transfer, then how many bits are actually transmitted on the bus before the transfer of one data byte is complete? Compare this with the number of bits transmitted on the UART interface to transfer one data byte. What is the reason for large overhead in case of I^2C interface compared to UART interface?

Exercise 11.8. In most cases, an I^2C bus will have external pull-up resistors installed as part of the bus interface. In this case, what will happen if the user enables the internal pull-ups as well? Will the two devices be able to communicate? Is there any scenario where the communication between the devices will be disrupted?

Exercise 11.9. What value should be configured to SCL_Period bit field of I^2C master time period register (I2CM_TIME_PERIOD R), when the desired I^2C clock frequency is 400 kHz and system clock frequency is 32 MHz?

Exercise 11.10. Write a C program to configure SPI mode to mode 3 and its clock frequency to 4 MHz, when the system clock frequency is 64 MHz.

Exercise 11.11. Let the system clock frequency be 48 MHz. We want to operate the CAN interface at a desired baud rate of 500 kbps. What values should be configured to registers CAN_BIT_TIMING_R and CAN_BAUD_PRE_EXT_R to achieve the desired baud rate?

Chapter 12

Analog Interfacing

Overview

To interact with natural phenomenon, embedded devices use different sensors/transducers and actuators. Many of these sensors and actuators use analog interface to connect to an embedded system. For proper interfacing between these sensors/actuators and an embedded system it is required to convert analog signals to digital and vice versa. This chapter describes the theory related to analog-to-digital conversion. Different methods employed for this conversion are introduced briefly. We then look at how Cortex-M performs analog to digital conversion by discussing its configuration details.

12.1 Need for Analog Interfacing

Most of the signals in the world are analog by nature. An analog signal is one that is continuous in both amplitude and time. Examples include speed, voltage, angle, position, force, pressure, temperature to name a few. An embedded system, being digital in nature, when interacts with the external world, it requires analog signals to be converted to their digital equivalent. For instance, a digital clock displaying temperature in addition to time, is required to convert the analog temperature signal, acquired from the temperature sensor, to digital equivalent before it can be processed and displayed. This can be achieved by using a device called Analog-to-Digital Converter (ADC). Lets look at some daily life examples where we encounter ADCs.

- When we use a scanner to scan a picture. The scanner performs an analog-to-digital conversion. It takes the analog information provided by the image sensor and converts it to digital equivalent.

- When we use a VoIP solution such as Skype on a computer, we are using an analog-to-digital converter to convert our voice, which is analog, to digital equivalent before its transmission.

- We use sensors to measure physical parameters, which mostly are analog signals. For their digital processing using a processor, we first convert those signals to digital by using an ADC.

Whenever we need to retrieve the analog signal back, an operation opposite to analog-to-digital conversion is performed. The device used for this purpose is called Digital-to-Analog Converter (DAC). For example, our friends when receive our voice in digital form over Skype, it is first converted to analog signal by DAC before feeding to headphones or speaker. In other words, our voice is heard once DAC has done its job. Similarly, when we play an audio CD, the CD player reads the digital information stored on the disk and converts it back to analog, so we can hear the music.

In this chapter, we will give an in-depth explanation about analog-to-digital conversion yet keeping an easy to follow language. We will also look at how the TM4C123 microcontroller provides the analog-to-digital conversion functionality. For this purpose, we will learn how we can program the ADC modules and will illustrate its working through examples.

There are some basic reasons to use digital signals instead of analog. One of the key reasons is the noise, which is an unwanted signal. As digital systems can understand only two numbers, zero and one, anything different from this is considered to be noise and can be easily discarded. For analog signals, any value can be the part of the signal itself and thus making the detection and removal of noise almost impossible. Another advantage of digital representation of signals over their analog counter part is the data compression capability. Unlike an analog signal, a digital signal is just a collection of binary numbers. This makes compression easier and simpler. The benefits of compression are reflected by the saving of storage space and reduction in bandwidth requirement.

12.2   Digital Representation of Analog Signals

This section describes how an analog signal is converted to its digital counterpart. For that purpose, we also define the quantization error and ADC resolution. Broadly speaking, analog to digital conversion can be decomposed into the following three steps, which are discussed in this section.

- Sampling

- Quantization

- Encoding

12.2.1   Sampling

The first step in the process of analog to digital conversion is *sampling*. To understand the concept of sampling, let's assume that two friends are communicating with each other over the old analog telephone. The human voice can be represented as an

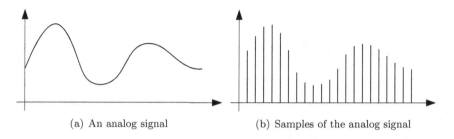

(a) An analog signal   (b) Samples of the analog signal

Figure 12.1: Sampling of an analog signal.

analog signal as shown in Figure 12.1(a). Along y-axis, we depict the voltage signal corresponding to voice, while the x-axis represents time.

In the process of signal digitization, an ADC periodically takes samples of the analog signal. Sampled signal is shown in Figure 12.1(b), which is now discrete time but continuous amplitude. These discrete time samples will be represented by numbers according to an individual sample's voltage level. The frequency at which the sampling is performed is also termed as *sampling rate*. For example, if sampling is done at 8 kHz, then 8000 samples of the analog signal are taken in one second. Thus, the time interval between two consecutive sampling points will be 1/8000 seconds or 125 μsecs. If we increase the sampling rate to 16 kHz, it means that 16,000 samples will be captured per second. In this case, two consecutive samples are separated by 1/16000 seconds or 62.5 μsecs.

To understand the impact of sampling rate, we need to look at the counter part of the ADC process. The digital-to-analog converter (DAC) does a job opposite to an ADC. During the conversion, it takes the numbers representing the digital signal and converts them back to voltages. Intuitively, the digital-to-analog converter should connect all the points captured by the analog-to-digital converter. Therefore, one can expect that the analog signal (in our example, the human voice between the two friends) reproduced won't be perfect as the DAC does not have all the points from the original analog signal. In Figure 12.2, we show this effect. In comparison to this figure, we see that the original waveform shown in Figure 12.1(a) is more continuous and *rounded* in shape.

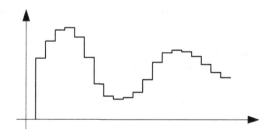

Figure 12.2: Analog signal generated by DAC from the digital signal.

So, if the DAC had more sampling points, less information would have been lost and as a result better approximation of the original analog signal would be obtained at the DAC output. Obviously, this can be achieved by increasing the sampling rate. However, capturing more samples requires more storage space to store the digital data. For example, an analog-to-digital conversion with a 16 kHz sampling rate will capture twice the samples from the original waveform as compared to the sampling performed at 8 kHz. This will generate twice the data.

We encounter a dilemma here: a high sampling rate generates a waveform close to perfection at the output of the DAC, but that requires a lot of storage space to hold the data. On the other hand, if the sampling rate is too low, the output quality will be bad. There is a trade-off between quality of the signal and the storage requirement. How can we determine the best sampling rate for an analog-to-digital conversion so that a good balance between storage and quality can be maintained? The Nyquist theorem provides answer to this question. According to Nyquist, the sampling rate must be at least two times the value of the highest frequency you want to capture. This will be different for different systems. For example, over the analog phone system human voice up to 4 kHz is transmitted and a sampling rate of 8 kHz is employed. Similarly, when the human ear listens to music sounds with frequencies up to 20 kHz, a sampling rate of at least 40,000 Hz is chosen.

12.2.2   Quantization

An analog signal is converted to a discrete time signal when samples are taken from it. Each sample is then represented by a numerical value according to its amplitude level. The ADC *resolution* is specified by the number of bits used to represent the analog value. While performing this conversion, an ADC divides the total signal span between its maximum and minimum values into m distinct levels. Then we encode the analog sample by one of the available m levels. For an 8-bit ADC, there will be $2^8 = 256$ levels. For unipolar signals, levels range from 0 to 255 while for bipolar signals the values are from -128 to 127. Similarly, if we have a 12-bit ADC, then it can encode an analog signal using $2^{12} = 4095$ different levels.

Considering a too small value for the number of levels can result in the same digital representation for two sampling points close to each other. Thus, generating a signal of poor quality at DAC's output. Using a too high value for the number of levels generates a good quality signal but requires more storage space and also complicates the ADC design. So, a 16-bit representation will require twice the storage space as compared to an 8-bit representation, but the quality will be far better.

Now assume that the analog signal can take any values between 0 V to 3 V. When we use an 8-bit ADC for this signal, an arbitrary signal sample is mapped to one of the 256 different levels. The process of mapping a signal sample to the nearest level is called *quantization*, which introduces *quantization noise*. A change from one level to the next level requires a change of $\Delta = 3/256 = 11.72$ mV in the analog signal. When a signal sample lies exactly at the mid point of two levels, it is approximated by one of those two levels. In this case, the approximation of the sampled value by the nearest level introduces maximum quantization noise, which is equal to $\Delta/2$ and is 5.86 mV for above mentioned case. If the ADC resolution is changed from 8-bit to 9-bit (for the analog signal between 0 to 3 V) the maximum quantization noise is reduced from 5.86 mV to 2.93 mV. In other words, increasing ADC resolution by one bit reduces the quantization noise by one-half or equivalently by -6 dB. In general, the signal to quantization noise ratio (SQNR) can be used to measure the signal quality after analog to digital conversion. The general expression for SQNR for is given by (12.1).

$$SQNR = 10\log_{10}\frac{P_s}{P_q} \quad \text{dB} \tag{12.1}$$

In (12.1), P_s and P_q, respectively, represent the signal and noise powers. Since quantization noise is uniformly distributed in the interval $[-\Delta/2, +\Delta/2]$, the corresponding noise power can be evaluated by $P_q = \frac{\Delta^2}{12}$. Now let us consider a sinusoidal signal, for which peak-to-peak swing has been normalized to unity. The power of this normalized sinusoidal signal is given by 1/8. For this signal the SQNR is given by (12.2).

$$SQNR = 1.76 + 6.02N \quad \text{dB}, \tag{12.2}$$

where N is the number of bits. The tolerable noise level can be different for different applications. The desired noise level can help in deciding about the number of bits required by an ADC. Higher the SNR, the better would be signal quality. A 12-bit ADC will provide a higher SNR as compared to an 8-bit ADC. These days, 24-bit resolution ADCs are also in use. The storage space required for storing the data generated by an ADC can be calculated by using the sampling rate and resolution. Consider our phone system example. It uses an 8000 Hz sampling rate and each sample is represented with 8 bits. In case, a phone conversation needs to be recorded in raw form (i.e., no compression technique is used), then the space of 8000 bytes per second is required.

12.2.3 Encoding

Once the samples are quantized, they can be encoded to the corresponding level values using binary codes. For instance, when a sample is quantized to level 47 (using an 8-bit ADC), the corresponding binary coded value of this sample is 00101111. It is not necessary to always use binary codes to encode quantized samples, rather any arbitrary encoding can also be used. However, most of the ADCs use binary encoding in the process of analog to digital conversion.

Example 12.1 (Evaluating ADC resolution for a given signal quality.). It is required to select an ADC for converting a sinusoidal signal of 0-3 V peak-to-peak value. The SQNR for the signal should be at least 90 dB. What should be the ADC resolution?

Using (12.2) the required ADC resolution for 90 dB SQNR is 15 bit.

12.3   ADC Types

So far our treatment about ADC has been similar to a black box as shown in Figure 12.3. V_{in} is the analog input and D_0 through D_n are the digital outputs. What is inside the box, is the question that we answer in this section. There are several ways to perform analog-to-digital conversion. Broadly, the ADC design can be divided into four groups.

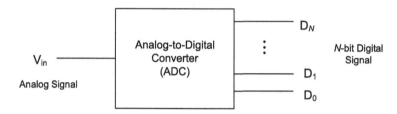

Figure 12.3: The ADC block diagram.

- Parallel design,

- Digital-to-analog converter (DAC) based design,

- Integrator based design,

- Sigma delta design.

12.3.1   Parallel Design

Figure 12.4 shows a parallel ADC also known as *flash* ADC. The main components include a number of operational amplifiers (op-amps) working as voltage comparators and a priority encoder. A flash ADC compares the voltage of an input analog signal V_{in} against a reference voltage V_{ref} and generates the output accordingly. The ADC shown in Figure 12.4 is a 3-bit ADC and the V_{ref} is set to 5 V. The comparisons are done through op-amps, while all the resistors are of same value. Each digital number represents $5/2^3 = 625$ mV. When the analog input is 0 V, it will generate 000, 0.625 V will be equal to 001, 1.250 V will be equal to 010 and so on, up to 5 V, for which the digital equivalent is 111.

The maximum value of the analog signal should not be greater than the reference voltage. For example, if the reference voltage is 5 V, then an analog signal with a peak value of 5 V should be applied. Flash ADC has a simple design and is the fastest ADC type available but requires $2^N - 1$ comparators, where N is the number of output bits. For example, for an 8-bit flash ADC, 255 comparators are needed while for a 16-bit flash ADC, 65,535 comparators are required. The priority encoder can be implemented as a single chip (e.g., 74148, which is 3-to-8 priority encoder) or with the help of XOR gates along with diodes and resistors.

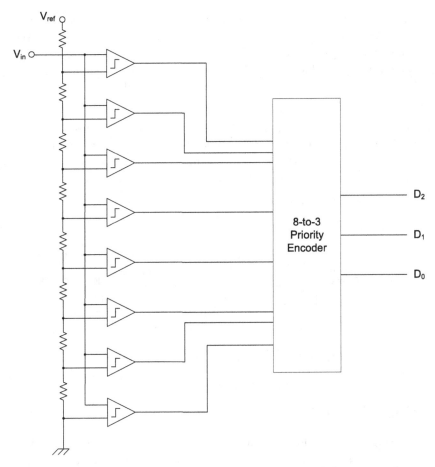

Figure 12.4: The flash type ADC. All resistors are of the same value.

12.3.2 DAC Based ADC Designs

Now, we describe an ADC design that employs a digital-to-analog converter in its comparison circuit. Two well-known DAC based ADC designs are listed below.

- Ramp counter based ADC
- Successive approximation based ADC.

Ramp Counter Based ADC

Ramp counter based ADC, also known as digital ramp ADC, is shown in Figure 12.5. It consists of a voltage comparator, counter and a DAC. The counter has a control input, which turns on the counter when activated, while the counter is stopped when it is deactivated. The START pulse starts the conversion of an analog input signal. The END signal indicates the end of conversion and digital equivalent for the signal sample is available at D_0 through D_n outputs. The CLOCK signal is used to increment the counter.

While performing conversion, the ramp counter increments at each clock pulse till the analog equivalent of counter value, found at the output of the DAC, matches the value of the analog input signal. In other words, the counter starts counting from 0 toward the maximum possible value of $2^N - 1$, until it finds the correct digital equivalent for V_{in}. At this point, the value on the counter is the digital equivalent of the analog signal.

The main issue with this technique is that it is really slow. At most, it may require up to $2^N - 1$ clock cycles to convert a sample. For instance, an 8-bit ADC may take up to 255 clock cycles to convert a single sample while 65,535 clock cycles can be required, in worst case, by a 16-bit ADC to perform the conversion.

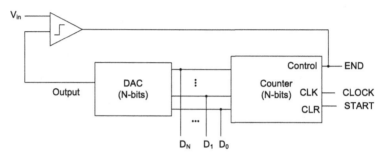

Figure 12.5: The ramp counter based ADC.

Successive Approximation Based ADC

Successive approximation based ADC is one of the most widely used DAC based ADC designs and is shown in Figure 12.6. The main building blocks of successive approximation ADC include an output buffer, successive approximation register (SAR) and a DAC. The buffer is responsible to make the digital data available, while the DAC converter is processing the next sample. The use of START, CLOCK and END control signals in successive approximation ADC is same as that of ramp counter ADC.

To understand the working of a successive approximation ADC, let's take the example of an 8-bit ADC. The conversion starts by setting the most significant bit D_7 of the 8-bit ADC output. A comparison between the V_{in} and DAC output (corresponding to D_7 set to 1) is made in order to determine whether the sample value is greater or lower than $2^7 = 128$. The op-amp output will update the control unit if this bit should remain set to 1 or should be updated to 0. Then D_6 output is set to 1 and the comparison made by the op amp reveals whether or not this bit should remain set or not and updates the control unit accordingly. This procedure continues until we reach the least significant bit D_0, at which point the conversion has completed.

Successive approximation ADC is more efficient as compared to ramp counter based ADC in terms of speed. It requires N clock cycles to find the correct digital value for a given sample, where N is the number of bits used. For example, an 8-bit ADC will be able to find the digital value for a sample in 8 clock cycles while for a 16-bit ADC,

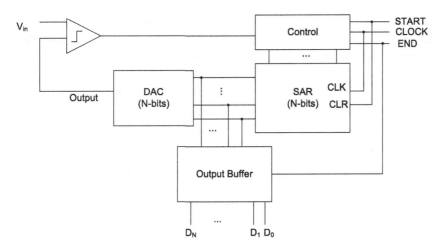

Figure 12.6: The successive approximation based ADC design.

it will take up to 16 clock cycles. Successive approximation ADCs are also suitable for converting multiple analog signals using multiplexing.

12.3.3 Integrator Based ADC Designs

An ADC converter of this type can achieve high resolution at the expense of speed and can provide high resolution at moderate sampling rates. There are two widely used integrator based techniques for an ADC.

- Single slope ADC

- Dual slope ADC

Single Slope ADC

Single slope ADC block diagram is shown in Figure 12.7. A comparison of this circuit against the ramp counter ADC reveals the difference between the two circuits. Instead of using a DAC for generating the comparison voltage, a single-slope ADC uses an integrator. An integrator consists of a capacitor, a resistor and an operational amplifier. For a constant input signal, the integrator produces a ramp signal at its output. The integrator output is zero at reset and increases linearly with a constant slope that is proportional to $-V_{ref}$. We use -ve V_{ref} due to inverted output from integrating amplifier. The integrator output, not only requires its input $-V_{ref}$ to be stable, but is also sensitive to the capacitor and resistor tolerance values. In addition, any change in the capacitor value overtime leads to drift in the ADC output.

When the analog input sample is applied at the comparator input V_{in}, it is compared against the integrator output. Since the integrator is reset at the beginning of conversion, its output is zero at that instant. As a result the comparator output is low at the start of conversion and counter starts counting from 0 to $2^N - 1$, where N is the number of bits of the ADC. When the integrator output becomes equal to the analog input, V_{in}, the comparator output goes high and stores the last value produced by the counter to the output buffer. This is the digital equivalent value of the analog sample being converted. At the same time, the comparator output resets the counter and the integrator, allowing to start the conversion for the next sample.

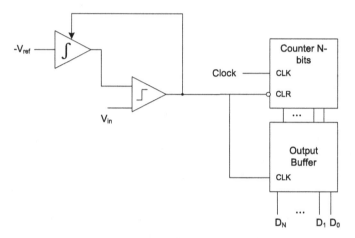

Figure 12.7: The single slope integrator based ADC.

The presence of an output buffer allows the ADC to start the new sample conversion while the last converted value is available. This is like the successive approximation ADC. However, like ramp counter design, it is still based on a counter and suffers from the same speed issue. The single-slope ADC may require up to $2^N - 1$ clock cycles to convert each sample. This means that in worst case, up to 255 clock cycles may be required for the conversion of a single sample in the case of an 8-bit ADC.

Dual Slope ADC

Dual slope ADC provides an improvement over single slope ADC by making the analog to digital conversion result insensitive to any errors in the component values. In addition, it overcomes the sensitivity, of single slope ADC, to any drift in the component values over long time durations. Figure 12.8 shows the block diagram of a dual-slope ADC.

The dual-slope ADC starts the conversion by first connecting V_{in} signal, to the integrator input using an analog switch, for fixed time interval, t_1. With this connectivity, the integrator generates a ramp signal output with a slope proportional to V_{in}. At the end of fixed time interval, the analog switch is moved to allow $-V_{ref}$ to be applied at the input of the integrator. As soon as -V_{ref} is connected at the integrator input, the binary counter is also started to count the clock cycles by the control logic. Since V_{ref} is negative, the slope of the ramp signal during this phase is also negative. As a result, the integrator output reduces and becomes zero after an arbitrary time interval t_2 and triggers the latching of counter value to the output buffer and resets the counter to start the conversion for next sample. At this point, the output buffer contains the binary equivalent (digital) value of the current sample. Increasing the frequency of the clock signal that is fed to the binary counter, one can increase the resolution of dual slope ADC.

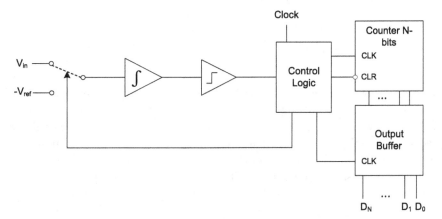

Figure 12.8: The dual slope ADC block diagram.

The waveform generated at the output of the integrator depicting this scenario is shown in Figure 12.9. Here, t_1 is fixed, while duration of t_2 is proportional to the value of V_{in}. The relationship among different parameters is given by

$$t_2 = t_1 \frac{V_{in}}{V_{ref}}.$$ (12.3)

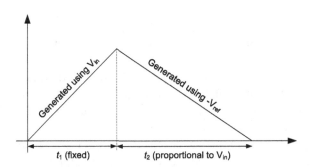

Figure 12.9: The waveform at the integrator output for dual slope ADC.

12.3.4 Sigma-Delta ADC

The sigma-delta ($\sum \Delta$) ADC, which is also known as delta-sigma, 1-bit ADC or oversampling ADC employs a different conversion strategy. The sigma-delta ADCs are primarily suitable for applications requiring low bandwidth and high resolution. Figure 12.10 shows its block diagram, from which we can see that it consists of two main components, namely an analog modulator and a digital filter. The purpose of the former is to take the analog signal and convert it into a stream of bits while the latter converts the output of the modulator into a digital number representing the equivalent of the input analog signal.

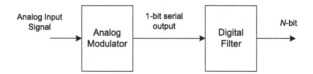

Figure 12.10: Block diagram of a sigma-delta ADC.

Figure 12.11 shows the basic sigma-delta modulator circuit. The input analog signal and the feedback signal from 1-bit DAC are combined to evaluate the difference between them, which can be considered as an error signal. This can be easily accomplished using a difference amplifier as well. This error signal is then applied to the input of an integrator. The output of the integrator is continuously compared against a reference value using a comparator. The comparator output is converted to 1-bit serial data stream at a certain high frequency using D-type flip-flop.

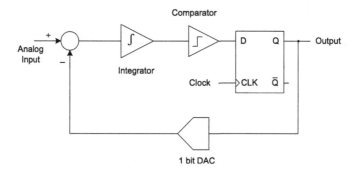

Figure 12.11: Sigma-delta ADC analog modulator basic design.

The serial data stream at the flip-flop output is converted to an analog equivalent signal using 1-bit DAC. The 1-bit DAC output is used as feedback signal as mentioned earlier. The 1-bit DAC simply converts the 0s or 1s generated at the flip-flop output to either positive or negative voltage signals, which are added to the input signal to generate the error signal as input to the integrator. This feedback based system is termed as first order sigma-delta modulator. The analog input signal average voltage is reflected in the bit stream average. The clock rate for the flip-flop is selected K times higher than the actual sampling rate. The parameter K is also referred to as oversampling ratio. It is important to mention that oversampling has an additional advantage of relaxing the analog anti-aliasing filter requirements.

Oversampling distributes the quantization noise over a larger bandwidth. This is also termed as noise shaping. Now using a low pass filter to recover the wanted signal, which has narrow bandwidth, we are able to remove a large part of the quantization noise in the high frequency band, which improves the SQNR. This is the job of the digital filtering stage. The SQNR for sigma-delta ADC can be further improved by using a higher order analog modulator. The sigma-delta ADCs are not suitable for analog multiplexing, due to digital filter response adaptation requirement.

12.4 ADC Details on TM4C123 Microcontroller

The TM4C123 microcontroller contains two integrated ADC modules, which share 12 analog inputs and can be operated completely independent of each other. The two modules are successive approximation based ADCs. Each of the ADC modules has the following features.

- Maximum sampling rate of 1 Msps

- 12 bit sample resolution

- 12 external analog input channels and one internal temperature sensor

- Sampling can be started using one of the multiple triggering sources

- Single ended or differential analog inputs

- Hardware sample averaging capability up to 64 samples (using multiple samples from same analog input)

- Four programmable sample sequencers to control the sampling sequence of multiple analog inputs without processor intervention.

Usually an analog to digital conversion in a microcontroller requires processor intervention to handle each conversion. However, sample sequencer based architecture of TM4C123 microcontroller enables the software to configure up to four different sampling sequences with separate custom configurations. The use of sample sequencers in each ADC module makes the sampling of multiple analog inputs quite flexible and also reduces the processor overhead. The ADC module in TM4C123 has four sample sequencers, namely SS0, SS1, SS2, and SS3. The sample sequencer SS0 allows sampling of maximum 8 analog signals, while SS1, SS2, and SS3 allow sampling of 4, 4, and 1 analog signals, respectively. Each sequencer has an associated FIFO to store the conversion result of that sequencer. The FIFO size is same as the sequencer size.

The 12 analog inputs, labeled as ANI_0 to ANI_11, are multiplexed to GPIO pins. The mapping of these analog inputs to the corresponding GPIO pins is tabulated in Table 12.1.

Table 12.1: Mapping of analog inputs to the corresponding GPIO port pins.

Analog input channel	GPIO pin assignment
ANI_0	Port E pin 3 (PE3).
ANI_1	Port E pin 2 (PE2).
ANI_2	Port E pin 1 (PE1).
ANI_3	Port E pin 0 (PE0).
ANI_4	Port D pin 3 (PD3).
ANI_5	Port D pin 2 (PD2).
ANI_6	Port D pin 1 (PD1).
ANI_7	Port D pin 0 (PD0).
ANI_8	Port E pin 5 (PE5).
ANI_9	Port E pin 4 (PE4).
ANI_10	Port B pin 4 (PB4).
ANI_11	Port B pin 5 (PB5).

The ADC module in TM4C123 combines four sample sequencers with actual analog to digital converter as shown in Figure 12.12. The sequencer inputs can be configured for any arbitrary analog channels and it is not necessary to configure all the sequencer input channels. The sequencer outputs are multiplexed to the ADC input and have configurable priorities. In addition, each sequencer can have a different triggering source. The ADC block, as shown in Figure 12.12, is responsible for sampling rate as well as hardware averaging control. All the analog inputs connected to a specific sequencer will have same sampling rate. It is also important to mention that hardware averaging control is an ADC attribute that, once enabled becomes applicable to all the sequencers which are enabled. The digital output, after conversion is stored to the associated FIFOs corresponding to each sequencer.

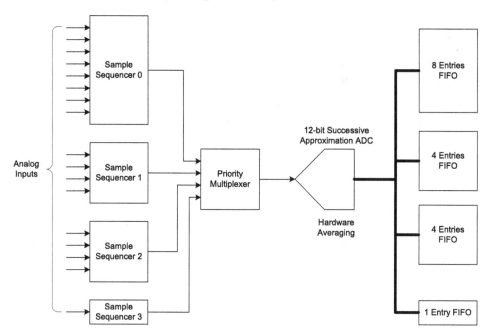

Figure 12.12: ADC block diagram showing the four sequencers and corresponding FIFOs.

The ADC module configuration involves two key configuration steps. One configuration step is related to sample sequencer, while second step is responsible for ADC module configuration. In addition, just like any other peripheral, ADC module also requires module clock, GPIO and interrupt related configurations. Below we list the configuration steps for an ADC module.

- Step 1: ADC clock gating control configuration

- Step 2: GPIO configuration as ADC input

- Step 3: ADC module configuration

- Step 4: ADC sample sequencer configuration

- Step 5: ADC interrupt configuration

12.4.1 ADC Clock Gating Control Configuration

There are two ADC peripheral modules integrated in TM4C123 microcontroller. The clock for each ADC module can be configured separately using RCGC_ADC_R register. The register address information along with brief description is provided in Table 12.2. Bit 0 of RCGC_ADC_R correspond to ADC0 module, while bit 1 correspond to ADC1. Writing 1 to these bits, enables and provides clock to corresponding ADC module. After enabling the clock to an ADC module, there must be a 3 clock cycle delay before accessing the ADC registers. Example 12.2 implements the clock enabling for ADC module 0.

Table 12.2: ADC clock gating control register address and usage.

Register label	Address	Reset value	Brief description
RCGC_ADC_R	0x400FE638	0x00000000	ADC clock gating control.

Example 12.2 (Enabling clock to ADC module 0.).
```
#define RCGC_ADC_R          *((volatile unsigned long *)0x400FE638)
#define ADC0_CLOCK_ENABLE   0x00000001 // ADC0 Clock Gating Control

// Enable the clock for ADC0
  RCGC_ADC_R |= ADC0_CLOCK_ENABLE;
```

12.4.2 GPIO Configuration as ADC Input

To configure analog functionality on a GPIO pin, the digital function should be disabled by clearing the corresponding bits in GPIO_DEN_R register. In addition, the corresponding bit in the GPIO_AMSEL_R register should be set to enable the analog functionality on that pin. The GPIO_AMSEL_R register offset address with respect to GPIO port base address is 0x528. Bit 0 to 7 of this register correspond to port pins 0 to 7, which can be configured individually for analog functionality. The mapping of different analog input channels to various GPIO port pins is summarized in Tables 8.2 and 8.3.

12.4.3 ADC Module Configuration

An ADC module requires multiple configurations, before it can be used for digitization of the desired analog signal. We have decomposed ADC related configurations into two groups. First set of configurations is performed for the entire ADC module and is discussed in this subsection. The second set of configurations are ADC sequencer related and are discussed in the following subsection. The key configuration steps for an ADC module are listed below.

- ADC module sampling rate configuration is performed to select the desired sampling rate.

- The triggering source, for each of the four sequencers in an ADC module, can be configured independently. This capability allows different sampling rates for each of the sequencers in an ADC module. Different sources that can be configured to trigger a sample sequencer, include the processor, analog comparators, timers, PWM unit or GPIO.

- It is possible to assign priorities to each of the four sample sequencers in an ADC module. However, same priority cannot be assigned to two different sequencers. The priority assignment allows to handle cases where multiple sequences are triggered by the same triggering source or they are triggered simultaneously.

- Hardware sample averaging can be enabled to increase the ADC accuracy or resolution. When it is enabled all the samples for different channels are over sampled by the same factor. Enabling hardware averaging reduces the effective sampling rate by the averaging factor. Hardware averaging is a global configuration and can not be enabled for a selected sequencer.

- Analog voltage reference configuration selects the positive and negative voltage reference signals for analog conversions.

The ADC registers used to perform the above mentioned configuration steps are listed in Table 12.3, while the base address information for the two ADC modules is provided in Table 12.4. Using ADC register definitions, the implementation details regarding each of these configuration steps are discussed next.

Table 12.3: ADC module configuration register labels, addresses, reset value, and brief description.

Register label	Offset	Reset value	Brief description
ADC_ACTIVE_SS_R	0x000	0x00000000	ADC active sample sequencer register.
ADC_TRIGGER_MUX_R	0x014	0x00000000	Triggering source multiplexer register.
ADC_PWM_TRIGGER_R	0x01C	0x00000000	PWM trigger source select register.
ADC_PRI_SS_R	0x020	0x00003210	ADC sample sequencer priority register.
ADC_PROC_INIT_SS_R	0x028	-	Processor initiate sequencer for sampling.
ADC_SAMPLE_AVE_R	0x030	0x00000000	Sample averaging control register.
ADC_CONTROL_R	0x038	0x00000000	ADC control register.
ADC_PERI_CONFIG_R	0xFC4	0x00000000	ADC sample rate configuration register.
ADC_CLK_CONFIG_R	0xFC8	0x00000000	Clock source configuration register.

Table 12.4: Base addresses for the two ADC modules.

Start address	End address	Description
0x40038000	0x40038FFF	ADC 0 module
0x40039000	0x40039FFF	ADC 1 module

ADC Clock Source and Sampling Rate Configuration

ADC clock source selection can be configured using ADC_CLK_CONFIG_R register. The least significant four bits (bit 0 to 3) of this register can be configured with a value of 0 or 1. When configured with a value of 0, the system clock is used as the clock source and a signal of 16 MHz frequency is derived from the system clock.

Configuring a value of 1 selects the precision internal oscillator (PIOSC) running at 16 MHz as the clock source for ADC.

The ADC sampling rate is one of the most critical parameters and is configured using ADC peripheral configuration register, ADC_PERI_CONFIG_R. The bit field allocations and their descriptions for ADC_PERI_CONFIG_R register are provided in Table 12.5.

Table 12.5: ADC peripheral configuration register and its bit field descriptions.

31	23	15	7	0
		reserved		Sample Rate

Bit field	Description
Sample Rate	This bit field configures the ADC sampling rate to four different values. Setting a value of 0x1 configures 125 ksps sample rate. Also a value of 0x3, 0x5 and 0x7 configures a corresponding sampling rate of 250 ksps, 500 ksps and 1000 ksps, respectively. The default value is 1 Msps.

When an attempt, to configure ADC to a sample rate other than supported by the device, is made then ADC either continues with the default sample rate or with the recently configured sample rate value. Example 12.3 describes the ADC configuration for a sampling rate of 500 ksps.

Example 12.3 (Configuring ADC sampling rate to 500 ksps.).

```
#define ADC_PERI_CONFIG_R   *((volatile unsigned long *)0x40038FC4)
#define SAMPLE_RATE_500KSPS 0x00000005 // 500ksps sample rate

// Configure sampling rate to 500 Ksps
ADC_PERI_CONFIG_R   |= SAMPLE_RATE_500KSPS;
```

ADC Sequencer Trigger Source Configuration

Each ADC sequencer can be triggered for sampling by different sources, including processor, GPIO, timer, PWM, or comparator. The triggering source selection is configured using ADC_TRIGGER_MUX_R register. Using two different triggering sources at different rates for two sequencers allows to sample analog inputs (connected to those sequencers) at different sampling rates. The bit field allocations and their descriptions for ADC_TRIGGER_MUX_R register are provided in Table 12.6.

As mentioned above, there are multiple sources that can be used to trigger the sampling of a sequencer. When the processor is configured as the sequencer trigger source, it can trigger a sequencer for sampling using ADC_PROC_INIT_SS_R register. The least significant four bits of this register are used to trigger the corresponding sequencer for sampling by the processor. When PWM is used as the triggering source

Table 12.6: ADC sample sequencer triggering source configuration register and its bit field descriptions.

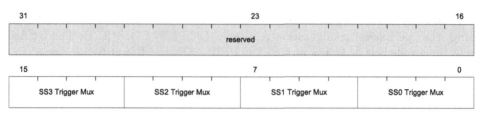

Bit field	Description
SS0 Trigger Mux, SS1 Trigger Mux, SS2 Trigger Mux, SS3 Trigger Mux	Each of these bit fields selects the triggering source for the corresponding sample sequencer. The configurable values for these bit fields and the resulting selections for ADC triggering are listed below. • Configuring a value of 0 makes the processor as ADC triggering source. • Configuring a value of 1 or 2 makes the analog comparator 0 or 1 as the trigger source using analog comparator control configuration. • Configuring a value of 4 makes the GPIO pin as the trigger source. • Configuring a value of 5 makes the timer as the trigger source. To use timer as trigger source timer output trigger enable bit must be set. • Configuring a value of 6, 7, 8 or 9 makes the PWM generator 0, 1, 2 or 3 as trigger source. • Configuring a value of 0xF enables continuous sampling mode, which does not require triggering.

then ADC_PWM_TRIGGER_R register is used to select the appropriate PWM module as the triggering source.

ADC Sequencer Priority Configuration

When multiple sequencers are triggered simultaneously, then the order of sampling among the sequencers is decided based on the assigned priority. The active sequencers should be assigned a unique priority using ADC sample sequencer priority configuration register, ADC_PRI_SS_R. The bit field allocations and their descriptions for ADC_PRI_SS_R register are provided in Table 12.7.

ADC Hardware Averaging Configuration

Hardware averaging functionality for improved resolution, can be configured using ADC_SAMPLE_AVE_R register. When hardware averaging is enabled, the ADC performs over sampling to average multiple samples. The bit field allocations and their descriptions for ADC_SAMPLE_AVE_R register are provided in Table 12.8.

Table 12.7: ADC sample sequencer priority configuration register and its bit field descriptions.

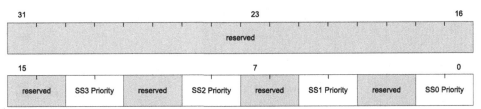

Bit field	Description
SS0 Priority, SS1 Priority, SS2 Priority, SS3 Priority	All the sequencer priority bit fields follow a common definition. Configuring a value of 0 makes the corresponding sequencer highest priority, while a value of 3 corresponds to lowest priority.

Table 12.8: ADC hardware averaging configuration register and its bit field descriptions.

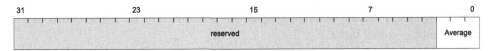

Bit field	Description
Average	This bit field specifies the magnitude of hardware averaging, which will be applied to the ADC samples. When averaging is enabled the ADC hardware over samples the signals. Configuring a value of 0 turns off averaging. Configuring a value of 1 over samples by a factor of 2 and then averages. Similarly configuring values of 2, 3, 4, 5 or 6 corresponds to over sampling by factors of 4, 8, 16, 32 or 64 and then averaging those samples.

ADC Control Configuration

Two different configurations are performed as part of ADC control. One configuration is related to the choice of ADC voltage reference source using ADC_CONTROL_R register. There is only one possible configuration which uses the voltages at the analog supply pins (VddA and GndA) and can be configured using bit 0 of ADC_CONTROL_R register. The second ADC control is used to enable or disable each of the sample sequencers individually using ADC_ACTIVE_SS_R register. The bit field allocations and their descriptions for ADC_ACTIVE_SS_R register are provided in Table 12.9.

12.4.4   ADC Sample Sequencer Configuration

Each sample sequencer can be configured independently using its associated set of registers. The set of configurations, which can be performed independently for each sample sequencer, are listed below.

- The selection of an analog input channel for each sequencer is completely independent and can be performed separately. It is also possible to configure same

Table 12.9: ADC active sample sequencer configuration register and its bit field descriptions.

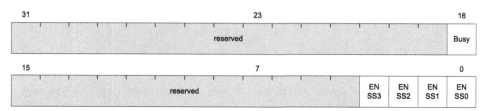

Bit field	Description
Busy	This bit indicates the ADC module status. When set to 1 it indicates that the ADC module is busy.
EN SS0, EN SS1, EN SS2, EN SS3	Each of these bit fields is used to enable or disable the corresponding sample sequencer. Writing a value of 1 enables the sequencer.

analog input for multiple channels of the same sequencer.

- Each sample sequencer analog input can be configured for single ended or differential type of input.

- When using a sequencer, it is not necessary to use all of its channels. Rather any arbitrary channel, of the sequencer, can be configured for end of sequence. For instance, any of the channels from 0 to 7 can be configured as end of sequence for sequencer 0. When a sequencer is triggered, sampling is started at the programmed sampling rate and continues until end of sequence has been reached for that sequencer.

Before modifying ADC configuration parameters, it is a good practice to disable the appropriate sample sequencer. The ADC registers used for sample sequencer configuration are listed in Table 12.10 and their use for different sequencer related configurations will be discussed below.

ADC Sequencer Input Multiplexing Configuration

Analog inputs to ADC sequencer can be configured using ADC sample sequencer input multiplexer register (ADC_SSx_IN_MUX_R, where x represents the sequencer number and can be 0 to 3). The bit field allocations and their descriptions for sequencer 0 register ADC_SS0_IN_MUX_R are provided in Table 12.11. The bit field definitions in Table 12.11 are equally applicable for other sequencer registers as well, with the exception that only four input channels can be configured for sequencers 1 and 2, while sequencer 3 can have one analog input.

ADC Sequencer Control Configuration

Each sample sequencer in an ADC module has a separate sequencer control configuration register ADC_SSx_CONTROL_R for sequencer x. We discuss the control configu-

Table 12.10: ADC sample sequencer related configuration register labels, addresses, reset value, and brief description.

Register label	Offset	Reset value	Brief description
ADC_SS0_IN_MUX_R	0x040	0x00000000	SS0 analog input selection register.
ADC_SS0_CONTROL_R	0x044	0x00000000	SS0 control register.
ADC_SS0_FIFO_DATA_R	0x048	-	SS0 FIFO data register.
ADC_SS0_FIFO_STATUS_R	0x04C	0x00000100	SS0 FIFO status register.
ADC_SS1_IN_MUX_R	0x060	0x00000000	SS1 analog input selection register.
ADC_SS1_CONTROL_R	0x064	0x00000000	SS1 control register.
ADC_SS1_FIFO_DATA_R	0x068	-	SS1 FIFO data register.
ADC_SS1_FIFO_STATUS_R	0x06C	0x00000100	SS1 FIFO status register.
ADC_SS2_IN_MUX_R	0x080	0x00000000	SS2 analog input selection register.
ADC_SS2_CONTROL_R	0x084	0x00000000	SS2 control register.
ADC_SS2_FIFO_DATA_R	0x088	-	SS2 FIFO data register.
ADC_SS2_FIFO_STATUS_R	0x08C	0x00000100	SS2 FIFO status register.
ADC_SS3_IN_MUX_R	0x0A0	0x00000000	SS3 analog input selection register.
ADC_SS3_CONTROL_R	0x0A4	0x00000000	SS3 control register.
ADC_SS3_FIFO_DATA_R	0x0A8	-	SS3 FIFO data register.
ADC_SS3_FIFO_STATUS_R	0x0AC	0x00000100	SS3 FIFO status register.

Table 12.11: ADC sample sequencer input multiplexing configuration register (for sequencer 0) and its bit field descriptions.

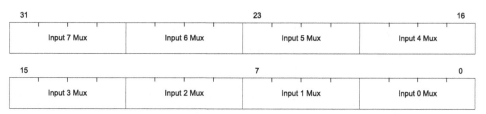

Bit field	Description
Input7 Mux - Input0 Mux	Each of these bit fields is used to configure an analog input channel for the corresponding sequencer. Configuring a value of 2 to bit field Input0 Mux, assigns analog input channel 2 as the first input to the sequencer 0. The possible values that can be configured are 0 to 11, which correspond to the 12 analog input channels.

ration for sequencer 1 here, which is equally applicable for the other sequencers as well. The only difference is that sequencer 0 will have eight such configuration sets, while sequencer 3 will have only one such configuration set. The bit field allocations and their descriptions for sample sequencer 1 control register, ADC_SS1_CONTROL_R, are provided in Table 12.12.

12.4.5 ADC Interrupt Configuration

Interrupt configuration for each analog input channel can be performed separately as mentioned in the description of ADC sequencer control configuration register. In addition, each sequencer can be a source of interrupt as well, which can be configured using the ADC interrupt configuration registers listed in Table 12.13. The least significant four bits of these registers are used for interrupt masking (using ADC_INT_MASK_R),

Table 12.12: ADC sample sequencer 1 control configuration register and its bit field descriptions.

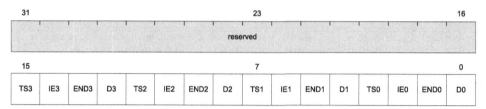

Bit field	Description
TS3-TS0	Each of these bit fields are used to configure the internal temperature sensor as the analog input source for the corresponding analog input to the sequencer.
IE3-IE0	Setting any of these bit fields to 1, enables the interrupt for the corresponding analog input. When enabled, the interrupt condition is generated at the end of conversion for the corresponding analog input. Multiple samples can be configured to generate interrupts in one sample sequence.
END3-END0	When any of these bits is set to 1, it indicates the end of sequence.
D3-D0	Each these bit fields indicate whether the corresponding analog input is differential or not. Setting it to 1 makes the corresponding analog input to be sampled differentially.

status (using ADC_RIS_R) and clearing (using ADC_INT_STATUS_CLR_R) of interrupts from the four sequencers. The bit 0 of these registers correspond to sequencer 0 while bit 3 corresponds to sequencer 3.

Table 12.13: ADC interrupt configuration register labels, addresses, reset value, and brief description.

Register label	Offset	Reset value	Brief description
ADC_INT_MASK_R	0x008	0x00000000	Interrupt mask register.
ADC_RIS_R	0x004	0x00000000	Interrupt raw status register.
ADC_INT_STATUS_CLR_R	0x00C	0x00000000	Interrupt status and clear register.

Example 12.4 (ADC configuration for reading analog temperature sensor data.). This example uses ADC module 0 for illustration. An external temperature sensor (LM35) is connected to Port E pin 2, which is also multiplexed for analog input 1 (AIN1) for the ADC. The LM35 sensor gives 0.25V output at 25 °C and changes its output value at a rate of 10mV/°C.

```
#define SYSCTL_RCGC_ADC_R   *((volatile unsigned long *)0x400FE638)
#define SYSCTL_RCGC_GPIO_R  *((volatile unsigned long *)0x400FE608)
#define ADC0_CLOCK_ENABLE   0x00000001 // ADC0 Clock Gating Control
#define CLOCK_GPIOE         0x00000010   //Port D clock control

// ADC module 0 base address
#define ADC_BASE            0x40038000

// ADC module configuration registers
#define ADC_ACTIVE_SS_R     *(volatile long *)(ADC_BASE + 0x000)
```

```
#define ADC_INT_MASK_R          *(volatile long *)(ADC_BASE + 0x008)
#define ADC_TRIGGER_MUX_R       *(volatile long *)(ADC_BASE + 0x014)
#define ADC_PROC_INIT_SS_R      *(volatile long *)(ADC_BASE + 0x028)
#define ADC_PERI_CONFIG_R       *(volatile long *)(ADC_BASE + 0xFC4)
#define ADC_INT_STATUS_CLR_R    *(volatile long *)(ADC_BASE + 0x00C)
#define ADC_SAMPLE_AVE_R        *(volatile long *)(ADC_BASE + 0x030)

// ADC sequencer 3 configuration registers
#define ADC_SS3_IN_MUX_R        *(volatile long *)(ADC_BASE + 0x0A0)
#define ADC_SS3_CONTROL_R       *(volatile long *)(ADC_BASE + 0x0A4)
#define ADC_SS3_FIFO_DATA_R     *(volatile long *)(ADC_BASE + 0x0A8)

// GPIO Port E analog function configuration
#define GPIO_PORTE_AMSEL_R *((volatile unsigned long *)0x40024528)
#define GPIO_PORTE_DEN_R   *((volatile unsigned long *)0x4002451C)

#define NVIC_EN0_R         *((volatile unsigned long *)0xE000E100)
#define NVIC_EN0_INT17     0x00020000 //Interrupt 17 for ADC0 Seq3

// Function prototypes and variables
void ADC_Init(void);
void ADC_Start_Sampling(void);
void Delay(volatile unsigned int delay);

unsigned char Temperature = 0;

// ADC 0 interrupt service routine for sequencer 3
void ADC0Seq3_Handler(void){
unsigned int adc_data = 0;

  // Read the temperature raw value from ADC FIFO
  adc_data = (ADC_SS3_FIFO_DATA_R & 0xFFF);

  // Convert the raw value to Celsius scale
  Temperature = (unsigned char) (((adc_data * 3.3)/4096) * 100);

  // Clear the interrupt from ADC0 sequencer 3
  ADC_INT_STATUS_CLR_R |= 0x08;

}

// Intialization routine for setting up the required ports
void ADC_Init(void)
{
  // GPIO configuration for ADC 0 analog input 1
  // as a temperature sensor is connected to GPIO pin PE2
  SYSCTL_RCGC_GPIO_R |=  CLOCK_GPIOE;

  GPIO_PORTE_DEN_R &= ~(0x04);
  GPIO_PORTE_AMSEL_R |= 0x04;

  // Enable the clock for ADC0
  SYSCTL_RCGC_ADC_R |= ADC0_CLOCK_ENABLE;
  Delay(3);

  ADC_PERI_CONFIG_R |= 0x03;         // 250 Ksps

  // Select AN1 (PE2) as the analog input
  ADC_SS3_IN_MUX_R = 0x01;
```

```
    // 1st sample is end of sequence and source of interrupt
    ADC_SS3_CONTROL_R |= 0x06;

    // 16x oversampling and then averaged
    ADC_SAMPLE_AVE_R |= 0x04;

    // Unmask ADC0 sequence 3 interrupt
    ADC_INT_MASK_R |= 0x08;

    // Enable ADC0 sequencer 3 interrupt in NVIC
    NVIC_EN0_R = NVIC_EN0_INT17;

    // Configure ADC0 module for sequencer 3
    ADC_ACTIVE_SS_R = 0x00000008;
}

void ADC_Start_Sampling(void)
{
    // Processor sample sequencer initiate for sequencer 3
    ADC_PROC_INIT_SS_R |= 0x08;
}


// Application main function
int main(void){

    ADC_Init();                    // Initialize the timer

    while(1){
        ADC_Start_Sampling();      // Start sampling of a new sample
        Delay(100);
    }
}

/* This function generates the delay. */
void Delay(volatile unsigned int delay)
{
    volatile unsigned int i, j;
    for(i = 0; i < delay; i++)
    {  //introduces a delay of about 10us at 16MHz
        for(j = 0; j < 12; j++);
    }
}
```

12.5 Summary of Key Concepts

In this chapter, we have introduced the basic concepts of analog to digital conversion
and the use of ADC module available on TM4C123. Specifically, the following key
concepts were discussed.

- We first answer the question why analog interfacing is needed. As signals are
  analog in nature, they have to be converted to digital equivalent so that they
  can be processed using a digital processor.

- There is a systematic way of converting analog signals to their digital equivalent.

- The first step in the process of analog to digital conversion is sampling. In addition, the rate at which sampling is performed is an important ADC parameter.

- The minimum sampling rate for an analog signal of bandwidth B is equal to $2B$ based on Nyquist sampling theorem.

- The next step after sampling is called quantization, which is the process of mapping the analog sample magnitude to one of the nearest levels. The number of quantization levels depends on the ADC bit resolution.

- The last step in analog to digital conversion is encoding. Binary encoding is the most widely used encoding scheme.

- There are four basic types of ADCs.

- The parallel or flash type ADC can perform fastest analog to digital conversion, but its complexity increases exponentially with an increase in the number of ADC bits.

- The second type of ADC uses digital-to-analog converter (DAC) based design. The successive approximation based ADC is one of the most widely used approaches.

- The third type of ADC is an integrator based design. This type of ADC can provide improved resolution but at the expense of slow sampling rate.

- The fourth type of ADC is based on sigma delta design, which can provide very high resolution and also provides improved signal to quantization noise ratio.

- The TM4C123 microcontroller contains two integrated ADC modules with 12 bit sample resolution and maximum sampling rate of 1 Msps.

- There are typically 5 steps in configuring an ADC module. ADC clock gating control configuration, GPIO configuration as ADC pin, ADC module configuration, ADC sample sequencer configuration, and ADC interrupt configuration.

- ADC sample sequencer reduces the processor overhead by taking care of sample prioritization and sequencing in case of multiple analog inputs.

Review Questions

Question 12.1. What parameters of analog to digital converter (ADC) are important when considering conversion of analog signals to their digital equivalent?

Question 12.2. If an analog signal of bandwidth B is to be converted to its digital equivalent, then what is the minimum sampling rate that should be used?

Question 12.3. What is quantization noise? How can it be reduced?

Question 12.4. If we double the number of quantization levels, then how much reduction, in the quantization noise, can be achieved?

Question 12.5. What type of ADC is suitable for fast analog to digital conversions?

Question 12.6. How does dual slope ADC improve the conversion compared to single slope ADC?

Question 12.7. What is the key benefit of delta-sigma analog to digital converter?

Question 12.8. What type of ADCs are suitable for converting multiple analog input signals using analog multiplexing?

Question 12.9. Is it possible to control the order in which analog signals are converted to their digital counterpart using sample sequencer of TM4C123 ADC module?

Question 12.10. How can we assign priorities to different analog to digital conversions?

Question 12.11. Is it possible to assign the same priority to two different sample sequencers of ADC module?

Question 12.12. What is the advantage of oversampling and sample averaging feature provided by the TM4C123 ADC module?

Question 12.13. Can we enable oversampling selectively for some of the sample sequencers?

Exercises

Exercise 12.1. Find a suitable ADC type for each of the following requirements.

- Fast sampling rate and low resolution
- Slow sampling rate but high resolution
- Slow sampling rate and moderate resolution

Exercise 12.2. Consider a bipolar tone signal taking values between -4V and +6V. If a 12-bit ADC is used for its digitization, what will be the resulting signal to quantization noise ratio?

Exercise 12.3. Write a C program that configures the ADC sampling rate to 500 ksps.

Exercise 12.4. Suppose we are interested in data acquisition from two different sensors, a temperature sensor and a light sensor. The data from the temperature sensor should be sampled at a rate of 1 Hz, while the sampling rate from the light sensor should be 100 Hz. Configure ADC module 0, using two of its sequencers, to implement different sampling rates for each sequencer. Clearly mention your choice of triggering source for each sequencer.

Appendices

Appendix A

System Startup and Configuration

A.1 System Startup

The assembly code for the system startup is provided below, which should be used with all C programs. The assembly code lines are numbered in the Listing A.1 for easy reference. The following key components are implemented by the startup file.

- Allocating memory area segment for Stack. This is implemented by the code listed in lines 19-22 of the system startup file provided in Listing A.1.

- Similar to stack memory region allocation, a memory region for heap can be allocated using the code listed in lines 27-31.

- The interrupt vector table for all the exceptions and peripheral interrupts is implemented by the code listed in lines 48-204.

- Following the interrupt vector table, the first interrupt service routine labeled, Reset_Handler, corresponding to reset interrupt is implemented using code listed in lines 210-235.

- Following the reset handler, all other interrupt service routines are listed in lines 242-526. However, all these interrupt service routines are empty. Specific functionality for each service routine can be implemented using these empty service routines.

- Following the interrupt service routines, the interrupt enabling and disabling functions are implemented (see the code listed in lines 552-587).

```
1 ;*****************************************************************
2 ; startup.S - Startup code for use with Keil's uVision.
3 ;
4 ; Copyright (c) 2012 Texas Instruments Incorporated.
5 ; All rights ;reserved.
6 ;*****************************************************************
```

```
 7 ;****************************************************************
 8 ;   Stack Size (in Bytes) <0x0-0xFFFFFFFF:8>
 9 ;****************************************************************
10 Stack   EQU     0x00000400
11
12 ;****************************************************************
13 ;   Heap Size (in Bytes) <0x0-0xFFFFFFFF:8>
14 ;****************************************************************
15 Heap    EQU     0x00000000
16
17 ;****************************************************************
18 ; Allocate space for the stack.
19 ;****************************************************************
20        AREA     STACK, NOINIT, READWRITE, ALIGN=3
21 StackMem
22        SPACE    Stack
23 __initial_sp
24
25 ;****************************************************************
26 ; Allocate space for the heap.
27 ;****************************************************************
28        AREA     HEAP, NOINIT, READWRITE, ALIGN=3
29 __heap_base
30 HeapMem
31        SPACE    Heap
32 __heap_limit
33
34 ;****************************************************************
35 ; Indicate that the code in this file preserves 8-byte alignment
36 ; of the stack.
37 ;****************************************************************
38        PRESERVE8
39
40 ;****************************************************************
41 ; Place code into the reset code section.
42 ;****************************************************************
43        AREA     RESET, CODE, READONLY
44        THUMB
45
46 ;****************************************************************
47 ; The interrupt vector table.
48 ;****************************************************************
49        EXPORT   __Vectors
50 __Vectors
51        DCD      StackMem + Stack        ; Top of Stack
52        DCD      Reset_Handler           ; Reset Handler
53        DCD      NMI_Handler             ; NMI Handler
54        DCD      HardFault_Handler       ; Hard Fault Handler
55        DCD      MemManage_Handler       ; MPU Fault Handler
56        DCD      BusFault_Handler        ; Bus Fault Handler
57        DCD      UsageFault_Handler      ; Usage Fault Handler
58        DCD      0                       ; Reserved
59        DCD      0                       ; Reserved
60        DCD      0                       ; Reserved
61        DCD      0                       ; Reserved
62        DCD      SVC_Handler             ; SVCall Handler
63        DCD      DebugMon_Handler        ; Debug Monitor Handler
64        DCD      0                       ; Reserved
65        DCD      PendSV_Handler          ; PendSV Handler
```

```
66       DCD      SysTick_Handler              ; SysTick Handler
67       DCD      GPIOPortA_Handler            ; GPIO Port A
68       DCD      GPIOPortB_Handler            ; GPIO Port B
69       DCD      GPIOPortC_Handler            ; GPIO Port C
70       DCD      GPIOPortD_Handler            ; GPIO Port D
71       DCD      GPIOPortE_Handler            ; GPIO Port E
72       DCD      UART0_Handler                ; UART0 Rx and Tx
73       DCD      UART1_Handler                ; UART1 Rx and Tx
74       DCD      SSI0_Handler                 ; SSI0 Rx and Tx
75       DCD      I2C0_Handler                 ; I2C0 Master and Slave
76       DCD      PWM0Fault_Handler            ; PWM 0 Fault
77       DCD      PWM0Generator0_Handler       ; PWM 0 Generator 0
78       DCD      PWM0Generator1_Handler       ; PWM 0 Generator 1
79       DCD      PWM0Generator2_Handler       ; PWM 0 Generator 2
80       DCD      Quadrature0_Handler          ; Quadrature Encoder 0
81       DCD      ADC0Seq0_Handler             ; ADC0 Sequence 0
82       DCD      ADC0Seq1_Handler             ; ADC0 Sequence 1
83       DCD      ADC0Seq2_Handler             ; ADC0 Sequence 2
84       DCD      ADC0Seq3_Handler             ; ADC0 Sequence 3
85       DCD      WDT_Handler                  ; Watchdog
86       DCD      Timer0A_Handler              ; Timer 0 subtimer A
87       DCD      Timer0B_Handler              ; Timer 0 subtimer B
88       DCD      Timer1A_Handler              ; Timer 1 subtimer A
89       DCD      Timer1B_Handler              ; Timer 1 subtimer B
90       DCD      Timer2A_Handler              ; Timer 2 subtimer A
91       DCD      Timer2B_Handler              ; Timer 2 subtimer B
92       DCD      Comp0_Handler                ; Analog Comp 0
93       DCD      Comp1_Handler                ; Analog Comp 1
94       DCD      Comp2_Handler                ; Analog Comp 2
95       DCD      SysCtl_Handler               ; System Control
96       DCD      FlashCtl_Handler             ; Flash Control
97       DCD      GPIOPortF_Handler            ; GPIO Port F
98       DCD      GPIOPortG_Handler            ; GPIO Port G
99       DCD      GPIOPortH_Handler            ; GPIO Port H
100      DCD      UART2_Handler                ; UART2 Rx and Tx
101      DCD      SSI1_Handler                 ; SSI1 Rx and Tx
102      DCD      Timer3A_Handler              ; Timer 3 subtimer A
103      DCD      Timer3B_Handler              ; Timer 3 subtimer B
104      DCD      I2C1_Handler                 ; I2C1 Master and Slave
105      DCD      Quadrature1_Handler          ; Quadrature Encoder 1
106      DCD      CAN0_Handler                 ; CAN0
107      DCD      CAN1_Handler                 ; CAN1
108      DCD      CAN2_Handler                 ; CAN2
109      DCD      Ethernet_Handler             ; Ethernet
110      DCD      Hibernate_Handler            ; Hibernate
111      DCD      USB0_Handler                 ; USB0
112      DCD      PWM0Generator3_Handler       ; PWM 0 Generator 3
113      DCD      uDMA_Handler                 ; uDMA Software Transfer
114      DCD      uDMA_Error                   ; uDMA Error
115      DCD      ADC1Seq0_Handler             ; ADC1 Sequence 0
116      DCD      ADC1Seq1_Handler             ; ADC1 Sequence 1
117      DCD      ADC1Seq2_Handler             ; ADC1 Sequence 2
118      DCD      ADC1Seq3_Handler             ; ADC1 Sequence 3
119      DCD      I2S0_Handler                 ; I2S0
120      DCD      ExtBus_Handler               ; External Bus Interf 0
121      DCD      GPIOPortJ_Handler            ; GPIO Port J
122      DCD      GPIOPortK_Handler            ; GPIO Port K
123      DCD      GPIOPortL_Handler            ; GPIO Port L
124      DCD      SSI2_Handler                 ; SSI2 Rx and Tx
```

```
125    DCD    SSI3_Handler              ; SSI3 Rx and Tx
126    DCD    UART3_Handler             ; UART3 Rx and Tx
127    DCD    UART4_Handler             ; UART4 Rx and Tx
128    DCD    UART5_Handler             ; UART5 Rx and Tx
129    DCD    UART6_Handler             ; UART6 Rx and Tx
130    DCD    UART7_Handler             ; UART7 Rx and Tx
131    DCD    0                         ; Reserved
132    DCD    0                         ; Reserved
133    DCD    0                         ; Reserved
134    DCD    0                         ; Reserved
135    DCD    I2C2_Handler              ; I2C2 Master and Slave
136    DCD    I2C3_Handler              ; I2C3 Master and Slave
137    DCD    Timer4A_Handler           ; Timer 4 subtimer A
138    DCD    Timer4B_Handler           ; Timer 4 subtimer B
139    DCD    0                         ; Reserved
140    DCD    0                         ; Reserved
141    DCD    0                         ; Reserved
142    DCD    0                         ; Reserved
143    DCD    0                         ; Reserved
144    DCD    0                         ; Reserved
145    DCD    0                         ; Reserved
146    DCD    0                         ; Reserved
147    DCD    0                         ; Reserved
148    DCD    0                         ; Reserved
149    DCD    0                         ; Reserved
150    DCD    0                         ; Reserved
151    DCD    0                         ; Reserved
152    DCD    0                         ; Reserved
153    DCD    0                         ; Reserved
154    DCD    0                         ; Reserved
155    DCD    0                         ; Reserved
156    DCD    0                         ; Reserved
157    DCD    0                         ; Reserved
158    DCD    0                         ; Reserved
159    DCD    Timer5A_Handler           ; Timer 5 subtimer A
160    DCD    Timer5B_Handler           ; Timer 5 subtimer B
161    DCD    WideTimer0A_Handler       ; WideTimer 0 subtimer A
162    DCD    WideTimer0B_Handler       ; WideTimer 0 subtimer B
163    DCD    WideTimer1A_Handler       ; WideTimer 1 subtimer A
164    DCD    WideTimer1B_Handler       ; WideTimer 1 subtimer B
165    DCD    WideTimer2A_Handler       ; WideTimer 2 subtimer A
166    DCD    WideTimer2B_Handler       ; WideTimer 2 subtimer B
167    DCD    WideTimer3A_Handler       ; WideTimer 3 subtimer A
168    DCD    WideTimer3B_Handler       ; WideTimer 3 subtimer B
169    DCD    WideTimer4A_Handler       ; WideTimer 4 subtimer A
170    DCD    WideTimer4B_Handler       ; WideTimer 4 subtimer B
171    DCD    WideTimer5A_Handler       ; WideTimer 5 subtimer A
172    DCD    WideTimer5B_Handler       ; WideTimer 5 subtimer B
173    DCD    FPU_Handler               ; FPU
174    DCD    PECI0_Handler             ; PECI 0
175    DCD    LPC0_Handler              ; LPC 0
176    DCD    I2C4_Handler              ; I2C4 Master and Slave
177    DCD    I2C5_Handler              ; I2C5 Master and Slave
178    DCD    GPIOPortM_Handler         ; GPIO Port M
179    DCD    GPIOPortN_Handler         ; GPIO Port N
180    DCD    Quadrature2_Handler       ; Quadrature Encoder 2
181    DCD    Fan0_Handler              ; Fan 0
182    DCD    0                         ; Reserved
183    DCD    GPIOPortP_Handler         ; GPIO Port P
```

```
184        DCD      GPIOPortP1_Handler              ; GPIO Port P1
185        DCD      GPIOPortP2_Handler              ; GPIO Port P2
186        DCD      GPIOPortP3_Handler              ; GPIO Port P3
187        DCD      GPIOPortP4_Handler              ; GPIO Port P4
188        DCD      GPIOPortP5_Handler              ; GPIO Port P5
189        DCD      GPIOPortP6_Handler              ; GPIO Port P6
190        DCD      GPIOPortP7_Handler              ; GPIO Port P7
191        DCD      GPIOPortQ_Handler               ; GPIO Port Q
192        DCD      GPIOPortQ1_Handler              ; GPIO Port Q1
193        DCD      GPIOPortQ2_Handler              ; GPIO Port Q2
194        DCD      GPIOPortQ3_Handler              ; GPIO Port Q3
195        DCD      GPIOPortQ4_Handler              ; GPIO Port Q4
196        DCD      GPIOPortQ5_Handler              ; GPIO Port Q5
197        DCD      GPIOPortQ6_Handler              ; GPIO Port Q6
198        DCD      GPIOPortQ7_Handler              ; GPIO Port Q7
199        DCD      GPIOPortR_Handler               ; GPIO Port R
200        DCD      GPIOPortS_Handler               ; GPIO Port S
201        DCD      PWM1Generator0_Handler          ; PWM 1 Generator 0
202        DCD      PWM1Generator1_Handler          ; PWM 1 Generator 1
203        DCD      PWM1Generator2_Handler          ; PWM 1 Generator 2
204        DCD      PWM1Generator3_Handler          ; PWM 1 Generator 3
205        DCD      PWM1Fault_Handler               ; PWM 1 Fault
206
207 ;*********************************************************************
208 ; This is the code that gets called when the processor first starts
209 ; execution following a reset event.
210 ;*********************************************************************
211        EXPORT   Reset_Handler
212 Reset_Handler
213    ;
214    ; Enable floating-point unit. This must be done here to handle
215    ; the case where main() uses floating-point and function prologue
216    ; saves floating-point registers (which will fault if floating-
217    ; point is not enabled). Any configuration of the floating-point
218    ; unit using DriverLib APIs must be done here prior to the
219    ; floating-point unit being enabled.
220    ;
221    ; Note that this does not use DriverLib since it might not be
222    ; included in this project.
223    ;
224        MOVW     R0, #0xED88
225        MOVT     R0, #0xE000
226        LDR      R1, [R0]
227        ORR      R1, #0x00F00000
228        STR      R1, [R0]
229
230    ;
231    ; Call the C library enty point that handles startup.  This will copy
232    ; the .data section initializers from flash to SRAM and zero fill the
233    ; .bss section.
234    ;
235        IMPORT   __main
236        B        __main
237
238 ;*********************************************************************
239 ; This is the code that gets called when the processor receives a
240 ; NMI. This simply enters an infinite loop, preserving the system
241 ; state for examination by a debugger.
242 ;*********************************************************************
```

```
243 NMI_Handler        PROC
244                    EXPORT   NMI_Handler            [WEAK]
245                    B        .
246                    ENDP
247
248 ;*********************************************************************
249 ;
250 ; This is the code that gets called when the processor receives a
251 ; fault interrupt.  This simply enters an infinite loop, preserving
252 ; the system state for examination by a debugger.
253 ;
254 ;*********************************************************************
255 HardFault_Handler\
256                    PROC
257                    EXPORT   HardFault_Handler      [WEAK]
258                    B        .
259                    ENDP
260
261 MemManage_Handler\
262                    PROC
263                    EXPORT   MemManage_Handler      [WEAK]
264                    B        .
265                    ENDP
266 BusFault_Handler\
267                    PROC
268                    EXPORT   BusFault_Handler       [WEAK]
269                    B        .
270                    ENDP
271 UsageFault_Handler\
272                    PROC
273                    EXPORT   UsageFault_Handler     [WEAK]
274                    B        .
275                    ENDP
276 SVC_Handler        PROC
277                    EXPORT   SVC_Handler            [WEAK]
278                    B        .
279                    ENDP
280 DebugMon_Handler\
281                    PROC
282                    EXPORT   DebugMon_Handler       [WEAK]
283                    B        .
284                    ENDP
285 PendSV_Handler     PROC
286                    EXPORT   PendSV_Handler         [WEAK]
287                    B        .
288                    ENDP
289 SysTick_Handler    PROC
290                    EXPORT   SysTick_Handler        [WEAK]
291                    B        .
292                    ENDP
293 IntDefaultHandler\
294                    PROC
295
296                    EXPORT   GPIOPortA_Handler      [WEAK]
297                    EXPORT   GPIOPortB_Handler      [WEAK]
298                    EXPORT   GPIOPortC_Handler      [WEAK]
299                    EXPORT   GPIOPortD_Handler      [WEAK]
300                    EXPORT   GPIOPortE_Handler      [WEAK]
301                    EXPORT   UART0_Handler          [WEAK]
```

```
302              EXPORT   UART1_Handler              [WEAK]
303              EXPORT   SSI0_Handler               [WEAK]
304              EXPORT   I2C0_Handler               [WEAK]
305              EXPORT   PWM0Fault_Handler          [WEAK]
306              EXPORT   PWM0Generator0_Handler     [WEAK]
307              EXPORT   PWM0Generator1_Handler     [WEAK]
308              EXPORT   PWM0Generator2_Handler     [WEAK]
309              EXPORT   Quadrature0_Handler        [WEAK]
310              EXPORT   ADC0Seq0_Handler           [WEAK]
311              EXPORT   ADC0Seq1_Handler           [WEAK]
312              EXPORT   ADC0Seq2_Handler           [WEAK]
313              EXPORT   ADC0Seq3_Handler           [WEAK]
314              EXPORT   WDT_Handler                [WEAK]
315              EXPORT   Timer0A_Handler            [WEAK]
316              EXPORT   Timer0B_Handler            [WEAK]
317              EXPORT   Timer1A_Handler            [WEAK]
318              EXPORT   Timer1B_Handler            [WEAK]
319              EXPORT   Timer2A_Handler            [WEAK]
320              EXPORT   Timer2B_Handler            [WEAK]
321              EXPORT   Comp0_Handler              [WEAK]
322              EXPORT   Comp1_Handler              [WEAK]
323              EXPORT   Comp2_Handler              [WEAK]
324              EXPORT   SysCtl_Handler             [WEAK]
325              EXPORT   FlashCtl_Handler           [WEAK]
326              EXPORT   GPIOPortF_Handler          [WEAK]
327              EXPORT   GPIOPortG_Handler          [WEAK]
328              EXPORT   GPIOPortH_Handler          [WEAK]
329              EXPORT   UART2_Handler              [WEAK]
330              EXPORT   SSI1_Handler               [WEAK]
331              EXPORT   Timer3A_Handler            [WEAK]
332              EXPORT   Timer3B_Handler            [WEAK]
333              EXPORT   I2C1_Handler               [WEAK]
334              EXPORT   Quadrature1_Handler        [WEAK]
335              EXPORT   CAN0_Handler               [WEAK]
336              EXPORT   CAN1_Handler               [WEAK]
337              EXPORT   CAN2_Handler               [WEAK]
338              EXPORT   Ethernet_Handler           [WEAK]
339              EXPORT   Hibernate_Handler          [WEAK]
340              EXPORT   USB0_Handler               [WEAK]
341              EXPORT   PWM0Generator3_Handler     [WEAK]
342              EXPORT   uDMA_Handler               [WEAK]
343              EXPORT   uDMA_Error                 [WEAK]
344              EXPORT   ADC1Seq0_Handler           [WEAK]
345              EXPORT   ADC1Seq1_Handler           [WEAK]
346              EXPORT   ADC1Seq2_Handler           [WEAK]
347              EXPORT   ADC1Seq3_Handler           [WEAK]
348              EXPORT   I2S0_Handler               [WEAK]
349              EXPORT   ExtBus_Handler             [WEAK]
350              EXPORT   GPIOPortJ_Handler          [WEAK]
351              EXPORT   GPIOPortK_Handler          [WEAK]
352              EXPORT   GPIOPortL_Handler          [WEAK]
353              EXPORT   SSI2_Handler               [WEAK]
354              EXPORT   SSI3_Handler               [WEAK]
355              EXPORT   UART3_Handler              [WEAK]
356              EXPORT   UART4_Handler              [WEAK]
357              EXPORT   UART5_Handler              [WEAK]
358              EXPORT   UART6_Handler              [WEAK]
359              EXPORT   UART7_Handler              [WEAK]
360              EXPORT   I2C2_Handler               [WEAK]
```

```
361                 EXPORT    I2C3_Handler            [WEAK]
362                 EXPORT    Timer4A_Handler         [WEAK]
363                 EXPORT    Timer4B_Handler         [WEAK]
364                 EXPORT    Timer5A_Handler         [WEAK]
365                 EXPORT    Timer5B_Handler         [WEAK]
366                 EXPORT    WideTimer0A_Handler     [WEAK]
367                 EXPORT    WideTimer0B_Handler     [WEAK]
368                 EXPORT    WideTimer1A_Handler     [WEAK]
369                 EXPORT    WideTimer1B_Handler     [WEAK]
370                 EXPORT    WideTimer2A_Handler     [WEAK]
371                 EXPORT    WideTimer2B_Handler     [WEAK]
372                 EXPORT    WideTimer3A_Handler     [WEAK]
373                 EXPORT    WideTimer3B_Handler     [WEAK]
374                 EXPORT    WideTimer4A_Handler     [WEAK]
375                 EXPORT    WideTimer4B_Handler     [WEAK]
376                 EXPORT    WideTimer5A_Handler     [WEAK]
377                 EXPORT    WideTimer5B_Handler     [WEAK]
378                 EXPORT    FPU_Handler             [WEAK]
379                 EXPORT    PECI0_Handler           [WEAK]
380                 EXPORT    LPC0_Handler            [WEAK]
381                 EXPORT    I2C4_Handler            [WEAK]
382                 EXPORT    I2C5_Handler            [WEAK]
383                 EXPORT    GPIOPortM_Handler       [WEAK]
384                 EXPORT    GPIOPortN_Handler       [WEAK]
385                 EXPORT    Quadrature2_Handler     [WEAK]
386                 EXPORT    Fan0_Handler            [WEAK]
387                 EXPORT    GPIOPortP_Handler       [WEAK]
388                 EXPORT    GPIOPortP1_Handler      [WEAK]
389                 EXPORT    GPIOPortP2_Handler      [WEAK]
390                 EXPORT    GPIOPortP3_Handler      [WEAK]
391                 EXPORT    GPIOPortP4_Handler      [WEAK]
392                 EXPORT    GPIOPortP5_Handler      [WEAK]
393                 EXPORT    GPIOPortP6_Handler      [WEAK]
394                 EXPORT    GPIOPortP7_Handler      [WEAK]
395                 EXPORT    GPIOPortQ_Handler       [WEAK]
396                 EXPORT    GPIOPortQ1_Handler      [WEAK]
397                 EXPORT    GPIOPortQ2_Handler      [WEAK]
398                 EXPORT    GPIOPortQ3_Handler      [WEAK]
399                 EXPORT    GPIOPortQ4_Handler      [WEAK]
400                 EXPORT    GPIOPortQ5_Handler      [WEAK]
401                 EXPORT    GPIOPortQ6_Handler      [WEAK]
402                 EXPORT    GPIOPortQ7_Handler      [WEAK]
403                 EXPORT    GPIOPortR_Handler       [WEAK]
404                 EXPORT    GPIOPortS_Handler       [WEAK]
405                 EXPORT    PWM1Generator0_Handler  [WEAK]
406                 EXPORT    PWM1Generator1_Handler  [WEAK]
407                 EXPORT    PWM1Generator2_Handler  [WEAK]
408                 EXPORT    PWM1Generator3_Handler  [WEAK]
409                 EXPORT    PWM1Fault_Handler       [WEAK]
410
411 GPIOPortA_Handler
412 GPIOPortB_Handler
413 GPIOPortC_Handler
414 GPIOPortD_Handler
415 GPIOPortE_Handler
416 UART0_Handler
417 UART1_Handler
418 SSI0_Handler
419 I2C0_Handler
```

```
420  PWM0Fault_Handler
421  PWM0Generator0_Handler
422  PWM0Generator1_Handler
423  PWM0Generator2_Handler
424  Quadrature0_Handler
425  ADC0Seq0_Handler
426  ADC0Seq1_Handler
427  ADC0Seq2_Handler
428  ADC0Seq3_Handler
429  WDT_Handler
430  Timer0A_Handler
431  Timer0B_Handler
432  Timer1A_Handler
433  Timer1B_Handler
434  Timer2A_Handler
435  Timer2B_Handler
436  Comp0_Handler
437  Comp1_Handler
438  Comp2_Handler
439  SysCtl_Handler
440  FlashCtl_Handler
441  GPIOPortF_Handler
442  GPIOPortG_Handler
443  GPIOPortH_Handler
444  UART2_Handler
445  SSI1_Handler
446  Timer3A_Handler
447  Timer3B_Handler
448  I2C1_Handler
449  Quadrature1_Handler
450  CAN0_Handler
451  CAN1_Handler
452  CAN2_Handler
453  Ethernet_Handler
454  Hibernate_Handler
455  USB0_Handler
456  PWM0Generator3_Handler
457  uDMA_Handler
458  uDMA_Error
459  ADC1Seq0_Handler
460  ADC1Seq1_Handler
461  ADC1Seq2_Handler
462  ADC1Seq3_Handler
463  I2S0_Handler
464  ExtBus_Handler
465  GPIOPortJ_Handler
466  GPIOPortK_Handler
467  GPIOPortL_Handler
468  SSI2_Handler
469  SSI3_Handler
470  UART3_Handler
471  UART4_Handler
472  UART5_Handler
473  UART6_Handler
474  UART7_Handler
475  I2C2_Handler
476  I2C3_Handler
477  Timer4A_Handler
478  Timer4B_Handler
```

```
479   Timer5A_Handler
480   Timer5B_Handler
481   WideTimer0A_Handler
482   WideTimer0B_Handler
483   WideTimer1A_Handler
484   WideTimer1B_Handler
485   WideTimer2A_Handler
486   WideTimer2B_Handler
487   WideTimer3A_Handler
488   WideTimer3B_Handler
489   WideTimer4A_Handler
490   WideTimer4B_Handler
491   WideTimer5A_Handler
492   WideTimer5B_Handler
493   FPU_Handler
494   PECI0_Handler
495   LPC0_Handler
496   I2C4_Handler
497   I2C5_Handler
498   GPIOPortM_Handler
499   GPIOPortN_Handler
500   Quadrature2_Handler
501   Fan0_Handler
502   GPIOPortP_Handler
503   GPIOPortP1_Handler
504   GPIOPortP2_Handler
505   GPIOPortP3_Handler
506   GPIOPortP4_Handler
507   GPIOPortP5_Handler
508   GPIOPortP6_Handler
509   GPIOPortP7_Handler
510   GPIOPortQ_Handler
511   GPIOPortQ1_Handler
512   GPIOPortQ2_Handler
513   GPIOPortQ3_Handler
514   GPIOPortQ4_Handler
515   GPIOPortQ5_Handler
516   GPIOPortQ6_Handler
517   GPIOPortQ7_Handler
518   GPIOPortR_Handler
519   GPIOPortS_Handler
520   PWM1Generator0_Handler
521   PWM1Generator1_Handler
522   PWM1Generator2_Handler
523   PWM1Generator3_Handler
524   PWM1Fault_Handler
525
526                   B           .
527                   ENDP
528
529   ;****************************************************************
530   ; Make sure the end of this section is aligned.
531   ;****************************************************************
532           ALIGN
533
534   ;****************************************************************
535   ; Some code in the normal code section for initializing the heap
536   ; and stack.
537   ;****************************************************************
```

```
538         AREA    |.text|, CODE, READONLY
539
540 ;*********************************************************************
541 ; Useful functions.
542 ;*********************************************************************
543         EXPORT  DisableInterrupts
544         EXPORT  EnableInterrupts
545         EXPORT  StartCritical
546         EXPORT  EndCritical
547         EXPORT  WaitForInterrupt
548
549 ;********** DisableInterrupts **************
550 ; disable interrupts
551 ; inputs:  none
552 ; outputs: none
553 DisableInterrupts
554         CPSID   I
555         BX      LR
556
557 ;********** EnableInterrupts **************
558 ; disable interrupts
559 ; inputs:  none
560 ; outputs: none
561 EnableInterrupts
562         CPSIE   I
563         BX      LR
564
565 ;********** StartCritical ********************
566 ; make a copy of previous I bit, disable interrupts
567 ; inputs:  none
568 ; outputs: previous I bit
569 StartCritical
570         MRS     R0, PRIMASK  ; save old status
571         CPSID   I            ; mask all (except faults)
572         BX      LR
573
574 ;********** EndCritical ********************
575 ; using the copy of previous I bit, restore I bit to previous value
576 ; inputs:  previous I bit
577 ; outputs: none
578 EndCritical
579         MSR     PRIMASK, R0
580         BX      LR
581
582 ;********** WaitForInterrupt ********************
583 ; go to low power mode while waiting for the next interrupt
584 ; inputs:  none
585 ; outputs: none
586 WaitForInterrupt
587         WFI
588         BX      LR
589
590 ;*********************************************************************
591 ; The function expected of the C library startup code for defining
592 ; the stack and heap memory locations.  For the C library version
593 ; of the startup code, provide this function so that the C library
594 ; initialization code can find out the location of stack and heap.
595 ;*********************************************************************
596     IF :DEF: __MICROLIB
```

```
597            EXPORT    __initial_sp
598            EXPORT    __heap_base
599            EXPORT    __heap_limit
600        ELSE
601    ;         IMPORT   __use_two_region_memory
602            EXPORT    __user_initial_stackheap
603    __user_initial_stackheap
604            LDR       R0, =HeapMem
605            LDR       R1, =(StackMem + Stack)
606            LDR       R2, =(HeapMem + Heap)
607            LDR       R3, =StackMem
608            BX        LR
609        ENDIF
610
611    ;****************************************************************
612    ; Make sure the end of this section is aligned.
613    ;****************************************************************
614            ALIGN
615
616    ;****************************************************************
617    ; Tell the assembler that we're done.
618    ;****************************************************************
619            END
```

Listing A.1: Startup code to be used with all user programs.

A.2   Linker Script File

For a properly configured project in Keil tools, linker script file is generated automatically. The linker script file is also termed as scatter-loading description file and enables the application developer to specify the device memory map for building an executable image. Scatter-loading is best utilized when the memory map at load time and the memory map at execution time are different. Each region or segment in the memory map can have different load time and execution time address.

Scatter-loading is in general used for developing an executable image for an embedded system, which integrate both ROM and RAM type of memories as well as memory mapped peripherals. Scatter loading is quite useful in the following situations.

- When the memory map is quite complex

- The system uses different types of memories

- The system has memory mapped peripherals

- When a function is required to be placed at a fixed address location

An example linker script file is provided in Listing A.2. This liker script file is generated for TM4C123 microcontroller from Texas Instruments, which has 256 KB Flash memory and 32 KB RAM memory. From Listing A.2, we can observe that the RESET section is put first in the flash memory region using '+First' option. This ensure that the interrupt vector table is located at the starting address of flash memory region. When this linker script file is used, a section named RESET should always be implemented with at least first two appropriate entries for proper initialization of SP and PC registers after reset. This memory section is implemented as part of the system startup file, as discussed in the previous section.

```
1  ;  ***************************************************************
2  ;  *** Scatter-Loading Description File generated by uVision ***
3  ;  ***************************************************************
4  LR_IROM1 0x00000000 0x00040000     ; load region size_region
5  {                                  ; load address = execution
6    ER_IROM1 0x00000000 0x00040000   ; address
7    { *.o (RESET, +First)
8     *(InRoot$ $Sections)
9     .ANY (+RO)
10   }
11   RW_IRAM1 0x20000000 0x00008000  {  ; RW data
12     .ANY (+RW +ZI)
13   }
14 }
```

Listing A.2: Linker script or scatter loading file used with first assembly program.

A.3 Application Startup

As discussed in Chapter 4, when a microcontroller based system is started, it follows a reset sequence. The reset sequence is a collection of steps performed at the very beginning of system startup and involves the loading of stack pointer and program counter. The program counter, on reset, is initialized with the Reset_Handler and begins the execution of reset interrupt service routine. The system startup related tasks are performed as part of the reset interrupt service routine. At the end of reset interrupt service routine, a function labeled __main is called and marks the beginning of application startup. The __main function is implemented as part of C library provided by the tools vendor. The C library from ARM contains precompiled code, which is used for application startup. When generating the executable, the linker includes all the necessary code from the C library to create application specific startup code. The key steps performed during application startup are listed below.

- Calling __main: The application startup code is managed by this tools library function.

- Calling __scatterload: This function is responsible for copying any code and data sections to a different memory location. It also copies or decompresses the RW (read write) data from code memory region to data memory region and initializes ZI (zero initialize) data to zeros. This is also depicted pictorially in Figure A.1. The __main always calls this function before calling __rt_entry.

- Calling __rt_entry: The __main function calls __rt_entry to initialize the stack, heap as well as any other C library related functions. The __rt_entry makes calls to other initialization functions and finally calls the user defined function main().

A simple scenario illustrating the use of scatter-load functionality of ARM C library, to copy read-write (RW) region from code memory to data memory during application startup, is illustrated in Figure A.1.

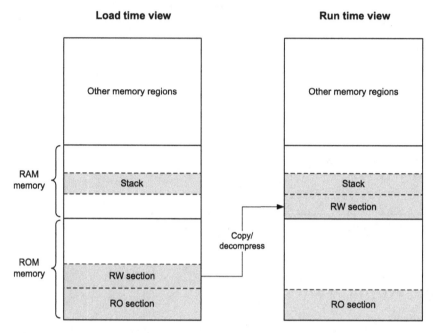

Figure A.1: Read write data region initialization in RAM memory during application initialization.

Appendix B

C Programming Review

During the early years of microprocessor based systems, the user programs were developed using assemblers and fused into EPROM memories. There used to be no mechanism in place to find out what actually had the program been doing. Different types of input and output devices such as LEDs, switches, etc. were used to check correct execution of the program. Some of the developers who were fortunate enough used to have an access to In-Circuit Emulators (ICEs), but they were too costly and were not quite reliable as well. As time progressed, use of assembly as the programming language reduced and embedded systems moved to C as the programming language of choice. Since then the C language has been the most widely used programming language for embedded systems. The assembly language is also used but primarily to implement those parts of the code that require higher execution efficiency and timing accuracy along with code size optimality.

B.1 Embedded Systems Programming

Embedded programs must work closely with the specialized components and custom circuitry that makes up the hardware. Unlike programming on top of an operating system, where the hardware details are pretty much hidden from the programmer's perspective, most embedded programs interact directly with the hardware. This includes not only the hardware of the CPU, but also the hardware which makes up all the peripherals (both on-chip and off-chip) of the system. Thus an embedded programmer must have a good knowledge of hardware, at least as it pertains to writing software that correctly interfaces with and manipulates that hardware. This knowledge will often extend to specifying key components of the hardware (microcontroller, memory devices, I/O devices, etc.) and in smaller organizations will often go as far as designing and laying out the hardware as a printed circuit board. An embedded programmer will also need to have a good understanding of debugging equipment such as multimeters, oscilloscopes and logic analyzers.

Another difference from more general purpose computers is that most embedded systems are quite limited as compared to the former. The microcomputers used in embedded systems may have program memory sizes of a few thousand to a few hundred thousand bytes rather than the gigabytes in the desktop machine, and will typically have even less data (RAM) memory than program memory.

There are many factors to consider when developing a program for embedded systems. Some of those are listed below.

- Efficiency, programs must be as short as possible and memory must be used efficiently

- Execution speed, programs must run as fast as possible

- Ease of implementation

- Maintainability

- Readability

B.2   C Programming for Embedded Systems

Embedded systems are commonly programmed using C, assembly and to some extent BASIC. C is a very flexible and powerful programming language, yet it is fairly simple to learn. C gives embedded programmers an extraordinary degree of direct hardware control without sacrificing the benefits of high-level languages so its compilers are available for almost every processor in use today and there is a very large body of experienced C programmers.

Programs that are developed for embedded systems are usually expected to monitor and control external devices and directly manipulate and use the internal architecture of the processor such as interrupt handling, timers, serial communications and other available features. C compilers for embedded systems must provide ways to examine and utilize various features of the microcontroller's internal and external architecture; this includes interrupt service routines, reading from and writing to internal and external memories, bit manipulation, implementation of timers/counters and examination of internal registers etc. Standard application program, communicates with the hardware components via the operating system of the machine but an application for an embedded system must communicate directly with the processor and its components. For example, consider the C program in Listing B.1.

```
printf("C Programming for Embedded Systems\n");

c = getchar();
```

Listing B.1: Information printing in C.

In standard C running on a PC platform, the *printf* statement causes the string inside the quotation to be displayed on the screen. The same statement in an embedded system causes the string to be transmitted via the serial port transmit pin (i.e., Tx pin) of the microcontroller provided the serial port has been initialized and enabled. Similarly, in standard C running on a PC platform *getchar()* causes a character to be read from the keyboard on a PC. In an embedded system the instruction causes a character to be read from the serial pin (i.e., RxD) of the microcontroller.

B.3 Template for Embedded C Program

```
#include "TM4C1233H6PM.h"

void main(void)
{
// body of the program goes here
}
```

Listing B.2: Template for embedded C program.

- First line of the template is the C directive. This tells compiler that during compilation, it should look into this file for symbols not defined in the program.

- The next line in the template declares the beginning of the body of the main part of the program. The main part of the program is treated as any other function in C program. Every C program should have a main function.

- Within the curly brackets we write the code for the application.

B.4 Type Casting

Type casting is a way to convert a variable from one data type to another data type. For example, if you want to store a long value into a simple integer then you can type cast long to int. You can convert values from one type to another explicitly using the cast operator as follows:

```
(type_name) expression
```

Listing B.3: Syntax to typecast a variable.

Consider the Example B.1 where the cast operator causes the division of one integer variable by another to be performed as a floating-point operation:

Example B.1 (Type casting illustration.).

```
#include <stdio.h>

main()
{
    int sum = 17, count = 5;
    double mean;

    mean = (double) sum / count;
    printf("Value of mean : %f\n", mean );
}
```

When the above code is compiled and executed, it produces the following result:

```
Value of mean : 3.400000
```

It should be noted here that the cast operator has precedence over division, so the value of sum is first converted to type double and finally it gets divided by count yielding a double value. Type conversions can be implicit which is performed by the compiler automatically, or it can be specified explicitly through the use of the cast operator. It is considered good programming practice to use the cast operator whenever type conversions are necessary.

B.5   Preprocessor Directives

Preprocessor directives are lines included in the code of our programs that are not program statements but directives for the preprocessor. Preprocessor directives begin with a hash symbol (#) in the first column. As the name implies, preprocessor commands are processed first, i.e., the compiler parses through the program handling the preprocessor directives.

These preprocessor directives extend only across a single line of code. As soon as a newline character is found, the preprocessor directive is considered to end. No semicolon (;) is expected at the end of a preprocessor directive. The only way a preprocessor directive can extend through more than one line is by preceding the newline character at the end of the line by a backslash (\).

The preprocessor provides the ability for the inclusion of header files, macro expansions, conditional compilation, and line control etc. Here we discuss only two important preprocessor directives.

B.5.1   Macro Definitions (#define)

To define preprocessor macros we can use #define. Its syntax is illustrated below.

```
#define identifier replacement
```

Listing B.4: Syntax to define macros in C.

When the preprocessor encounters this directive, it replaces any occurrence of *identifier* in the rest of the code by *replacement*. This *replacement* can be an expression, a statement, a block or simply anything. The preprocessor does not understand C, it simply replaces any occurrence of the *identifier* by the corresponding *replacement* value.

```
#define DELAY 20000
```

During compilation, wherever the DELAY is found as a label, it is replaced with the value 20000.

Advantages of Using A Macro

- The speed of the execution of the program is the major advantage of using a macro.

- It saves a lot of time that is spent by the compiler for invoking/calling the functions.

- In some cases macros can reduce the size of the program.

B.5.2 Including Files (#include)

To include the contents of a definition header file in a source code file we use the #include compiler directive. There are two slightly different ways to specify a file that is to be included.

```
#include "TM4C1233H6PM.h"
#include <stdio.h>
```

Listing B.5: Two different ways to specify the file name to be included.

In first illustration, the include directive will include file named 'TM4C1233H6PM.h', which is a user defined file and defines all the I/O port names for TM4C123 microcontroller. In the second illustration, the 'stdio' is a system header file.

B.6 Bitwise Operations in C

The byte is the lowest level at which we can access data, as there is no 'bit' type variable, and we cannot access individual bits while performing memory read/write operations (with the exception of bit-banding memory region). In fact, we cannot perform operations on a single bit. Every bitwise operator will be applied, at a minimum, to an entire byte. This means we will be considering the entire representation of a number whenever we talk about applying a bitwise operator. Table B.1 summarizes the bitwise operators available in C.

Table B.1: Bitwise operators in C.

Operator	Description	Example	Result
&	Bitwise AND	0x88 & 0x0F	0x08
∧	Bitwise XOR	0x0F ∧ 0xFF	0xF0
|	Bitwise OR	0xCC | 0x0F	0xCF
≪	Left Shift	0x01 ≪ 4	0x10
≫	Right Shift	0x80 ≫ 6	0x02

B.6.1   Bit Masking

Bitwise operators treat every bit in a word as a Boolean (two-value) variable, apply a column-wise Boolean operator, and generate the result. Unlike binary arithmetic, there is no carry or borrow and every column is operated independently.

Clearing Bits

Bit masking uses the bits in one operand to 'mask' or select any arbitrary bits of the second operand, using the bitwise Boolean AND operator. The 1 bits in the 'mask' select which bits we want to retain from the second operand, and the 0 bits in the mask clear all the corresponding bits to zeroes. In other words, the bits set to 1's are the 'holes' in the mask that let the corresponding bits in the second operand flow through to the result.

Setting Bits

The opposite of masking (turning bits OFF) is setting bits, where we use the bitwise Boolean OR operator to turn ON one or more bits in an operand. We select an appropriate value which when ORed with the second operand, turns ON selected bits and leaves the other bits unchanged.

Toggling Bits

Sometimes it does not really matter what is the bit value. Rather we are interested in changing the current bit value to opposite. This operation is called bit toggling and can be achieved using the XOR (Exclusive OR) operation. XOR operation returns 0 if and only if the two bits operated on are of same binary value. Therefore, if two corresponding bits are 1, the result will be a 0, but if only one of them is 1, the result will be 1. As a result, bit inversion is achieved by XORing the selected bits with a 1. If the original bit was 1, it returns 0.

B.7   Type Qualifiers

The variables in C have two attributes associated with them. One of them is called *type specifier*, while the other is called *type qualifier*. Table B.2 provides a list of various type specifiers along with information regarding their memory allocation. Notice that the keywords 'volatile' and 'const' are not present. This is because these are type qualifiers and not type specifiers. We will now look at these two type qualifiers in some detail.

Table B.2: List of type specifiers.

Variable type	Keyword	Size	Range
Character	char	1	-128 to 127
Unsigned Character	unsigned char	1	0 to 255
Integer	int	2	-32678 to 32767
Short Integer	short int	2	-32678 to 32767
Long integer	long int	4	-2147483648 to 2147483647
Unsigned Integer	unsigned int	2	0 to 65535
Unsigned Short Integer	unsigned short int	2	0 to 65535
Unsigned Long Integer	unsigned long int	4	0 to 4294967295
Float	float	4	1.2E-38 to 3.4E38
Double	double	8	2.2E-308 to 1.8E308
Long Double	long double	10	3.4E-4932 to 1.1E+4932

B.7.1   Const Keyword

A constant data object can be declared using 'const' type qualifier, when defining a variable. Handling of constants by the compiler is similar to regular variables with the exception that they cannot be modified during execution to a different value. To be useful, a variable with 'const' qualifier should be initialized with an appropriate value.

B.7.2   Volatile Keyword

The reason for having this type qualifier is mainly to do with the problems that are encountered in real-time or embedded systems programming using C. What volatile keyword does is that it tells the compiler that the object is subject to sudden change for reasons which cannot be predicted from a study of the program itself, and forces every reference to such an object to be a genuine reference. It is a qualifier that is applied to a variable at the time of its declaration. It tells the compiler that the value of the variable may change at any time, without any action taken due to the program execution.

To declare a variable volatile, include the keyword volatile before or after the data type in the variable definition. For instance, both of the declarations given next, will declare foo to be a volatile integer.

```
volatile int foo;
int volatile foo;
```

Now, it turns out that pointers to volatile variables are very common, especially with memory-mapped I/O registers. Both of the declarations provided next declare pReg to be a pointer to a volatile unsigned 8-bit integer.

```
volatile uint8_t * pReg;
uint8_t volatile * pReg;
```

Proper Use of Volatile

A variable should be declared volatile whenever its value could change unexpectedly. In practice, this can happen in the following three scenarios:

- Memory-mapped peripheral registers

- Global variables modified by an interrupt service routine

- Global variables accessed by multiple tasks within a multi-threaded application

B.8   Pointers

Pointers are extremely powerful tool, which can simplify things and help improve the program efficiency. For example, using pointers is one way to have a function modify a variable passed to it. It is also possible to use pointers to dynamically allocate memory, which means that you can write programs that can handle nearly unlimited amount of data on the fly and you do not need to know, at the time of writing the program, how much memory is required. As we know, each variable effectively is a memory location and each memory location has an associated address that can be obtained using the ampersand (&) operator, which provides a memory address. Consider the following illustration, which will print the variable addresses.

Example B.2.
```
#include <stdio.h>

int main ()
{
   int   varA;
   char varB[8];

   printf("Address of varA variable: %x\n", &varA  );
   printf("Address of varB variable: %x\n", &varB  );

   return 0;
}
```

When we compile and execute the above code, it gives a result something similar to the following listing.

```
Address of varA variable: 0x20001000
Address of varB variable: 0x20001004
```

B.8.1   Declaring a Pointer

A pointer is a variable whose value is the address of another variable, i.e., direct address of the memory location. Like any variable or constant, you must declare a

pointer before you can use it to store any variable address. The general form of a
pointer variable declaration is provided below.

```
type *var_name;
```

Here, type is the pointer's base type, which must be a valid C data type and var_name
is the name of the pointer variable. In this statement, the asterisk is used to explicitly
specify that the variable is of type pointer. Following are few examples of valid pointer
type variable declarations.

```
int    *p_iv;   /* pointer to an integer variable */
double *p_dv;   /* pointer to a double variable */
char   *p_cv;   /* pointer to a character variable */
float  *p_fv;   /* pointer to a float variable */
```

The actual data type of all pointers, irrespective of the type of the variable being integer, float, character, or otherwise, is the same, which is a four byte long hexadecimal
number (for 32-bit address bus) representing a memory address. The key difference
among different pointers is the data type of the variable or constant that the pointer
points to.

B.8.2 How To Use Pointers

There are some important operations, which are performed quite frequently using
pointers. When using a pointer the first step is to define a pointer variable. The
second step is to assign the address of a variable to the pointer and the final step is
to access the value at the address available in the pointer variable. This is done by
using the * operator which provides the variable value stored at the specified address.
The next example illustrates the use of these steps.

Example B.3.

```
#include <stdio.h>

int main ()
{
    int  varA = 0x22;   /* declaring an initialized variable */
    int  *p_iv;         /* declaration of a pointer variable  */

    p_iv = &varA;   /* assign address of varA to pointer variable */

    printf("Address of varA is: %x\n", &varA );

    /* address stored in pointer variable */
    printf("Address stored in p_iv variable: %x\n", p_iv );

    /* accessing the variable value using pointer */
    printf("Value of *p_iv variable: %d\n", *p_iv );

    return 0;
}
```

After compilation when this program is executed, it gives the following output.

```
Address of var variable: 0x20001010
Address stored in ip variable: 0x20001010
Value of *ip variable: 0x22
```

B.9 Structures

Structures are user defined variables which allow to combine multiple variables of different data types. The use of structures allows to manage data in a compact manner. Structures are generally useful whenever different data types need to be grouped together. For instance, they can be used to hold records from a database or to store information about contacts in an address book. In the contacts example, a structure can be used that holds all of the information about a single contact, including name, address, phone number, etc.

B.9.1 Defining a Structure

The structure can be defined using the C statement 'struct'. The listing below illustrates the generic syntax to define a structure.

```
struct [structure tag]
{
   member definition;
   member definition;
   ...
   member definition;
} [one or more structure variables];
```

Listing B.6: Syntax to define structure in C.

The 'structure tag' is an optional attribute, while each 'member definition' can follow any arbitrary normal variable definition, for instance int x, or float y etc. In the structure definition, before the final semicolon, one can optionally declare single or multiple structure variables of this type of structure. The listing below illustrates the definition of an example Book structure.

Example B.4.

```
struct Books
{
   char    title[50];
   char    author[50];
   char    subject[100];
   int     year;
   int     book_id;
} book;
```

B.9.2 Accessing Members of a Structure

To access an arbitrary member of a structure, the member access operator (.) is used. The member access operator is a period operator used between the structure name and the structure member name that is to be accessed. The next example illustrates the usage of structure.

Example B.5.

```
struct Books
{
   char   title[50];
   char   author[50];
   char   subject[100];
   int    book_id;
};

int main( )
{
   struct Books Book1;        /* Declare Book1 of type Book */

   strcpy( Book1.title, "ARM Architecture");
   strcpy( Book1.author, "Liu");
   strcpy( Book1.subject, "Embedded Systems");
   Book1.book_id = 123456;


   printf( "Book 1 title : %s\n", Book1.title);
   printf( "Book 1 author : %s\n", Book1.author);
   printf( "Book 1 subject : %s\n", Book1.subject);
   printf( "Book 1 book_id : %d\n", Book1.book_id);


   return 0;
}
```

When the above user program is compiled and then executed, it gives the following output.

```
Book 1 title : ARM Architecture
Book 1 author : Liu
Book 1 subject : Embedded Systems
Book 1 book_id : 123456
```

Just like any other variable, we can declare pointers to structures as well. A structure pointer can be initialized with a structure variable of same type. To access a member of a structure using the structure pointer we need to use '-¿' operator. This is illustrated using the following example.

Example B.6.

```
int main( )
{
   struct Books Book1;        /* Declare Book1 of type Book */
   struct Books *book_ptr;
```

```
    book_ptr = &Book1;

    strcpy( Book1.title, "ARM Architecture");
    printf( "Book title : %s\n", book_ptr->title);
    return 0;
}
```

B.10   Functions

A function is a group of statements that collectively perform a certain task. Almost all C programs have functions. One such function, which is almost present in every C program, is the function main(). The user is free to many additional functions based on its application requirements. A function prototype is used define the function declaration, which informs the compiler about the function name, its return type, and its parameters. A function definition/body contains the actual implementation of the functionality. There different names used to refer to a function, including method, sub-routine, procedure etc. The general syntax for function definition in C programming is illustrated below.

```
return_type function_name(parameter list)
{
   body of the function
}
```

Listing B.7: Syntax to define functions in C.

Following is the example C source code for min() function. This function accepts two parameters named numA and numB, and returns the minimum of the two.

Example B.7.

```
/* function returning the minimum of two numbers */
int min(int numA, int numB)
{   /* local variable declaration */
   int result;

   if (numA < numB)
      result = numA;
   else
      result = numB;

   return result;
}
```

B.10.1   Passing Parameters by Value

There are different ways to pass parameters to functions. One simple possibility is to pass parameters by value. When parameters are passed by value, a copy of the parameter is actually passed to the function. Therefore, any changes made to the passed parameter by the function called will have no effect on the corresponding actual parameter value. To illustrate this let us consider the following example program.

Example B.8.

```
#include<stdio.h>

void foo(int y)
{
    printf("y = %d\n", y);
}

int main()
{
    foo(5); // first call

    int x = 6;
    foo(x); // second call
    foo(x+1); // third call

    return 0;
}
```

In the first call to foo(), the argument is the literal 5. When foo() is called, variable y is created, and the value of 5 is copied into y. Variable y is then destroyed when the program exits from foo(). In the second call to foo(), the argument is the variable x, which is initialized with a value 6. When foo() is called for the second time, variable y is created again, and the value of 6 is copied to y. Variable y is then destroyed when foo() is exited. In the third call to foo(), the argument is the expression x+1. For this case the expression x+1 is evaluated first to produce a value 7, which is copied to variable y. Variable y is once again destroyed as soon as foo() exits. The output of this program is given in the following listing.

```
y = 5
y = 6
y = 7
```

Since a copy of the variable is passed to the function as parameter, the original parameter value cannot be modified by the function. This is verified in the following example.

Example B.9.

```
#include<stdio.h>

void foo(int y)
{
    printf("y = %d\n", y);

    y = 6;
    printf("y = %d\n", y);
} // y is destroyed here
```

```
int main()
{
    int x = 5;
    printf("x = %d\n", x);

    foo(x);
    printf("x = %d\n", x);

    return 0;
}
```

The output of this program can be seen in the following listing.

```
x = 5
y = 5
y = 6
x = 5
```

At first, x is 5. When foo() is called, the value of x equal to 5 is passed as parameter y to function foo(). The variable y is first printed then modified to a value of 6 and printed again. Finally, the variable y is destroyed. The value of x remains unchanged, even though y was modified.

Arguments passed by value can be variables (e.g., x), literals (e.g., 6) or expressions (e.g., x+1). There are two reasons to call a function by value. When a parameter is passed by reference, it is possible to have unwanted side effects, which are usually caused by inadvertent changes made to original value of the parameter. Mostly data is required to be private and only someone calling the function, if permitted, can modify it. However, passing large structures or data by value can take a lot of time and memory to copy, and this can cause a performance penalty, especially if the function is called many times. So, it is better to use a call by value option when only a few parameters with moderate size are used by the function. One can use call by reference if data changes inside the function are expected.

B.10.2   Passing Parameters by Reference

When passing parameters by value, the only way to return a value back to the caller is using the return statement. While this is suitable in many cases, there are few cases where improved performance can be obtained by using parameter passing by reference. In parameter passing by reference, we pass the parameter reference or address rather its value. As a result no new copy of the parameters is created, rather the address of the original parameter is passed as parameter reference. One such scenario where this approach can be quite useful, is where a function needs to modify the values of multiple parameters (e.g., sorting an array). In this case, it is more efficient to have the function modify the actual array passed to it as reference. In pass by reference, we declare the function parameters as references rather than normal variables as illustrated by the following listing. Specifically, this is called parameter passing by pointers and different from true parameter passing by reference supported by C++ programming.

```
void AddOne(int *y) // y is a reference variable
{
    y = y + 1;
}
```

When the function is called, y will become a reference to the parameter. The following example illustrates this option.

Example B.10.

```
#include<stdio.h>

void foo(int *y) // y is now a reference
{
    printf("y = %d\n", y);

    *y = 6;
    printf("y = %d\n", y);
}

int main()
{
    int x = 5;
    printf("x = %d\n", x);

    foo(&x);
    printf("x = %d\n", x);

    return 0;
}
```

This program is the same as the one we used for the pass by value example, except the parameter value for function foo() is now a reference instead of a normal variable. When we call foo(x), y becomes a reference to x. This example gives the output given in the following listing. Note that the value of x was changed by the function foo in this case.

```
x = 5
y = 5
y = 6
x = 6
```

B.10.3 Function-Like Macro

Function-like macros were very popular before inline functions were introduced to C/C++, because they are often faster than regular function calls and without function call overhead. However, the fact that macros are processed in the preprocessing stage means they do not offer type safety and can result in very cryptic error messages. However, there are quite some old legacy that macros have that will stay at least in the near future. Some of them were introduced well before inline functions. There is also no problem if macros use very special name and do not conflict with very generic names that functions want to use.

Defining Function-Like Macro

A function-like macro definition declares the names of formal parameters within parentheses, separated by commas. An empty formal parameter list is legal, such a macro can be used to simulate a function that takes no parameters. C adds support for function-like macros with a variable number of arguments. The generic syntax for declaring an identifier as a function-like macro is given below.

```
#define <identifier>(<parameter list>) <replacement token list>
```

The function-like macro declaration must not have any whitespace between the identifier and the first, opening, parenthesis. If whitespace is present, the macro will be interpreted as object-like with everything starting from the first parenthesis added to the token list. An example of a function-like macro is given below.

```
#define RADTODEG(x) ((x) * 57.29578)
```

This defines a radians-to-degrees conversion which can be inserted in the code where required, i.e., RADTODEG(34). This is expanded in place, so that repeated multiplication by the constant is not shown throughout the code. The macro here is written as all uppercase to emphasize that it is a macro, not a compiled function. The second x is enclosed in its own pair of parentheses to avoid the possibility of incorrect order of operations when it is an expression instead of a single value.

Appendix C

Introduction to Keil Tools

Below we discuss step by step approach to use the Keil tools from ARM for software development.

C.1 Setup Keil μVision to Write Code

1. Run the software by clicking the icon on desktop, if available, or by clicking on **Start** $\to$ **All Programs** $\to$**Keil** μ**Vision**. An interface similar to one shown in Figure C.1 will open.

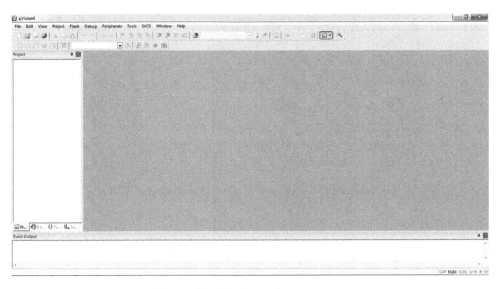

Figure C.1: Keil interface on start.

2. Click on *Project* tab and choose **New** μ**Vision Project** from the drop-down list as shown in Figure C.2

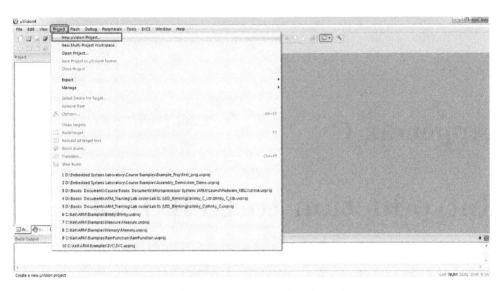

Figure C.2: Create new μVision project.

3. Select and create a directory, then assign a name to your project (project name can be different from folder name) then click on **Save**. **Do not make a directory, file or project name with a space in it**. A space will prevent simulation from working properly.

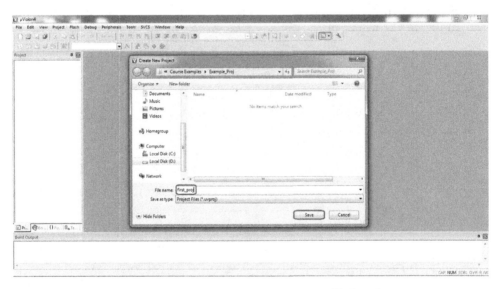

Figure C.3: Type the name of the project in Keil and save it.

4. To select a microcontroller double click on *Texas Instruments* and then select the **TM4C1233H6PM**. Click *OK*, (see Figure C.4 and C.5).

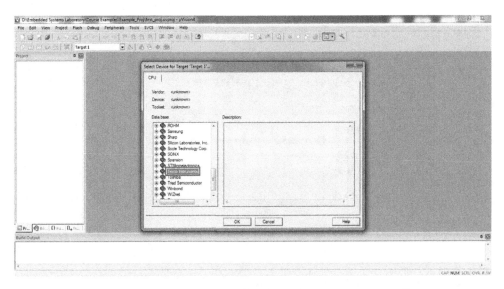

Figure C.4: Select the manufacturer of your microcontroller.

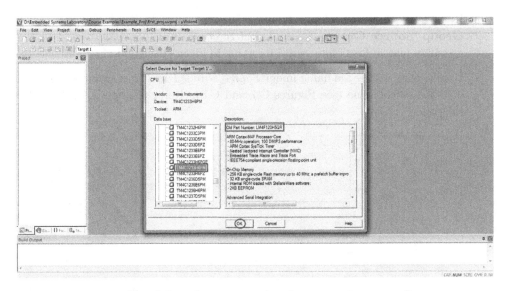

Figure C.5: Select the part number for your microcontroller.

5. When prompted to copy **'Startup_TM4C123.s to project folder'** click on *Yes* or *No* according to the requirement of your project (see Figure C.6).

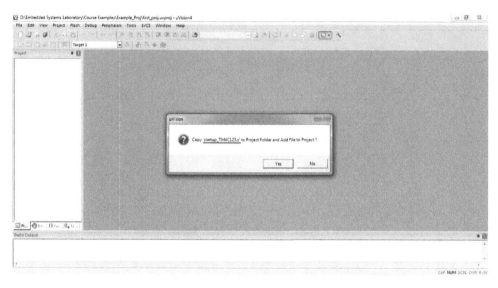

Figure C.6: Add or discard the startup file to the project.

6. Right click on *Source Group 1* under *Target 1*, click on **Add New Item to Group 'Source Group 1'...** and elect the type of file you want to add (**.s** for assembly and **.c** for C file), write its name in given space and click *OK*.

7. Double click on the file name under *Source Group 1* in *Project* window to open it in the editor pane (see Figures C.7 and C.8). Here, you can write and edit the code.

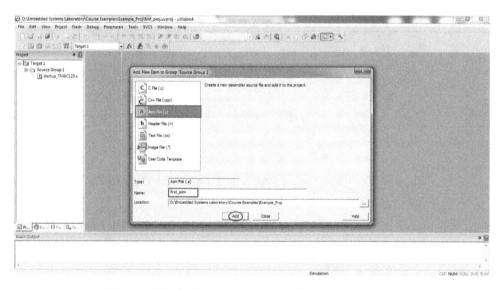

Figure C.7: Add and save a new file to the project.

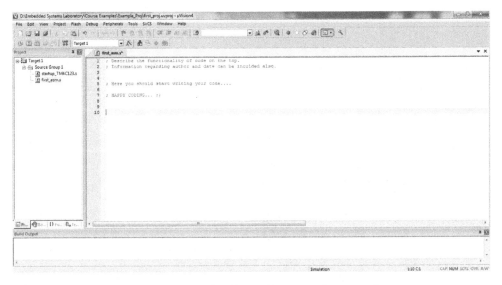

Figure C.8: Edit the file in the text editor window.

C.2 Compilation and Debugging Using KEIL Tools

1. Next step after writing the code is to build your code. Click on Project menu and select Build Target from the drop down list. You can also build your project by clicking Build button in Build bar. Build Output window at displays the errors, warning and build messages during build process. Double-click a message to open the corresponding source file. Build button translates modified or new source files and generates the executable file. The Rebuild command translates all source files regardless of modifications. Simulation highlighted at the bottom signifies that we are not downloading our code on hardware.

2. After successfully building the code, you can run the program through the Debug menu. As shown in Figure C.9 select the Start/Stop Debug Session option from the debug menu or press the debug button. Click on 'OK' for the pop up window showing 'EVALUATION MODE, Running with Code Size Limit: 32K'.

3. Open your uVision to full screen to have a better and complete view. In Figure C.10 the left-hand side window shows you the registers and the right side window shows the program code. There are some other windows open. You may adjust the size of them to see better. Run the program step by step as shown in, you can observe the change of the values in the registers. Click on the Start/Stop Debug Session again to stop executing the program.

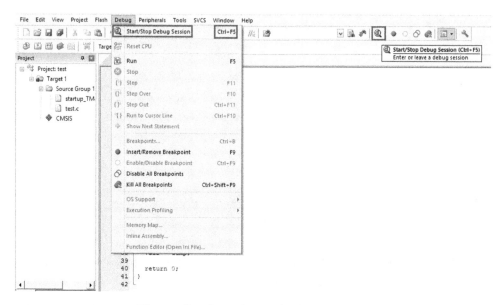

Figure C.9: Launching of debug session.

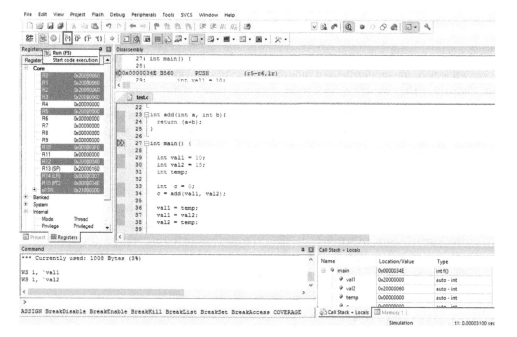

Figure C.10: Debugging of user program.

Appendix D

Introduction to Sourcery CodeBench Tools

Below we discuss step by step procedure to use Sourcery CodeBench tools for software development.

D.1 Sourcery CodeBench Setup for C Program

1. Double click on the Sourcery Codebench icon and start execution of code composer studio. The default location of where the projects are stored is given and can be changed to another location. Use the default location given.

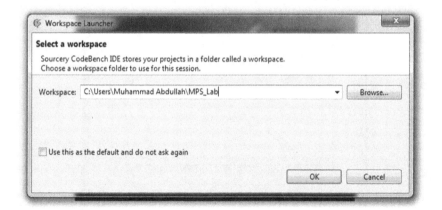

Figure D.1: Setup the project workspace.

2. To create a new project, select from the File menu, New and then C Project (File → Project → C Project).

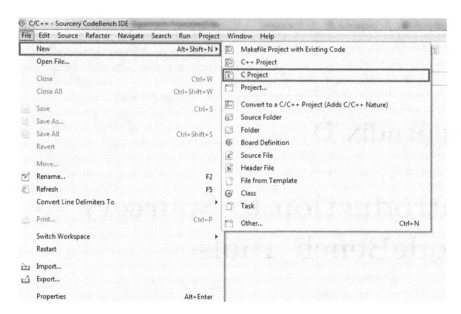

Figure D.2: Create new project.

3. Select the desired project name and the type of project as Empty Project and click next.

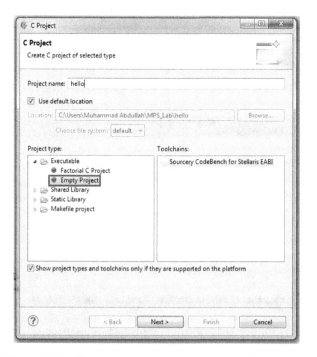

Figure D.3: Choose appropriate name for the project.

4. Select the board and processor you want to use. In this lab we will use TivaC

TM4C123/Stellaris LM4F120 LauncPad board.

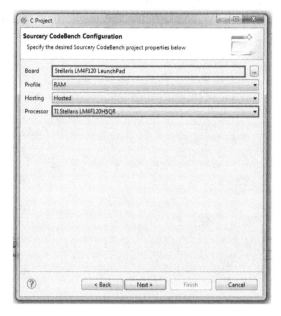

Figure D.4: Select the hardware development platform.

5. Select Sourcery Codebench Debug as Launch type and Stellaris USB as Debug interface to create a debug launch configuration and associate it with the project. Select Finish to create the project.

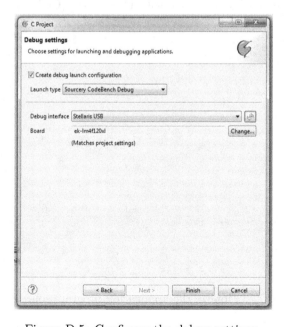

Figure D.5: Configure the debug settings.

6. At this point, the project exists, but there is no associated source code. So, the next step is to create the main program. Right-click on the project, and select New → Source File. Give the new file the name main.c and click the Finish button.

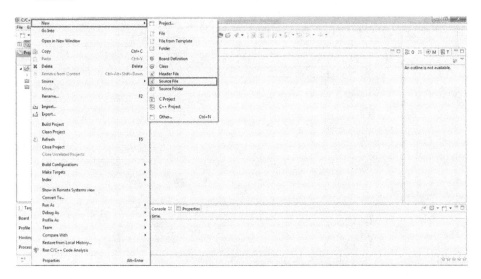

Figure D.6: Either include existing user program or write one.

7. The Sourcery CodeBench IDE now displays an editing window for you to use to create the program. Type a C program to print "Hello World!" into the editor. When you are done, save the file with File → Save(Ctrl+S).

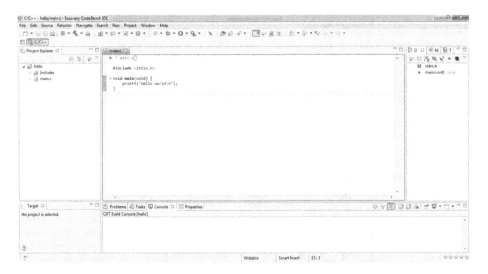

Figure D.7: Finalize the application program source code.

8. After you save the file, build your project by selecting it in the Project Explorer pane on the left, then going to the Project menu and selecting Build Project.

The output of the commands run by the IDE is displayed in the Console tab in the lower pane. If the project build is successful, the IDE prints statistics about the code size of the hello executable at the bottom of the console.

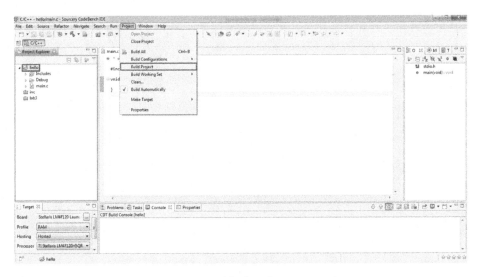

Figure D.8: Build the executable for the user application program.

9. To debug a project, the debug option must be selected under the Run menu and the experimenter board must be connected to the desktop. The Build Project and Debug options should be selected every time the program source code changes.

10. To execute the program, the Resume option under the Run menu is selected. The console window should now display the "Hello World" message.

11. To quit the program, the Terminate option under the Run menu is selected.

Bibliography

[ARM(2010)] ARM. *ARMv7-M architecture reference manual*, 2010. URL `http://infocenter.arm.com/help/index.jsp?topic=/com.arm.doc.ddi0403e.b/index.html`.

[ARM(2012)] ARM. *Procedure call standard for the ARM architecture*, 2012. URL `http://infocenter.arm.com/help/topic/com.arm.doc.ihi0042e/IHI0042E_aapcs.pdf`.

[ARM(2014)] ARM. *Illustration of the benefits of using conditional instructions*, 2014. URL `http://infocenter.arm.com/help/topic/com.arm.doc.dui0473c/BABJGFDD.html`.

[Intel(2014)] Intel. *ARK — Intel Core2 Duo Processor E8400 (6M Cache, 3.00 GHz, 1333 MHz FSB)*, 2014. URL `http://ark.intel.com/products/33910/Intel-Core2-Duo-Processor-E8400-6M-Cache-3_00-GHz-1333-MHz-FSB`.

[Keil(2014)] Keil. *MDK-ARM Primer, Keil MDK-ARM Microcontroller Development Kit*, 2014. URL `http://www.keil.com/support/man/docs/gsac/`.

[Leens(2009)] Frédéric Leens. An introduction to i2c and spi protocols. *IEEE Instrumentation & Measurement Magazine*, 12(1):8–13, 2009.

[Messmer(2001)] Hans-Peter Messmer. *The indispensable PC hardware book*. Addison-Wesley Longman Publishing Co., Inc., 2001.

[TI(2014)] TI. *TM4C123GH6PM, High performance 32-bit ARM Cortex-M4F based MCU*, 2014. URL `http://www.ti.com/product/tm4c123gh6pm`(Lastaccessed: Oct20,2015).

[Yiu(2013)] Joseph Yiu. *The Definitive Guide to ARM® Cortex®-M3 and Cortex®-M4 Processors*. Newnes, 2013.

Index

Printed in the United States
by Baker & Taylor Publisher Services